GUIDE TO Clinical and Diagnostic Virology

GUIDE TO
Clinical and Diagnostic Virology

REETI KHARE, PhD, D(ABMM)
Department of Pathology and Laboratory Medicine
Northwell Health Laboratories
Lake Success, New York

Washington, DC

Library of Congress Cataloging-in-Publication Data

Names: Khare, Reeti, author.
Title: Guide to clinical and diagnostic virology / Reeti Khare, PhD, D(ABMM), Department of Pathology and Laboratory Medicine, Northwell Health Laboratories, Lake Success, New York.
Description: Washington, DC : ASM Press, [2019] | Includes bibliographical references and index.
Identifiers: LCCN 2018055597 (print) | LCCN 2018057746 (ebook) | ISBN 9781555819941 (ebook) | ISBN 9781555819910 (softcover)
Subjects: LCSH: Medical virology—Handbooks, manuals, etc. | Diagnostic virology—Handbooks, manuals, etc.
Classification: LCC QR201.V55 (ebook) | LCC QR201.V55 K43 2019 (print) | DDC 616.9/101—dc23
LC record available at https://lccn.loc.gov/2018055597

Printed in Canada

10 9 8 7 6 5 4 3 2 1

Address editorial correspondence to
ASM Press, 1752 N St., N.W.,
Washington, DC 20036-2904, USA

Send orders to ASM Press, P.O. Box 605, Herndon, VA 20172, USA
Phone: 800-546-2416; 703-661-1593
Fax: 703-661-1501
E-mail: books@asmusa.org
Online: http://www.asmscience.org

To my husband, for being my best friend. You are the eye of the storm.

To my children. You are my bubbles of light.

To my parents, for the privilege of your unwavering support.

To my sister, for being my role model.

CONTENTS

PREFACE

The field of clinical and diagnostic virology is undergoing a period of remarkable change. Viruses and viral infections are becoming more relevant in the practice of modern medicine, but it is increasingly difficult to keep up with them. Newly discovered viruses with novel clinical presentations, a dizzying array of new diagnostic techniques with unique terminology, and even new antivirals are driving a gap in knowledge. Unlike bacteriology, which has matured over many decades, the explosion in molecular testing has rapidly placed viral etiologies of disease into a prominent light. Now clinicians and microbiologists need to develop a similar level of comfort with virology and its jargon, diagnostic techniques, clinical syndromes, and therapies.

The purpose of this guide is to provide a simple reference for medical and scientific professionals so that we may tread more comfortably in this new era. Its goals are to summarize essential concepts and to highlight important terms in clinical and diagnostic virology. It is designed for a wide range of professionals, including students in medical and laboratory fields (such as infectious disease fellows, pathology residents, medical students, microbiology fellows, medical technologists, technicians, and graduate students), scientists and virologists who need to understand the context for their fields of study, and practicing clinicians who need to understand changes that are occurring in the field.

Comprehensive material in clinical and diagnostic virology can be found in textbooks, primary literature, and on public health sites. This book is a "CliffsNotes" version of all this information. Here critical concepts are carefully selected and concisely summarized in bullet points to make things simple and easy to digest. Key terms are featured in bold, and notable "pearls" are emphasized in the margins. There are also questions at the end of each chapter to help reinforce important ideas.

I hope that this guidebook will be a useful reference and that it helps make clinical and diagnostic virology even more interesting and accessible. I welcome any suggestions, additions, or corrections that would improve the quality of this book.

Reeti Khare

ACKNOWLEDGMENTS

I thank the reviewers who contributed their time and expertise to this book:

- Neil Anderson, MD, D(ABMM), Assistant Medical Director of Clinical Microbiology, Washington University School of Medicine in St. Louis
- Esther Babady, PhD, D(ABMM), Director of Clinical Operations, Microbiology, Memorial Sloan Kettering Cancer Center
- Gregory J. Berry, PhD, D(ABMM), Director of Molecular Diagnostics, Northwell Health
- Matthew J. Binnicker, PhD, D(ABMM), Director of Clinical Virology, Mayo Clinic
- Scott Duong, MD, Assistant Director of Infectious Disease Diagnostics, Northwell Health
- Christine C. Ginocchio, PhD, MT(ASCP), Vice President of Microbiology, BioMérieux, Vice President of Scientific/Global Affairs, BioFire
- Aya Haghamad, PharmD, Clinical Solutions Specialist, Northwell Health
- Stefan Juretschko, PhD, D(ABMM), Senior Director of Infectious Disease Diagnostics, Northwell Health
- Michael J. Loeffelholz, PhD, D(ABMM), Senior Director of Medical Affairs, Cepheid
- Benjamin Pinsky, MD, PhD, Medical Director of Clinical Virology, Stanford Health Care and Stanford Children's Health
- Elitza Theel, PhD, D(ABMM), Director of Infectious Disease Serology, Mayo Clinic
- Xiomin Zheng, MD, PhD, Pathology Resident, Staten Island University Hospital, Northwell Health

I also thank Drs. Bobbi Pritt, James Crawford, and Megan Angelini for their support and encouragement throughout this project.

ABBREVIATIONS

Throughout the book, certain abbreviations will be used. The below list highlights some of the most important and frequently used.

(-) ssRNA	Negative-sense, single-stranded RNA
(+) ssRNA	Positive-sense, single-stranded RNA
ADE	Antibody-dependent enhancement
AIDS	Acquired immunodeficiency syndrome
BSL	Biosafety level
CAP	College of American Pathologists
cccDNA	Covalently closed circular DNA
CCHFV	Crimean-Congo hemorrhagic fever virus
CDC	Centers for Disease Control and Prevention
cDNA	Complementary DNA
CLSI	Clinical and Laboratory Standards Institute
CMV	Cytomegalovirus
CNS	Central nervous system
CoV	Coronavirus
CPE	Cytopathic effect
CSF	Cerebrospinal fluid
DAA	Direct acting antiviral
dNTP	Deoxynucleotide triphosphate
dsDNA	Double-stranded DNA
dsRNA	Double-stranded RNA
EBV	Epstein-Barr virus
EEEV	Eastern equine encephalitis virus
EIA	Enzyme immunoassay
ELISA	Enzyme-linked immunosorbent assay
FDA	Food and Drug Administration
HAV	Hepatitis A virus
HBV	Hepatitis B virus
HCV	Hepatitis C virus
HDV	Hepatitis D virus
HEV	Hepatitis E virus
HHV-1	Human herpesvirus 1; also known as herpes simplex virus 1
HHV-2	Human herpesvirus 2; also known as herpes simplex virus 2

HHV-3	Human herpesvirus 3; also known as varicella-zoster virus
HHV-4	Human herpesvirus 4; also known as Epstein-Barr virus
HHV-5	Human herpesvirus 5; also known as cytomegalovirus
HHV-6	Human herpesvirus 6
HHV-7	Human herpesvirus 7
HHV-8	Human herpesvirus 8; also known as Kaposi sarcoma-associated virus
HIV	Human immunodeficiency virus
HMPV	Human metapneumovirus
HPIV	Human parainfluenza virus
HPV	Human papillomavirus
HSV	Herpes simplex virus; may refer to either human herpesvirus 1 or 2
HTLV	Human T-cell lymphotropic virus
IHC	Immunohistochemistry
ISH	*In situ* hybridization
KSHV	Kaposi sarcoma-associated herpesvirus; also known as human herpesvirus 8
LCMV	Lymphocytic choriomeningitis virus
MERS	Middle East respiratory syndrome
MMR	Measles, mumps, rubella
NAAT	Nucleic acid amplification testing
NGS	Next-generation sequencing
NNPI	Non-nucleoside polymerase inhibitor
NNRTI	Non-nucleoside reverse transcriptase inhibitor
NPI	Nucleoside polymerase inhibitor
NPV	Negative predictive value
NRTI	Nucleoside reverse transcriptase inhibitor
NTP	Nucleotide triphosphate
PCR	Polymerase chain reaction
PI	Protease inhibitor
PPV	Positive predictive value
RIDT	Rapid influenza diagnostic tests
RSV	Respiratory syncytial virus
RT-PCR	Reverse transcription PCR (*not* real-time PCR)
SARS	Severe acute respiratory syndrome
ssDNA	Single-stranded DNA
ssRNA	Single-stranded RNA
VZV	Varicella-zoster virus; also known as human herpesvirus 3
WEEV	Western equine encephalitis virus
WNV	West Nile virus

ABOUT THE AUTHOR

Reeti Khare, PhD, D(ABMM), is the Director of Microbiology at Northwell Health Laboratories in New York. She received her PhD in Virology and Gene Therapy at Mayo Clinic and did a postdoctoral fellowship at the University of Washington. Her research involved reengineering viral vectors, developing adenoviruses for liver gene therapy, and creating viral vector vaccines against MRSA. She returned to Mayo Clinic for her clinical microbiology fellowship and is a diplomate of the American Board of Medical Microbiology. Reeti enjoys teaching and learning about microbiology and has authored numerous publications, chapters, and reviews. At Northwell Health Labs she continues to pursue clinical research and provide student education, and is responsible for laboratory oversight, improving efficiency, designing workflows, and diagnostic microbiology testing.

SECTION I

FOUNDATIONS OF CLINICAL VIROLOGY

1

CHAPTER 1

INTRODUCTION TO VIRUSES

I. OVERVIEW. Viruses are **obligate intracellular** parasites. Unlike all other organisms, they are not "alive" because they are metabolically inactive on their own. They are also not "dead" because they can metabolize and reproduce when associated with a host cell. Instead, they are referred to as being "active" or "inactive." Viruses are difficult to study because of their minuscule size, but they are even more abundant than bacteria. Most are part of normal environmental or human flora but some viruses are medically relevant and can cause infections that fall anywhere on the spectrum, from asymptomatic to fulminant. Several factors affect the pathogenicity of a virus.

1. **Virus-specific factors**
    - Virulence: Some viruses are more virulent than others. For example, rabies virus is highly pathogenic, while torque teno virus does not cause disease despite being ubiquitous in normal flora.
    - Persistence: Some viruses, like herpesviruses, cause mild disease but infect humans for life.
    - Indirect effects: Viruses like bacteriophages only infect bacteria but are still indirectly pathogenic to humans. For example, *Corynebacterium diphtheriae* does not generally cause clinically significant disease unless it is infected with the bacteriophage containing the diphtheria toxin gene.
2. **Host-specific factors:** Hereditary genetic mutations can allow viruses that are weakly pathogenic to cause significant disease. For example, a specific mutation in CCR5 (a host cell receptor for HIV) has been shown to prevent infection with this virus. On the other hand, other mutations can result in overgrowth of viruses. For example, human papillomavirus 2 (HPV2) typically causes benign warts, but individuals with genetic defects in cell-mediated immunity can demonstrate uncontrolled giant warty overgrowths ("tree man" disease).
3. **Immunosuppression:** Immunosuppressive drugs, virus-induced immunosuppression, and even pregnancy are all instances in which the immune system is depressed. This can leave patients vulnerable to unique viral infections.

II. VIRUS STRUCTURE. The structure of a virus defines its life cycle, mechanism of pathogenicity, and how it is detected by laboratory assays. A virus particle, or

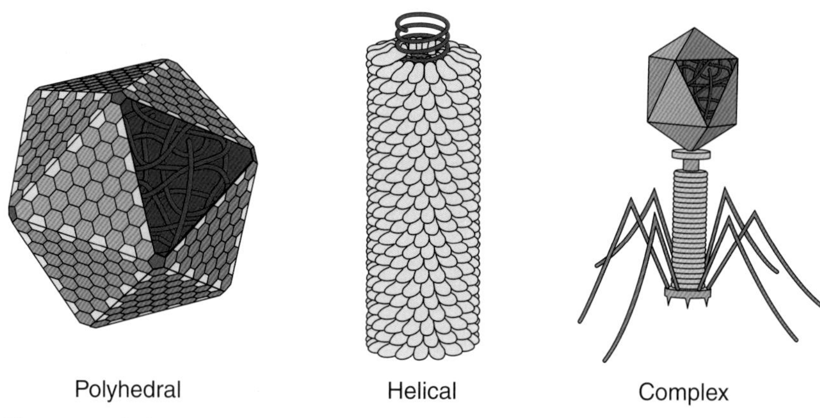

Figure 1.1. Viral capsids come in three main shapes.

virion, is composed of nucleic acid surrounded by a protective protein coat called a capsid. Together, the nucleic acid and capsid are called the **nucleocapsid**.

1. **Capsids:** Occur in three main shapes (Fig. 1.1).
   - Polyhedral capsids have multiple flat sides that form a rigid shell around the viral nucleic acid. Viruses with lots of flat sides can appear round. Viruses with these kinds of capsids are highly regular and have a rigid shape and size. Most DNA viruses are polyhedral.
   - Helical capsids wrap proteins around the strand of nucleic acid to form a spiral, elongated nucleocapsid. These viruses tend to be more variable in size and shape. Most RNA viruses are helical.
   - Complex capsids have other shapes or are a combination of helical and polyhedral capsids.

Most blood-borne viruses are enveloped because they need to evade the immune system efficiently.

2. **Envelopes:** Some viruses have an envelope, which is a lipid bilayer that surrounds the nucleocapsid.
   - Envelopes are acquired from cell membranes and act as a shield from the immune system. The disadvantage is that these viruses are susceptible to detergents, drying, and pH changes and typically do not survive for long on external surfaces.
   - **Nonenveloped**, or **naked**, viruses are more resistant to harsh conditions and tend to be more stable in the environment. They are relatively resistant to disinfectants (e.g., alcohol, dilute bleach, quaternary ammonium compounds, and even water disinfectants like chlorine). This means that they are difficult to eliminate from community and hospital environments.

Most gastrointestinal viruses are naked because they must be highly resistant to the acidic environment of the stomach.

Rule of thumb: Viruses are about 1/10 the size of a bacterial cell.

3. **Size:** Viruses cannot be seen using light microscopes. Medically important viruses range from ~20 to 500 nm in length (Fig. 1.2).

III. LIFE CYCLE. The life cycle of the virus is how it binds to a host cell, replicates its nucleic acid, and then spreads to new cells. Knowing each virus's life cycle is critical to understanding what part of the body will be affected, how long the infection will last, how it can be detected, and which antivirals will work.

Replication: making new genomes.

Transcription: making messenger RNA (mRNA) from the genome.

Translation: making proteins from mRNA.

1. **Incubation period:** Viruses infect target host cells and replicate. Due to the low level of virus at the beginning of infection, patients are typically asymptomatic.

2. **Spread:** Viruses spread through the host by infecting adjacent cells, traveling in migratory cells, disseminating through the bloodstream (**viremia**), and diffusing through body fluids. Viremia can be identified through detection of viral nucleic acids and/or antigens in the blood.
   - Viruses infect cells by binding to specific cellular receptors; cells that do not display the correct receptors will not be infected. **Tropism** is the affinity of a virus for some cell types and not others.
   - Once bound, virions enter cells by **endocytosis** or **fusion**. During endocytosis, the host cell membrane invaginates and engulfs the virus. During fusion, the viral envelope will fuse with the cell membrane in order to release the viral nucleocapsid into the cell.

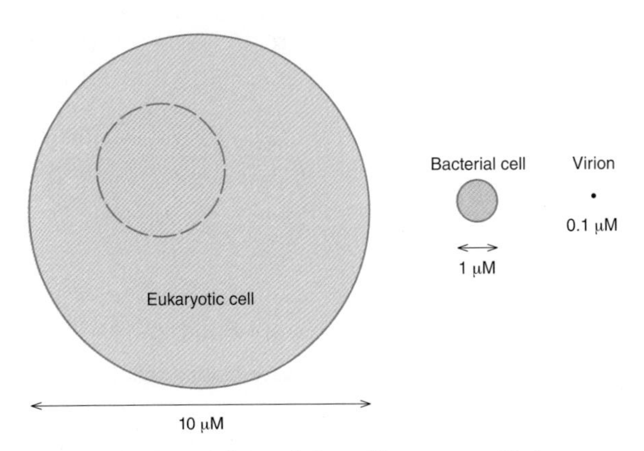

Figure 1.2. Comparison of sizes. Viruses are ~10 times smaller than a bacterium and ~100 times smaller than a eukaryotic cell.

3. **Prodromal phase:** Viruses may produce early, nonspecific symptoms (e.g., fever, aches, pain, and nausea) as they replicate.
   - Once inside the cell, viruses replicate, transcribe mRNA, and translate it into viral proteins. These proteins and nucleic acids are assembled into infectious viral particles in the nucleus or cytoplasm.
   - These viral aggregates can sometimes be seen as viral inclusions.

Latent viruses are dormant. They do not actively produce virions or trigger the immune system, and therefore result in a persistent viral reservoir.

Proviruses = integrated viral genomes
Episomes = nonintegrated, persistent viral genomes

4. **Active disease:** Viruses cause an immediate or long-lived infection in the cell.
   - **Lytic** viruses lyse (destroy) the host cell to get out immediately after replication. This manifests as an acute infection where the patient may show characteristic signs of the viral infection.
   - **Lysogenic** viruses cause a long-lived, latent infection by integrating into the host genome. These infections are often subclinical. Environmental triggers can cause the integrated viruses to excise and enter the lytic cycle. The integrated viral genome is called a **provirus**. Proviral DNA replicates passively with the cell every time the host cell replicates. This makes them very long-lived and largely undetected by the immune system.
   - **Pseudolysogenic** or **episomal** viruses persist for a long time in the host cell without integrating. Their nucleic acids remain separate as an **episome**. These episomes are diluted as cells divide and removed from circulation when cells die (Fig. 1.3).
   - Depending on the type of virus, new viral particles will exit the cell by four main methods (Fig. 1.4).
     - **Lysis:** bursting of the host cell
     - **Budding:** Viruses push into the perimeter of the cell and capture part of the cellular membrane. Many enveloped viruses surround their nucleocapsid with the host plasma membrane in order to evade detection by the immune system.
     - **Exocytosis:** Virions are enclosed in a cellular vesicle, which then fuses with the plasma membrane in order to release virus particles outside the cell.
     - **Cell-to-cell transport:** Some viruses can cause host cells to fuse together (syncytia). This allows the new viruses to directly enter neighboring cells without exposing them to the immune system.

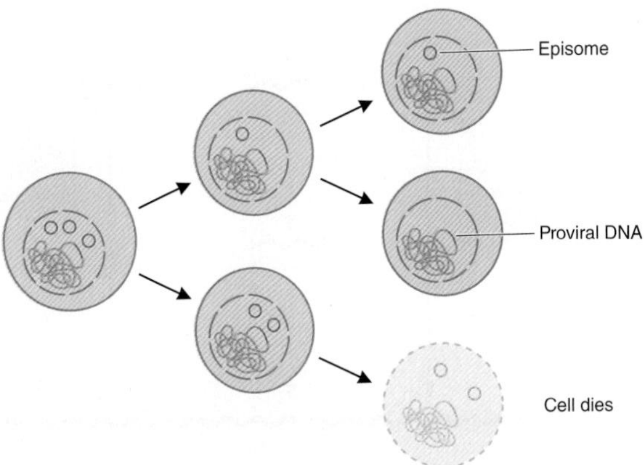

Figure 1.3. Integration versus episomal persistence. Episomes are diluted when the cell replicates while integrated viruses are multiplied when the cell replicates.

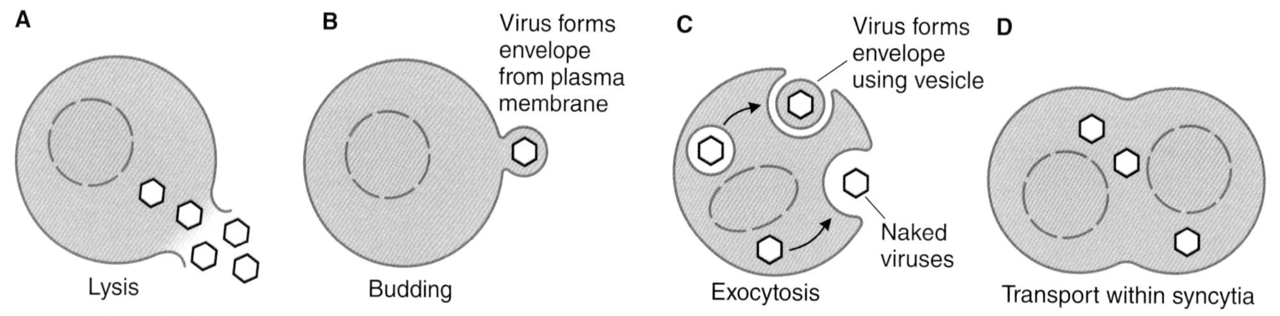

Figure 1.4. Methods of viral release from a cell. (A) Lysis; (B) budding; (C) exocytosis; (D) cell-to-cell transport within syncytia.

5. **Resolution of disease:** Both the innate and adaptive immune responses clear or suppress the viral infection.
   - Innate responses are able to suppress viral infections rapidly.
     - Antiviral cytokines, like **interferon,** are produced. They activate lymphocytes and upregulate proteins that inhibit viral replication.
     - **Natural killer (NK) cells** are specialized lymphocytes that kill virus-infected or cancerous cells.
     - $CD8^+$ killer T cells also destroy infected cells, and $CD4^+$ helper T cells activate the memory response.
   - Adaptive responses take several weeks to develop but produce immunoglobulins (**antibodies**) against pathogens that are highly effective, specific, and long-lived. There are 5 different classes of immunoglobulins: IgA, IgD, IgE, IgG, and IgM.
     - **IgM** is produced first, within ~1 to 2 weeks of exposure.
     - **IgG** is produced within 2 to 4 weeks of exposure and provides immunity for life.
     - IgA is localized to mucosal membranes and provides immunity against respiratory and gastrointestinal pathogens.

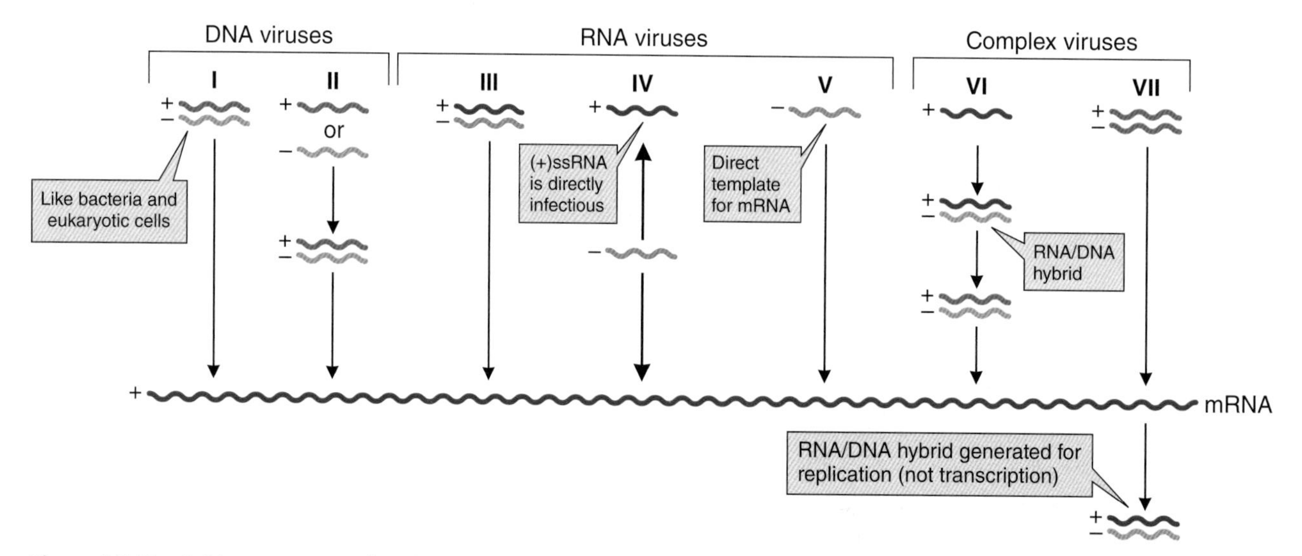

Figure 1.5. The Baltimore system classifies a virus based on the virus's nucleic acid and its method of mRNA synthesis. Blue indicates DNA, red indicates RNA.

- IgE provides immunity against parasites.
- IgD is present on B cells.

IV. VIRAL GENOMES. Viruses carry only a small number of essential genes that are necessary for infection and replication. Unlike prokaryotes and eukaryotes, the genome configurations of viruses vary widely—they have variable shapes, structures, copy numbers, and genome sequences. Most importantly, they can be composed of either deoxyribonucleic acid (DNA) or ribonucleic acid (RNA).

1. **The Baltimore system of classification:** This classification scheme categorizes viruses into classes I to VII according to the type of nucleic acids they contain and how they are transcribed into mRNA (Fig. 1.5 and Table 1.1). How viruses generate mRNA is important because the faster they make mRNA, the faster they can make proteins and the faster they can assemble these proteins into new virions.

"Be HAPPPPy!" for DNA viruses: hepatitis **B** virus, **h**erpesvirus group, **a**denovirus, **p**oxvirus, **p**arvovirus, **p**apillomavirus, **p**ol**y**omavirus

Table 1.1. Characteristics of viruses in each Baltimore class

BALTIMORE CLASS	NUCLEIC ACID	DESCRIPTION	EXAMPLE(S)
I	dsDNA	These viruses make DNA and mRNA like cells do, so they can use the cell's own machinery and redirect its resources into making more virus particles. Also, dsDNA genomes are **stable** and can **persist** in host cells for a long time.	Herpesviruses, adenovirus
II	ssDNA	Like class I viruses, these can also redirect and use the cell's own machinery to make more DNA and mRNA. Because they are single stranded, they can **mutate** and recombine more readily than dsDNA. Also, they have half the amount of DNA (only one, instead of two, strands), so they can be packed into very **small** particles.	Parvovirus B19
III	dsRNA	These viruses are mutagenic, but since cellular proteins are not designed to work on RNA templates, RNA viruses must encode and create their own polymerase.	Rotavirus
IV	(+) ssRNA	This class is significant in terms of biosafety because these genomes are equivalent to mRNA. So even if the viruses are not assembled, their genomes are **directly infectious**, as well as highly **mutagenic**. However, mammalian cells do not make polymerases that can replicate RNA, so these viruses need to encode and create their own viral RNA polymerases.	Poliovirus, West Nile virus, norovirus, hepatitis C virus
V	(–) ssRNA	This class is very **common** among medically relevant viruses because negative-sense ssRNA is a template for mRNA. So the genomes are not infectious but mRNA can be made directly from it. These viruses are also mutable but need to encode and create their own viral RNA polymerases.	Influenza virus, Ebola virus
VI	(+) ssRNA	Unlike class IV viruses, which also have (+) ssRNA, these viruses use **reverse transcriptase** to generate a **DNA intermediate**. Although this costs the virus time and resources, the DNA intermediate can **integrate** into the host genome and causes latent infection. Then more mRNA is transcribed from the integrated provirus. Retroviruses are classified within this group.	HIV
VII	dsDNA	These viruses are just like class I viruses when it comes to making mRNA, but during replication they produce an RNA intermediate. **Reverse transcriptase** converts it back to DNA. Again, this costs the virus time and resources but allows it to **integrate** into the host genome.	Hepatitis B virus

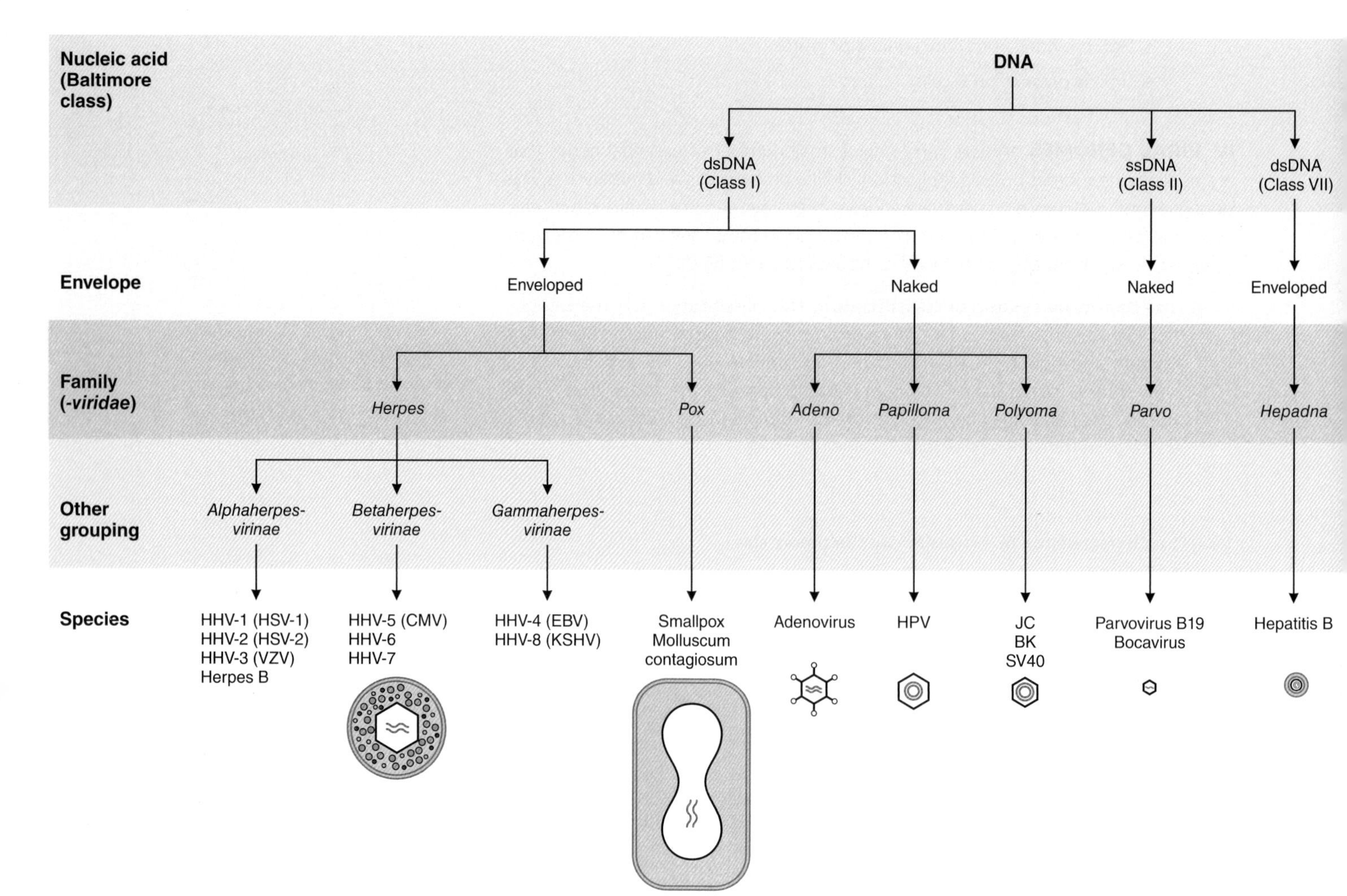

Figure 1.6. Organization of clinically relevant DNA viruses. Virus images: blue indicates DNA, black indicates capsid, and double brown line indicates an envelope.

A rapid mutation rate is so useful that the majority of medically relevant viruses contain RNA.

Negative-sense nucleic acid is the **template** for mRNA because they are complementary to mRNA.

Positive-sense nucleic acid is the same orientation as mRNA.

Proteins can be translated directly from (+) ssRNA (class IV) viral genomes. So class IV genomes are directly infectious, even when they are not packaged in a capsid.

2. **DNA viruses (Fig. 1.6):** Replication of DNA has high fidelity, which means that replication errors are rare. This is an advantage because viruses that contain DNA are highly stable and can persist in the host cell for a long time. They can also redirect the cell's own polymerases and enzymes towards replication of more viral particles.

3. **RNA viruses (Fig. 1.7):** RNA replication machinery is significantly more error prone and mutations are incorporated at a much higher rate. Because of this, viruses with RNA genomes mutate quickly, which is an advantage because it allows them to adapt to new environments. Over many rounds of replication there can be so many new mutations that there are almost distinct viral populations within a single patient. These quasispecies can evade memory immune responses and can be difficult to treat because they are so diverse.

4. **Strands and sense:** Viral nucleic acids are usually **single stranded** or **double stranded** (yielding, for example, ssRNA or dsDNA), but some are partially single and double stranded. Single-stranded genomes are either negative or positive sense.

   - **Negative-sense (−)** sequences are back to front (i.e., they are in 3′-to-5′ orientation, or the "non-sense" direction). This makes them the blueprints for mRNA production.

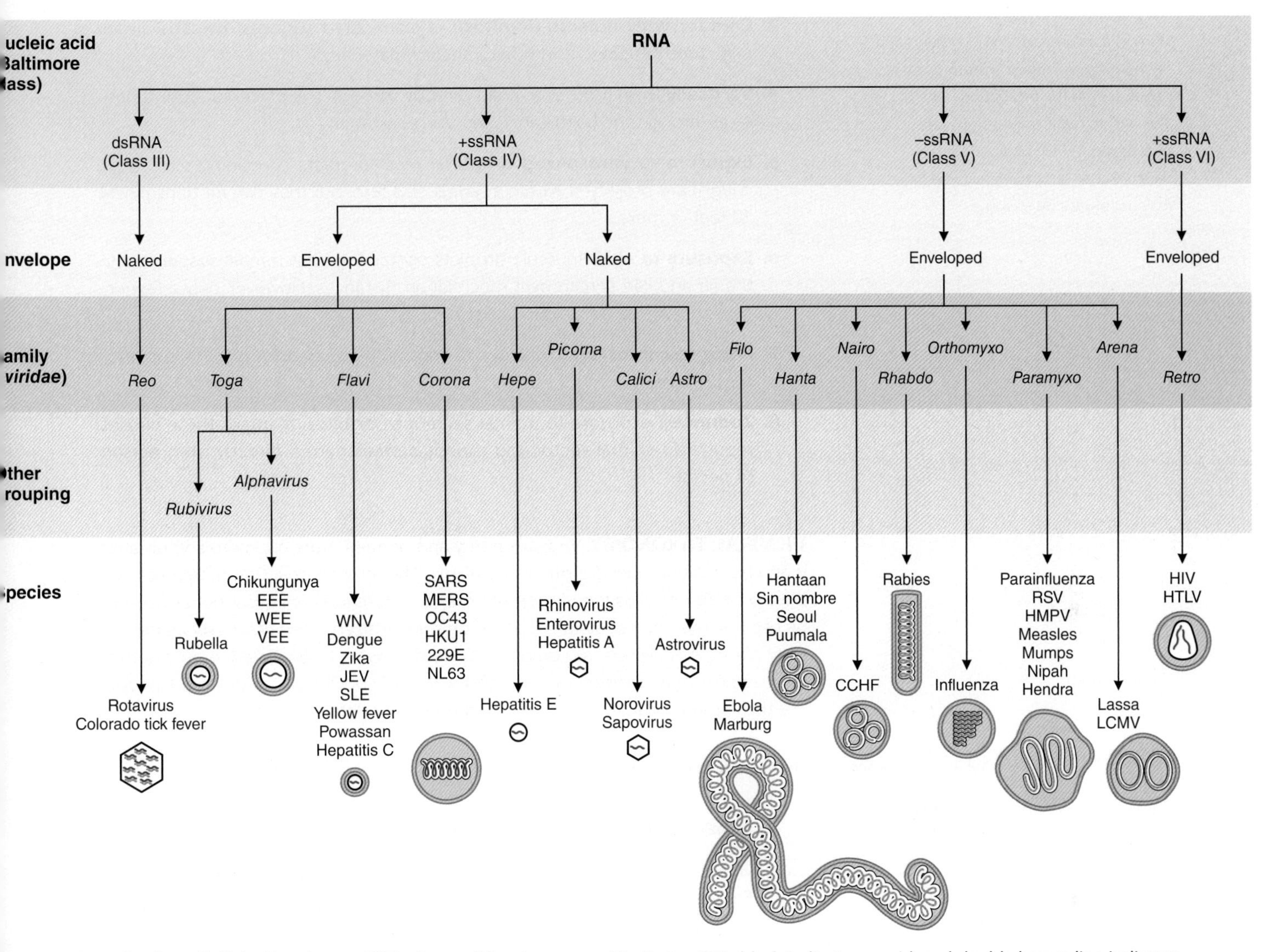

ure 1.7. Organization of clinically relevant RNA viruses. Virus images: red indicates RNA, black indicates capsid, and double brown line indicates envelope.

- **Positive-sense (+)** sequences are in the correct 5′-to-3′ orientation (i.e., they have the same orientation as mRNA).

5. **Structure:** Viral nucleic acids also have structure. They can be linear or circularized, can form hairpin loops, and can be **segmented**. Segmented genomes allow viruses to shuffle segments from multiple strains together (i.e., reassort) and make entirely new viral strains (*see* chapter 3).

V. VIRAL TRANSMISSION. Understanding the way different viruses spread helps prevent new exposures and contain outbreaks. Viruses can be spread through various routes.

1. **Direct contact:** infected tissue, contact with mucous membranes during sexual intercourse
2. **Contact with body fluids:** infected blood, saliva, respiratory secretions, seminal fluid, fecal materials

Many viruses are shed in body fluids and secretions even before symptoms have begun (i.e., in the incubation period). This has dramatic implications for transmission, since quarantines of patients after symptoms have begun may not always be effective.

3. **Contact with objects (fomites):** contaminated surfaces, personal items (e.g., toothbrushes), and other infected materials
4. **Vertically** from mother to child directly through the placenta, during passage through the birth canal, and via breast milk
5. **Exposure to aerosolized droplets:** Most droplets containing virus (e.g., cough) are projected within a radius of 3 feet, but they can be transmitted 10 feet or more.
6. **Exposure to air:** Minuscule droplets containing virus remain suspended in the air and can spread over much larger distances. However, only a few organisms can be transmitted by this route.
7. **Iatrogenic intervention:** organ transplant, blood transfusion, and immunosuppression
8. **Zoonoses:** exposure to animal secretions or bites. Many of these viruses require an animal vector and cannot be transmitted directly from person to person.

VI. VIRAL TAXONOMY. Viral taxonomy and nomenclature are more complicated than classification rules for other organisms. The International Committee on Taxonomy of Viruses classifies viruses into only 5 hierarchical ranks, but not all need to be used. Names are not assigned based on standardized criteria and may be named after places, discoverers, region of the body they were first isolated, or symptoms. Most confusingly, viruses are assigned a formal species name but most people use the common name, which may be the same or different (Table 1.2).

Table 1.2. Nomenclature for viral taxonomic categories

TAXONOMIC CLASSIFICATION	NOMENCLATURE	EXAMPLE 1	EXAMPLE 2
FORMAL CATEGORIES			
Order	Ends in -*virales*	*Herpesvirales*	NA[a]
Family	Ends in -*viridae*	*Herpesviridae*	*Orthomyxovirus*
Subfamily	Ends in -*virinae*	*Gammaherpesvirinae*	NA
Genus	Ends in -*virus*	*Lymphocryptovirus*	*Influenzavirus A*
Species	**Varies**; may contain multiple words and may or may not contain "virus."	*Human herpesvirus 4*	*Influenza virus A*
Subtypes	**Varies**; may contain the word "genotype," "serotype," or other label	NA	H1N1
OTHER DEMARCATIONS			
Common name (this is used much more often than the species name)	[Name] virus	Epstein-Barr virus	Influenza virus
Abbreviation of common name	**Varies**; often the first letter of each word	EBV	NA or "flu"
Other	Virus specific descriptors	NA	A/Michigan/45/2015

[a]NA, not applicable.

1. **Formal species names:** Formal names are usually used when referring specifically to a virus's taxonomy. This name is italicized and the first letter is capitalized (as well as other proper nouns), just like species names for bacteria (1). For example, the species *Human respiratory syncytial virus* belongs to the order *Paramyxoviridae*. All higher viral taxon names are also italicized (e.g., *Flaviviridae*).
2. **Common names:** These are used more often than formal species names. They are not italicized and are capitalized only if they are proper nouns. For example, Epstein-Barr virus is capitalized because it is named after people, while herpes simplex virus is not. The commonly used name is not uniform and may or may not be the same as the species name. For example, the common name for *Mumps virus* is mumps virus while the common name for *Hepatovirus A* is hepatitis A virus.
3. **Abbreviations:** Abbreviations for viruses are also not uniform. Common names are often abbreviated by capitalizing the first letter of each word, including the word "virus" (e.g., HSV for herpes simplex virus). Sometimes part of the name is used for more clarity (e.g., CHIKV for chikungunya virus), with or without capitalization (e.g., Ad or AdV for adenovirus).
4. **Subtypes:** Viral species can be further divided into subtypes. Serologic techniques were originally used to differentiate the strains into categories called **serotypes**. Newer techniques based on sequencing identify them as **genotypes**. Genotypes and serotypes tend to coincide, but not always.
5. **Other descriptors:** Some viruses, like influenza virus, have additional naming conventions that convey extra information (see chapter 3).

Most of the time people use common virus names. These are not italicized, and only the proper nouns should be capitalized.

Multiple-Choice Questions

1. **Some viruses have a lipid layer around them. What is it called?**
   a. A capsule
   b. A capsid
   c. An envelope
   d. A tegument
2. **What do all viruses contain?**
   a. Envelope and capsid
   b. Viral polymerase and nucleic acid
   c. Capsule and envelope
   d. Capsid and nucleic acid
3. **Which of the following is true about Baltimore class I viruses?**
   a. They contain DNA and are able to replicate outside of a cell.
   b. They contain DNA and are able to use the host cell's replication machinery.
   c. They contain RNA and are able to reassort.
   d. They contain RNA and are able to mutate rapidly.
4. **Which of the following is an advantage of Baltimore class IV viruses?**
   a. Their nucleic acid is directly infectious.
   b. Their nucleic acid is able to integrate.
   c. They do not need to carry any viral proteins.
   d. They are nonmutagenic.

5. Which viruses are most stable in the environment?

a. Enveloped viruses, due to the extra layer of protection

b. Naked viruses

c. RNA viruses

d. Integrating viruses

6. Many medically relevant viruses belong to Baltimore class V. This is because they

a. Can all reassort to create novel, highly virulent strains

b. Are highly stable and are able to integrate

c. Can be replicated rapidly and are highly mutable

d. Are transmitted by direct contact

7. Which of the following terminology is correct?

a. There are many human papillomaviruses.

b. There are many Human Papilloma Viruses.

c. There are many *Human Papillomaviruses*.

d. There are many *Human papilloma* Viruses.

8. Which of the following mechanisms does NOT hide viruses from the immune system?

a. Syncytia

b. Envelopes

c. Provirus

d. Lysis of the host cell

9. How do Baltimore class VI viruses differ from class IV viruses?

a. They are intrinsically more pathogenic.

b. They package both RNA and DNA within the capsid.

c. They encode reverse transcriptase.

d. They are transmitted by direct contact.

10. Which nucleic acid do most episomally persistent viruses contain?

a. dsDNA

b. dsRNA

c. (+) ssRNA

d. (–) ssRNA

11. Which of the following viruses are in the same family?

a. Dengue and chikungunya viruses

b. Parainfluenza and influenza viruses

c. Lassa and Marburg viruses

d. West Nile and hepatitis C viruses

12. Which of the following viruses are NOT in the same family?

a. Respiratory syncytial and mumps viruses

b. Hepatitis A and rhinovirus viruses

c. Hepatitis B and hepatitis C viruses

d. Herpes simplex and varicella-zoster viruses

Match the following. Use each answer only once.

13. Virus shape

Bullet shaped	A. Ebola virus
Dumbbell-shaped nucleocore	B. HIV
Filamentous	C. Rabies virus
Conical nucleocapsid	D. Smallpox virus

True or False

14. Only viruses that integrate can cause lifelong infection. **T F**

15. Immunosuppression exacerbates viral infections. **T F**

16. Syncytia protect viruses from exposure to the immune system. **T F**

2

CHAPTER 2

LABORATORY DIAGNOSIS OF VIRAL INFECTIONS

I. OVERVIEW. Viruses can cause a wide range of manifestations, from asymptomatic infection to latent, acute, localized, and systemic infections or even cancer. Different viruses are associated with different diseases (Fig. 2.1). To diagnose the etiology of an infection, it is essential that the correct specimen is collected and the right tests are ordered. The appropriate specimen type depends on the location of the infection, the type of patient, the collection device, the stage of infection, and the pretest probability of disease. It is also essential that the correct assay is chosen because it will affect how likely it is that the pathogen will be detected accurately, the turnaround time of results, and the type of specimen that needs to be collected.

II. SPECIMEN TYPE. Viruses are intracellular microbes, so the best specimens for virus identification depend on where the infection occurs. This is important because some viruses may be present and detectable in a sample but are not the cause of active symptoms.

1. **Tissue:** Viruses live in cells, so tissues and biopsy specimens are often preferred samples for viral detection using PCR, culture, shell vial, and histology. Formalin-fixed paraffin-embedded tissue is useful for histology but cannot be cultured because all the organisms are inactivated. Fixing also cross-links DNA, which inhibits PCR, although it can be done in some cases.
2. **Cerebrospinal fluid (CSF):** Used to diagnose central nervous system (CNS) infections, like meningitis. Viral meningitis is sometimes called "aseptic meningitis" to differentiate it from bacterial meningitis. However, this term is no longer used because there may be other non-bacterial causes of meningitis.
   - Overall, viral culture has poor sensitivity from CSF.
   - PCR or serology is often the preferred diagnostic approach.
   - Other lab findings can also indicate infection (Table 2.1).
3. **Blood:** It is important to consider the life cycle of viruses before trying to identify them in blood because not all viruses are blood borne.

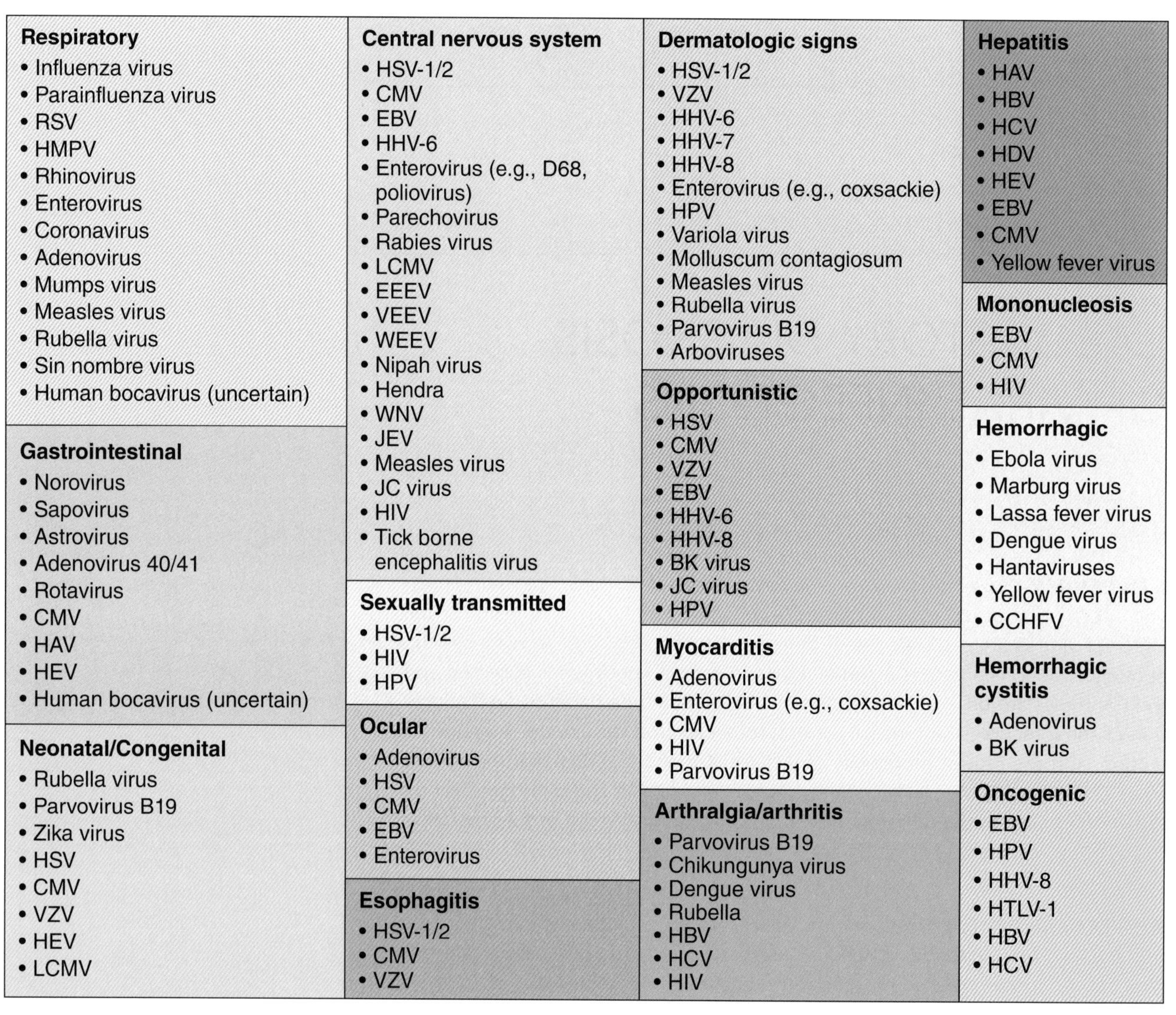

Figure 2.1. **Viruses associated with various clinical syndromes.** See abbreviations list at the front of this book.

- Blood is not always a good source for diagnosis because viruses are intracellular. Many viruses have only a short viremic period (e.g., arboviruses are viremic only for a few days), while others are spread from cell to cell instead of hematogenously (e.g., herpesvirus).
- Blood is a good source for diagnosing and **monitoring** certain bloodborne viral infections, for example, HIV, hepatitis B and C viruses (HBV and HCV), and systemic (instead of localized) cytomegalovirus (CMV) or Epstein-Barr virus (EBV) infection.
- Serum is typically the preferred specimen for serology, although plasma or whole blood can be used in some situations. Serology is not the preferred diagnostic test for many viruses that are common among humans. However, it is very useful for viruses that are rare (e.g., Zika virus) or if a positive result always necessitates some follow-up (e.g., HCV).

4. **Respiratory specimens:** Are used to detect and diagnose respiratory pathogens. Specimens include nasal, nasopharyngeal, buccal and throat swabs, as well as bronchoalveolar lavage fluid and sputum specimens.

Table 2.1. Lab findings in CSF associated with different types of pathogens

TYPE OF ORGANISM	GLUCOSE	PROTEIN	WHITE CELLS
Viruses	Normal	Normal to high	Normal to high
Bacteria	Low	High	High, mostly polymorphonuclear cells
Mycobacteria	Low	High	High, mostly lymphocytes
Fungi	Low	High	High

5. **Urine and stool:** Gastrointestinal (e.g., norovirus) and genitourinary viruses (e.g., BK virus) are often excreted in stool and urine specimens at very high titers. However, identification of viruses from these specimens can be difficult to interpret for several reasons.
   - Viruses that cause disease in other areas of the body may be excreted in the urine and/or stool (e.g., enteroviruses causing respiratory disease may be detected in the stool).
   - Urine and stool contain bacterial flora that overgrow in culture, components that are toxic to cell monolayers, and **PCR inhibitors.**
   - Several gastrointestinal or genitourinary pathogens cannot be identified by some methods (e.g., norovirus is difficult to grow in cell culture). On the other hand, these pathogens may be excreted into urine and stool even in healthy individuals (e.g., BK virus).
   - Some viruses may be shed for days to months after resolution of symptoms, so detection from these specimens may not correlate with disease.
6. **Other common specimens:** Includes body fluids (vaginal secretions, semen, saliva, and ocular, joint, and amniotic fluid) and cells from lesions.

III. COLLECTION DEVICES AND TRANSPORT.

1. **Swabs:** These are convenient collection devices, but they gather very little specimen.
   - Traditional swabs have long fibers wrapped around the end of a shaft.
   - Flocked swabs are made of nylon. They are brush-like and increase the surface area of collection. These are preferred for collection.
   - **Wooden-shafted swabs** and swabs with **calcium alginate or cotton** tips should not be used because they can inhibit viral recovery.
2. **Universal viral transport medium:** This is a pH-buffered medium containing antibiotics and antimycotics to inhibit bacterial and fungal overgrowth.
   - Specimens such as swabs, tissue, and scrapings are commonly collected into transport media to preserve the integrity of the sample.
   - Liquid specimens, such as blood and CSF, should not be added to transport media to prevent dilution of the specimen.
3. **Storage conditions:**
   - For transport and short-term storage (<30 days), specimens are typically placed at refrigerated temperature (2 to 8°C).
   - For long-term storage, specimens may be frozen at −70°C, depending on the assay's storage requirements. Nucleic acid extracts can be frozen at −70°C or refrigerated (~4°C).

Table 2.2. Comparison between commonly used viral identification techniques

	CULTURE BASED ASSAYS			IMMUNOLOGY BASED ASSAYS	
PARAMETER	**VIRAL CULTURE**	**SHELL VIAL**	**MOLECULAR, NUCLEIC ACID AMPLIFICATION ASSAYS**	**ANTIBODY DETECTION**	**RAPID ANTIGEN DETECTION**
Turnaround time	**Slow** (2–14 days)	Moderate (2–3 days)	Rapid (~minutes to hours)	Rapid (~1 day)	**Very rapid** (~seconds to minutes)
Target of detection	Changes in cell morphology (cytopathic effect)	Viral proteins produced after replication in cells	Nucleic acid	Host antibodies	Viral proteins
Labor	**High**	Moderate	Low (newer assays) or moderate (older assays)	Low (if automated) or moderate (if manual)	Low
Diagnostic sensitivity	Poor to moderate	Moderate	**High**	Variable (immunosuppressed persons may not mount adequate antibody responses)	Poor (exception: HIV testing)
Diagnostic specificity	Low (cytopathic effects between viruses can overlap)	Moderate. Monoclonal antibodies impart specificity but may cross-react with other proteins. Also, interpretation can be subjective.	High	Moderate (**Crossreactive antibodies** can decrease specificity)	Variable
Breadth of detection	Moderate (5–10 viruses)	Limited (1 per vial)	Extensive (newer assays detect >20 targets) or limited (older assays detect 1–3 targets)	Limited (some cross-reactive antibodies can detect closely related viruses)	Limited
Specimen type	Many	Many	Limited. Different specimen types must be validated separately.	Blood components or CSF	Varies
Cost of reagents	Low	Moderate	High	Low to moderate	Low
Risk of lab-acquired infection	Moderate. Other, more dangerous viruses may grow unexpectedly.	Low to moderate	Low	Low	Low

IV. COMPARISON OF ASSAYS USED FOR DIAGNOSIS OF VIRAL INFECTIONS. Currently, the most common methods of viral detection are antigen or antibody detection, serology, culture, and nucleic acid amplification. These tests have different advantages and limitations, such as differences in sensitivity/specificity, ease of use, and cost. Table 2.2 shows a general comparison of these tests, but they are covered in greater detail in section III of this book

Multiple-Choice Questions

1. Which of the following is true of diagnostic testing for viruses?

a. It is always necessary.

b. Testing can differentiate between viruses that cause similar diseases.

c. Testing does not depend on preanalytic factors.

d. All of the above.

2. A patient with respiratory symptoms is seen in the emergency room. An antigen test for influenza virus is negative but the culture is positive. Which of the following is true?

a. The patient is infected with influenza virus.

b. Serology should be performed for confirmation.

c. PCR should be performed for confirmation.

d. The patient should probably be tested for more viral and bacterial pathogens because of the negative antigen test.

3. Which of the following is most accurate regarding a patient that is shedding virus in their respiratory secretions?

a. The patient will be symptomatic for as long as they are shedding virus.

b. The patient is likely contagious.

c. The patient will not produce antibodies to the virus until the shedding stops.

d. Any patient shedding virus definitely has active disease.

4. Which of the following is a limitation of PCR testing?

a. It cannot be performed on serum.

b. It is inexpensive.

c. It is often falsely negative.

d. A positive result does not always mean a patient has active disease.

5. Which of the following is an advantage of viral culture?

a. It can detect multiple viruses from a single specimen.

b. It is highly sensitive for most specimens.

c. It is easy to perform.

d. All of the above

6. What is the danger of identifying respiratory viruses in stool?

a. Detection in stool may not correlate with the virus causing respiratory disease.

b. False positivity due to the presence of hemin

c. Respiratory viruses cannot be present in stool.

d. All of the above

True or False

7. Blood is the ideal specimen for testing of all viral pathogens. **T F**

8. Viral culture from CSF detects most viruses causing CNS disease. **T F**

9. A positive result from viral culture is indicative of disease (i.e., culture is clinically specific). **T F**

10. Swabs containing calcium alginate should not be used for viral testing. **T F**

SECTION II

VIRAL PATHOGENS AND CLINICAL PRESENTATION

3

CHAPTER 3

RESPIRATORY VIRUSES

I. OVERVIEW. Many viruses can cause respiratory symptoms, from mild cough and cold to severe lower respiratory tract infections.

1. **Background:** Most respiratory infections are caused by RNA viruses. Most of them are also enveloped (Box 3.1).

Most respiratory viruses contain RNA genomes.

2. **Transmission:** Respiratory viruses are highly contagious and are transmitted primarily by respiratory secretions. Some individuals shed viruses asymptomatically for a prolonged period and others (e.g., children) shed large amounts of virus. Common modes of transmission include the following.
   - Inhalation of respiratory droplets (e.g., cough or sneeze)
   - Direct contact with respiratory secretions
   - Contact with contaminated objects (fomites). Viruses can survive for days to weeks on environmental surfaces, especially hard, nonporous surfaces like metal and plastic (2).
   - Many respiratory viruses have a distinct pattern of seasonality (Fig. 3.1). Seasonality is affected by various factors, such as the amount of close contact, temperature, humidity, and precipitation. For example, seasonal influenza outbreaks in temperate climates occur in the winter months, likely because people spend more time indoors (e.g., transmission via close contact). However, in tropical climates, influenza can occur all year-round, or during the rainy season (3).
3. **Clinical presentation:** Respiratory viruses can cause overlapping clinical symptoms. Disease ranges from mild to severe upper and lower respiratory

Box 3.1. Common and/or important respiratory viruses

PRESENCE OF AN ENVELOPE	(–) ssRNA	(+) ssRNA	dsDNA
Enveloped	*Orthomyxoviridae* (influenza virus) *Paramyxoviridae* (parainfluenza virus, RSV, HMPV, measles, mumps, rubella viruses)	*Coronaviridae* (mild coronaviruses, SARS, MERS)	
Nonenveloped		Rhinoviruses	Adenoviruses

Winter	Spring	Summer	Fall
Influenza virus			
RSV			
Coronavirus			
HMPV			
	HPIV-3		HPIV-1, HPIV-2
Rhino, entero, and parechovirus			
Adenovirus			

Figure 3.1. Seasonality of respiratory viruses in temperate regions.

tract symptoms such as nasal discharge, cough, cold, fever, croup, bronchiolitis, pneumonia, and acute respiratory distress syndrome (4)

- Viruses have different incubation periods before time to onset of symptoms (Fig. 3.2).
- The duration of symptoms for most respiratory viruses is typically 7 to 14 days.
- Immunosuppressed patients (e.g., transplant recipients) or patients with underlying pulmonary disorders (e.g., individuals with chronic obstructive pulmonary disease or asthma) have a high risk of prolonged infection, serious respiratory disease, and persistent shedding.

Croup (laryngotracheobronchitis) exhibits a characteristic inspiratory stridor, or "**seal bark cough**."

4. **Diagnostic testing:** In cases where there are underlying risk factors or very severe infection, broad "syndromic" testing can be done to identify multiple pathogens that can cause overlapping symptoms, or narrow testing can be done for agents that are treatable (such as bacterial infection or infection with influenza virus or respiratory syncytial virus [RSV]).
   - No testing: Most uncomplicated respiratory infections are self-limited and testing is not necessary.
   - Nucleic acid amplification tests (NAATs): a commonly used method because it provides rapid results. Some newer PCR assays can be used to detect multiple respiratory pathogens simultaneously.
   - Culture: can detect several respiratory pathogens simultaneously, but it is labor-intensive and has a long turnaround time. This makes diagnosis of acute infections difficult. Shell vials are also labor-intensive but have a shorter turnaround time.
   - Antigen tests: Rapid antigen detection tests and direct fluorescence antigen (DFA) testing have relatively low sensitivity but are sometimes used to rapidly screen for some respiratory pathogens.
   - Serology: not typically used. Respiratory viruses cause acute illness and antibodies form only after ~1 to 2 weeks. Also, these viruses are common in the population, so it is difficult to differentiate between current and past exposures.
   - Specimen: The ideal samples for respiratory pathogens are respiratory secretions. Specimens should be collected during the time of greatest viral shedding (i.e., within 10 days of onset). Specimens include the following.

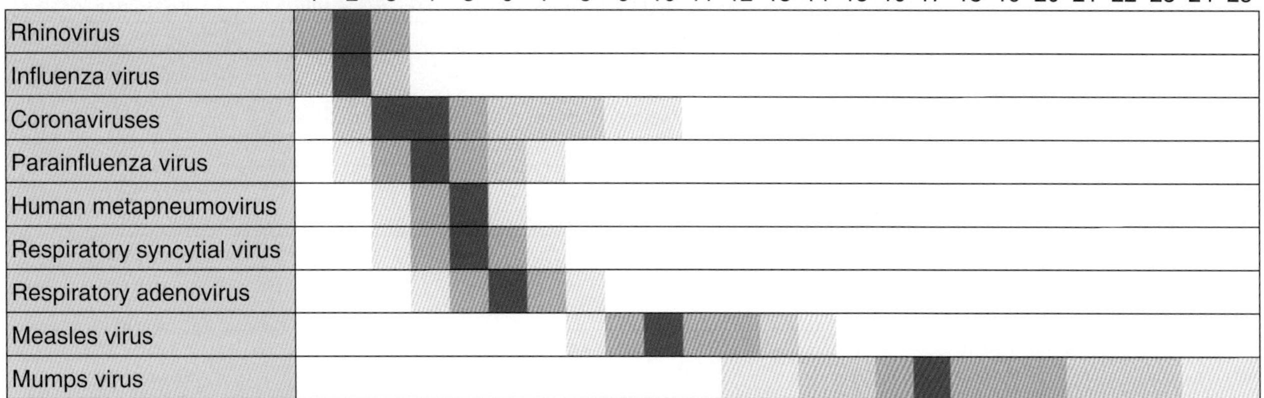

Figure 3.2. Respiratory viruses have different incubation periods (time before symptom onset). Colored squares represent the range; dark orange shows the most frequent incubation period.

- Nasal swabs, nasopharyngeal swabs, aspirates, or washes
- Tracheal aspirates, bronchoalveolar lavage fluid, sputum, throat swabs, and lung tissue

5. Prevention and treatment

- Prevention
  - Handwashing
  - Vaccines are available for some respiratory viruses.
  - Wearing a mask
  - Contact and/or droplet isolation of infected patients
  - Prophylactic treatment is available for RSV.
- Treatment
  - Most respiratory viruses do not need treatment. Serious infections are usually treated with supportive care, although broad antivirals are sometimes used with varying levels of success.
  - Only influenza virus and some RSV infections can be treated with therapeutic antivirals.

Most respiratory viruses are enveloped so handwashing with soap can inactivate them.

II. INFLUENZA VIRUS. Influenza virus is a tremendous burden on human health and productivity. Every year, outbreaks cause significant morbidity and mortality (~20,000 to 40,000 deaths in the United States alone) and increase absenteeism, hospital visits, and hospitalizations.

1. **Background:** Influenza viruses are moderately sized (~100 nm in diameter), round, and enveloped (Fig. 3.3). They are in the family *Orthomyxoviridae*.
   - Their genomes are (–) ssRNA (Baltimore class V) and are divided into 8 **segments**.
   - There are three main genera that infect humans: influenza viruses A, B, and C. These three viruses have different structures, host ranges, and clinical presentations (Table 3.1).
   - Influenza A virus can be categorized into subtypes based on the combination of two major surface proteins, **hemagglutinin** and **neuraminidase**.

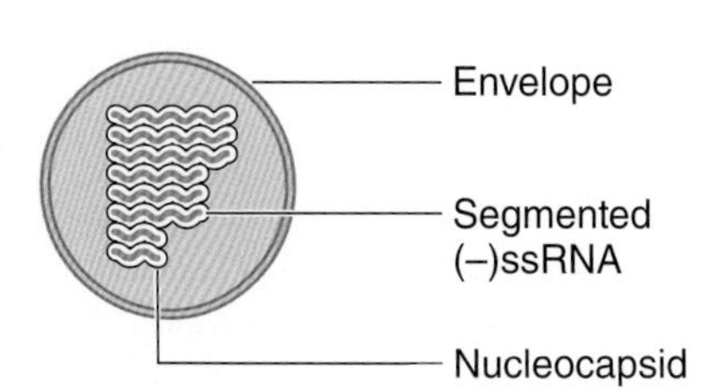

Figure 3.3. Influenza virus.

Table 3.1. Important proteins produced by influenza viruses

PROTEIN FUNCTION	INFLUENZA VIRUS A	INFLUENZA VIRUS B	INFLUENZA VIRUS C
Binds to sialic acid on host cells to initiate cell entry	Hemagglutinin	Hemagglutinin	Hemagglutinin-esterase-fusion
Releases virus particles as they bud from the cell	Neuraminidase	Neuraminidase	
Virus assembly and nuclear transport	Matrix 1		Matrix 1
Channel protein that helps uncoat the virus for replication	Matrix 2	BM2	

- There are 18 known hemagglutinin types. In humans, the prevalence is as follows: H1 > H3 > H2.
- There are 11 known neuraminidase types. In humans, the prevalence is as follows: N1 > N2.

- Strains of influenza A and influenza B viruses are named according to an influenza virus-specific naming convention (5):

  Genus/animal host, if applicable/geographical origin/strain number/year of isolation (hemagglutinin/neuraminidase type for influenza A)

  - Example 1, influenza A virus from duck: A/duck/Alberta/35/76 (H1N1)
  - Example 2, influenza A virus from human: A/Perth/16/2009 (H3N2)
  - Example 3, influenza B virus: B/Yamagata/16/88.

Influenza virus has an extended naming convention that is different from other viruses.

2. **Transmission:** Influenza virus is transmitted from human to human or animal to animal through respiratory droplets and secretions.
   - Virus starts shedding into the respiratory tract within 24 hours of exposure. As a result, patients are infectious even before symptoms become apparent.
   - Influenza A and B viruses are distinctly seasonal in temperate areas, causing yearly outbreaks in the winter months. (Fig. 3.1). Influenza C is not seasonal.
   - **Antigenic drift:** Seasonal infections are the result of gradual strain variations called antigenic drift. Influenza virus genomes can change by slowly accumulating mutations as they replicate (i.e., variations in the genome "drift" over time). Over time a variant is generated that looks sufficiently different to the immune system from previous strains. This process results in new, seasonal strains of influenza virus each year (Fig. 3.4).
   - **Antigenic shift:** influenza virus genomes can "shift" (i.e., change abruptly and substantially) by recombining with other strains so that they look completely novel to the human immune system. These new strains can cause significantly higher morbidity and mortality than slowly evolving strains (Fig. 3.4) due to the lack of immunity in the population. Pandemics are usually caused by antigenic shift (Table. 3.2).
     - Antigenic shift occurs only with **segmented** viruses. This is because segments from different strains can mix and match (**reassort**) when new virions are being assembled. This creates a completely novel strain of virus.
     - The risk of a completely novel virus is highest when human strains reassort with strains that normally infect a different species. This can

Influenza virus is distinctly seasonal and infection occurs in the winter.

Live poultry markets are a risk factor for interspecies influenza genome reassortment.

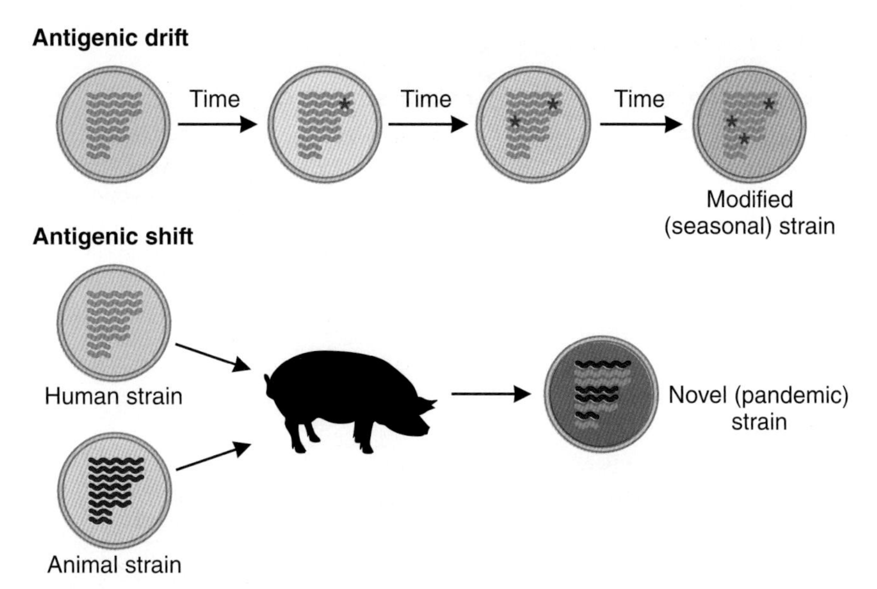

Figure 3.4. The difference between antigenic shift and antigenic drift. In antigenic drift (top), accumulation of mutations creates a novel virus strain over time. In antigenic shift (bottom), reassortment of viral genome segments suddenly results in a novel strain.

happen if an intermediate host (such as pigs or chickens) is infected by two different viral species. The risk of interspecies infection increases in areas of close animal and human contact.

- Influenza A virus is more likely to produce highly novel, pandemic viral strains because it can infect a wide range of species (humans, birds, swine, whales, horses, seals, and cats). Influenza B and C viruses are primarily restricted to humans (although influenza B virus can also infect seals and horses), so reassortment between unrelated virus strains is rare.
- Pandemic strains induce immunity in the surviving population, so subsequent outbreaks in an exposed region are much milder. In fact, after a pandemic the viral strain will often start circulating as a seasonal strain.

Pandemics are usually caused by antigenic shift in influenza A virus.

- Direct zoonotic transmission (i.e., without reassortment) from an animal to a human is extremely rare. This is because influenza virus strains usually bind to slightly different sites. For example, avian influenza virus binds to sialic acid that is linked to galactose with an **alpha (2,3) linkage**. Human influenza virus binds to sialic acid linked to galactose with an **alpha (2,6) linkage**.

3. Clinical presentation

- Incubation period: ~2 (range, 1 to 4) days
- Symptoms last for ~1 to 2 weeks. They are usually most severe in elderly and very young patients (>65 and <2 years old).

FACT: fever, aches, chills, tiredness

- Influenza A and B viruses cause **fever** (temperatures can be very high, up to 106°F) and chills lasting for 3 to 4 days, headache, malaise, **myalgia**, dry cough, severe **pharyngitis**, and nasal discharge. Some patients, especially children, can have vomiting or diarrhea. Influenza C virus

Table 3.2. Notable epidemics/pandemics caused by influenza A virus (6–8)

PANDEMIC	YEAR	SUBTYPE	SOURCE OF SPREAD	APPROXIMATE DEATHS WORLDWIDE	NOTES
Spanish flu	**1918**	**H1N1**	Humans (avian strain adapted to human spread)	>40 million	Highly virulent strain. Unlike most epidemics, **mortality was higher among young, healthy people** (20–50 years old) than among people at the extremes of age. It was often associated with bacterial superinfection.
Asian flu	1957	H2N2	Humans (reassortment between avian and human strains)	>4 million	Primarily affected people with underlying lung or heart disease
Hong Kong flu	1968	H3N2	Humans (reassortment between avian and human strains)	>1 million	High morbidity but low mortality compared to the previous viral pandemics
Avian flu	1997	H5N1	Birds. Limited human-to-human spread.	<10	Did not spread to the same degree as the other pandemic strains. However, it was unique because it could spread directly from birds to humans.
Swine flu	2009	H1N1	Humans (swine origin, but exposure to pigs was not a risk factor)	>500,000	Notable because it caused disease primarily in young, healthy persons (<25 years old)
Avian flu	2013–2016	H7N9	Birds (live poultry markets). Limited human-to-human spread.	~300	High mortality rate. It can cause severe pneumonia and death. It is currently restricted to China.

causes mild respiratory symptoms. Seropositivity studies show that most people have been exposed to it during childhood.

- **Viral pneumonia:** Influenza A or B virus infection may progress from classic to severe infection with significant morbidity and high mortality.
  - This is associated with underlying cardiovascular disease or pregnancy
  - This was also the cause of death for many young, healthy patients infected during the 1918 outbreak
- **Secondary bacterial pneumonia:** A bacterial superinfection is when bacterial disease occurs in addition to the viral infection. This can significantly increase mortality rate.
  - Bacterial pathogens commonly involved include *Streptococcus pneumoniae*, *Staphylococcus aureus*, and *Haemophilus influenzae*.
  - Associated with people ≥65 years of age and people with underlying (e.g., pulmonary) disease
  - Patients with bacterial superfection can be treated with antibiotics.
- **Reye's syndrome:** a rare condition resulting from influenza virus or varicella-zoster virus (VZV) infection with aspirin use. It affects the brain and liver and can cause mental status changes such as lethargy, seizures, delirium, and death. It occurs primarily in children and adolescents.

- Other: exacerbation of underlying pulmonary or cardiac conditions such as chronic pulmonary disease, asthma, cystic fibrosis, myositis, myocarditis, and pericarditis. Rare: neurologic complications.

4. **Diagnostic testing:** clinically useful because a confirmed diagnosis of influenza can affect management of the patient.
   - Antigen assays: These tests are simple to perform, can differentiate between influenza A and B viruses (depending on the test), and produce a rapid positive or negative result (<30 minutes). These rapid influenza diagnostic tests (RIDTs) can be performed at the point of care (9).
     - Important: The sensitivity of antigen-based RIDTs is very limited. As a result, negative results have to be confirmed by another laboratory method to ensure that cases are not missed. Specificity is moderate to good (ranges from 85 to 100%).
     - When the prevalence is high (i.e., when the patient has clear clinical symptoms in the middle of an outbreak), the positive predictive values of RIDTs are acceptable.
     - Note that older tests have sensitivity as low as 40 to 80%. In January 2018, the Food and Drug Administration (FDA) improved the standards for antigen-based RIDTs so that they must maintain sensitivities of at least 80% compared to NAATs.
   - NAAT: PCR and other NAATs are the most sensitive and specific method for detection of influenza virus, and they are relatively rapid. Some assays also detect and differentiate between several influenza A virus subtypes and influenza B virus.
     - A few molecular tests are available as point-of-care (≤15-minute) tests.
     - Most tests take 1 to 2 hours and are performed within the laboratory. When accounting for pre-analytic factors, like transportation of the specimen, these results are still available within 24 hours.
     - The sensitivity of NAAT-based assays may need to be checked against new influenza virus strains. Influenza virus mutates regularly, and mutations within the target sequence can cause false-negative results.
   - Culture/shell vials: this method of testing for influenza is not preferred because it is too slow.
     - In clinical labs cytopathic effect (CPE) for influenza virus usually occurs in ~2 to 5 days (*see* chapter 13).
     - CPE is not always present or distinctive, so **hemadsorption** can be done for confirmation. Hemadsorption of guinea pig red blood cells to rhesus monkey kidney (RMK) cells should be equally strong at 4°C and 20°C.

RIDTs have generally had poor sensitivity and should be used only when the prevalence is high.

Shedding continues only up to a week after symptoms begin. Specimens should ideally be collected in the first 4 days or tests may be falsely negative.

5. **Prevention and treatment**
   - Prevention:
     - Annual vaccination is recommended for everyone >6 months old.
     - Most vaccine strains are grown in eggs, so egg shortages can affect vaccine availability. Cell-based and recombinant subunit vaccine strains are also available.

Seasonal vaccination is recommended annually due to antigenic drift.

- Vaccine formulations: Vaccines in the United States contain strains that are predicted to circulate in the upcoming season and are often based on strains currently circulating in other regions of the world. Because of this, they may not cover the strain that actually circulates during the flu season (*see* Table 18.1).
  - **Trivalent vaccine:** two strains of influenza A virus and one strain of influenza B virus
    - **Inactivated virus**
    - Generally intramuscular
    - Standard dose: for use in most people
    - High dose: for people ≥65 years old
    - Recombinant subunit vaccine: for people ≥18 years old with egg allergy
  - **Quadrivalent vaccine:** two strains of influenza A virus and two strains of influenza B virus. Several formulations are produced and are administered by the following routes.
    - Intranasal spray
      - **Live attenuated** virus
      - Not given to pregnant or immunocompromised people. Is used for 2- to 49-year-olds.
    - Intramuscular injection
    - Intradermal injection: for 18- to 64-year-olds

- Treatment: antivirals are available for influenza (*see* Table 19.4).
  - **Oseltamivir** is the most commonly administered anti-influenza antiviral. It is most efficacious when administered within 48 hours of symptom onset. Treatment is recommended for severe cases (hospitalized patients or patients with complications).
  - Other anti-influenza antivirals are zanamivir, peramivir, and baloxavir marboxil.

Oseltamivir should be administered within 48 hours of symptom onset.

III. RESPIRATORY SYNCYTIAL VIRUS. RSV is a common virus, and most people are infected before they are 2 years old. The initial exposure is the major cause of severe lower respiratory tract infection in infants. However, immunity is not long-lasting, and humans are infected multiple times during their lifetime.

1. **Background:** enveloped, irregular shaped (–) ssRNA virus (Baltimore class V). It is in the family *Paramyxoviridae* (Fig. 3.5).
   - **F protein** is involved in virus binding, or fusing with, target cells. It also causes infected cells to fuse together (called syncytia). Syncytia allow the virus to pass from cell to cell without being exposed to the immune system.
   - There are two main groups: A and B. These can be categorized into subtypes, which are indicated by number. Both subgroups circulate simultaneously, but RSV A is often more prevalent (2:1 ratio) and may cause more severe disease (10).
   - Temperature-labile virus
2. **Transmission:** spreads easily via contact with respiratory secretions.
   - Primary exposure usually occurs before the age of 2.
   - RSV does not induce a memory response, so reinfections are common.

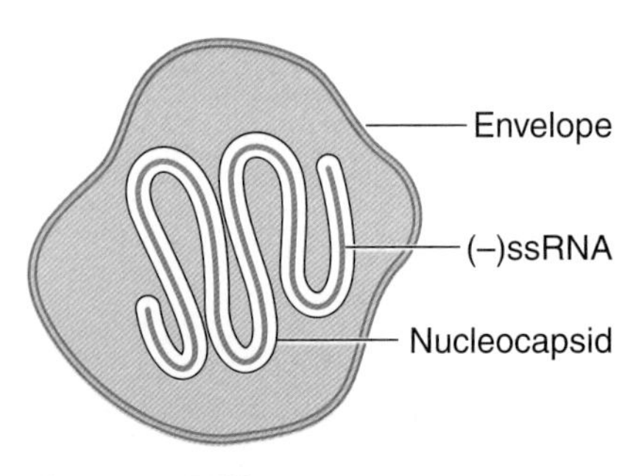

Figure 3.5. RSV.

- Important transmission risks
  - Nosocomial spread, especially among pediatric and elderly patients
  - Elderly people in long-term care facilities
- Shedding
  - From infants: up to 4 weeks
  - From older individuals: a few days
  - From immunocompromised hosts: up to several months

3. Clinical presentation

- Incubation period: ~2 to 6 days
- Infants: most people are exposed to RSV before the age of 2. Disease is usually mild and presents like the common cold. However in some infants it may cause severe lower respiratory tract disease that can require hospitalization.
  - Symptoms include difficulty breathing, **croup**, bronchitis, **bronchiolitis**, **pneumonia**, and **hypoxemia**.
  - Risks of severe disease include the following.
    - Age <2 years old, especially <6 months old
    - Prematurity
    - Lung, heart, and neuromuscular conditions
- Children and adults: Reinfections with RSV are common throughout life. They are generally milder than the primary infection and present in the upper respiratory tract. Symptoms are **common cold**-like, such as cough, nasal congestion, wheezing, fatigue, and low-grade fever.
- Other risk factors that can cause severe disease (such as shortness of breath and pneumonia) include the following.
  - Advanced age and association with long-term health care facilities
  - Underlying pulmonary disease, such as asthma and COPD
  - Chronic heart disease
  - Exposure to tobacco smoke

People can be reinfected multiple times with RSV.

RSV is the major cause of bronchiolitis and pneumonia in children **under 2 years** old.

4. Diagnosis

- Rapid direct antigen testing: similar to rapid influenza testing, a rapid diagnosis of RSV may affect management of the patient and prevent nosocomial spread.
  - Can be performed in <30 minutes
  - These tests have good specificity (>90%) but may have low sensitivity (50 to 90%), so patients with negative results must get additional testing.
- NAAT: the most sensitive and specific method
- Cell culture: RSV grows well in multiple cell lines. It produces characteristic CPE, with refractile cells and syncytia. Syncytia form when the growth medium contains calcium and glutamine.
- Shell vials: decreases culture-based turnaround time.

Respiratory **syncytial** virus produces **syncytia** (individual cells that fuse together and form a bigger, multinucleated cell). Other paramyxoviruses can also produce syncytia.

5. Prevention and treatment

- Prevention
  - Prophylaxis can be given for high-risk infants/patients (e.g., preterm infants)

There is no vaccine for RSV because protective immunity is not long-lasting.

- Intravenous immunoglobulin
- **Palivizumab:** a monoclonal antibody against the viral F protein. It is administered intravenously (*see* chapter 19).

- Treatment
  - Supportive care (oxygenation, fluids, nasal suctioning, etc.)
  - Corticosteroids and bronchodilators are sometimes used but have uncertain benefits.
  - Aerosolized and oral **ribavirin** is not recommended for serious lower respiratory tract infection (*see* chapter 19) in children because of unclear efficacy, expense, and risk to health care workers. However, it can be useful in treating immunocompromised adults with RSV infections.

IV. HUMAN PARAINFLUENZA VIRUS. Human parainfluenza virus (HPIV) is a common cause of upper and lower respiratory tract symptoms, like **croup**, in infants and children <5 years old.

Unlike influenza viruses, *Paramyxoviridae* are not segmented.

1. **Background:** enveloped, (–) ssRNA virus (Baltimore class V). It is in the family *Paramyxoviridae*. It is variable in size and shape (Fig. 3.6). There are 4 main types, HPIV-1 to HPIV-4.
   - Like influenza virus, HPIV also binds sialic acid.
   - HPIV has a combined hemagglutinin-neuraminidase protein on its surface. Like influenza hemagglutinin, it can cause red blood cells to hemagglutinate.
2. **Transmission:** spreads easily via **contact** with respiratory secretions
   - There are four HPIV types that have a seasonal pattern of infection.
     - **Fall:** HPIV-1, HPIV-2, often in alternating years (HPIV-1 in odd-numbered years and HPIV-2 in even-numbered years)
     - **Spring:** HPIV-3, although it can also occur all year-round. This strain is associated with a more severe infection.
     - Seasonality for HPIV-4 is not yet defined. It is traditionally not easily recognized because it does not grow in culture.
   - May be spread nosocomially (especially HPIV-3)
3. **Clinical presentation**
   - Human parainfluenza viruses typically cause a self-limited respiratory infection (fever, cough, pharyngitis, and rhinitis) that is milder than influenza. They may also cause otitis media.
   - May cause severe lower respiratory tract disease that is similar to severe RSV infection.
     - Bronchitis, bronchiolitis, and pneumonia
     - Causes croup in infants/children
   - Reinfections are common in children and adults. These are usually mild, cold-like upper respiratory infections.
   - Infants, immunosuppressed, or elderly individuals are at greater risk of severe disease.
4. **Diagnosis**
   - Human parainfluenza viruses are thermolabile, so specimens should be processed quickly

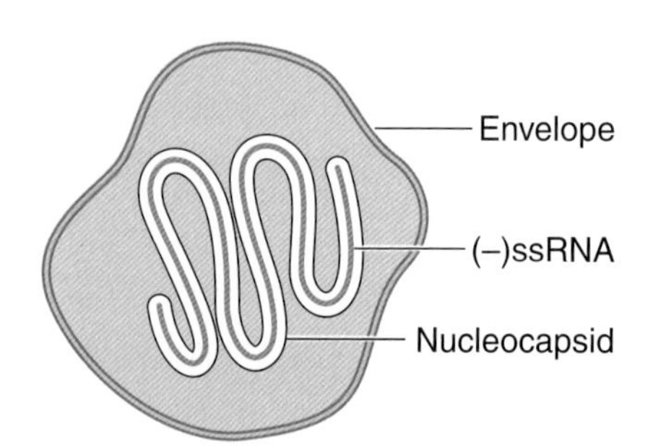

Figure 3.6. Parainfluenza virus.

- NAAT (usually PCR): Readily detects all four HPIV types with high sensitivity and specificity.
- Culture
  - HPIV-1, -2, and -3 grow well in primary cell culture lines (e.g., RMK). HPIV-4 does not. They may produce syncytia like RSV and other paramyxoviruses.
  - Parainfluenza viruses show a positive **hemadsorption** reaction, with greater binding to guinea pig red blood cells at 4°C.
- DFA: This test method is not as sensitive as other assays. If used, then testing from swabs of nose or throat should be performed within the first few days of symptoms when virus shedding is greatest.
- Serology: not useful for acute infection because exposure is common. However, recent infection may be detected by observing a 4-fold rise in IgG titer between acute- and convalescent-phase sera.

5. **Prevention and treatment**
   - There is no vaccine for human parainfluenza viruses.
   - The antiviral ribavirin has been used in severe infections. Some studies show benefit while others do not.

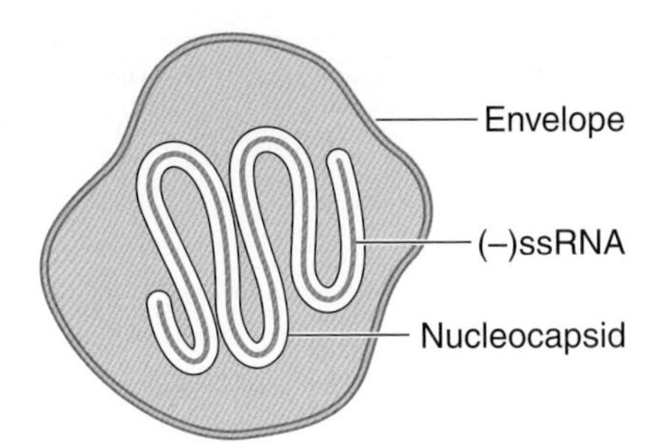

Figure 3.7. HMPV.

V. HUMAN METAPNEUMOVIRUS. Human metapneumovirus (HMPV) is an extremely common virus, with almost all people exposed during childhood (<10 years).

1. **Background:** enveloped, (–) ssRNA virus (Baltimore class V) in the family *Paramyxoviridae*. It is highly variable in size and shape (Fig. 3.7).
2. **Transmission:** spreads easily via contact with respiratory secretions
3. **Clinical presentation**
   - Like RSV and HPIV, HMPV causes upper respiratory tract symptoms such as fever, cough, pharyngitis, rhinitis, and wheezing. Reinfections are common.
   - Severe lower respiratory infections occur mainly in infants and children but also immunosuppressed and elderly patients.
   - HMPV is an important cause of **bronchiolitis.** It can also cause croup, bronchitis, and pneumonia.
4. **Diagnosis**
   - Culture: HMPV can grow in cell culture but is slow and difficult to detect because it does not produce CPE on typically used cell lines.
   - NAAT: The preferred test method. It should be designed to detect the 4 main lineages of HMPV.
5. **Prevention and treatment**
   - There are no vaccines or approved treatments.
   - Supportive care

VI. RHINOVIRUSES. Rhinoviruses are a species of *Enterovirus* that replicate in the nose ("rhino"). They are the most common cause of colds and are a huge burden on human health. However, infections are not usually severe and have a very low mortality rate.

Picornaviruses = tiny (pico) RNA (rna) viruses

Rhinoviruses are a subset of enteroviruses. Rhinoviruses are acid labile while other enteroviruses are acid stable.

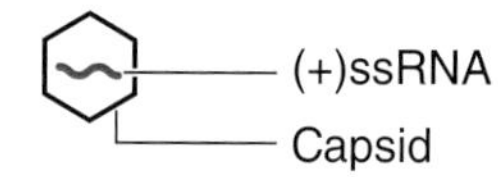

Figure 3.8. **Rhinovirus.**

1. **Background:** very small (~25-nm), nonenveloped, (+) ssRNA virus (Fig. 3.8).
    - Family *Picornaviridae*, genus *Enterovirus*, species *Rhinovirus A*, *Rhinovirus B*, and *Rhinovirus C*. There are more than 100 serotypes of rhinoviruses within the 3 species.
    - There is very little cross-protective immunity between the different serotypes, so reinfections with other strains are common.
    - Rhinoviruses are extremely similar in structure and genetic sequence to non-rhinoviral enteroviruses and are difficult to differentiate from them even with commonly used NAAT assays. However, rhinoviruses are susceptible to low pH (pH <6), so they do not infect the gastrointestinal tract. Non-rhinoviral enteroviruses are stable at low pH, can infect the gastrointestinal tract, and can be transmitted via the fecal-oral route (*see* chapter 6).
2. **Transmission:** Rhinoviruses are easily transmissible through transfer of nasal secretions to hands, eyes, and nasal mucosa. Transmission through droplets and aerosols can also occur (11).
3. **Clinical presentation**
    - Incubation period: ~2 to 3 days
    - Common cause of colds
    - Asymptomatic infection can occur, but upper respiratory tract symptoms are the most common.
        - **Profuse nasal discharge**, nasal congestion, sneezing, mild headache, pharyngitis, cough. Fever is uncommon.
        - Symptoms last for 10 to 14 days.
        - Shedding occurs during active symptoms (i.e., highest shedding during the first few days of symptoms).
    - Otitis media and sinusitis
    - Serious lower respiratory tract symptoms are less common and are usually associated with immune compromise, underlying pulmonary conditions, and young or elderly individuals.
4. **Diagnosis:** Clinically, rhinoviruses do not look different from other viruses that cause colds. Nevertheless, diagnostic testing is not necessary because the infection is primarily mild and self-limited.
    - Specimen: respiratory tract specimens such as nasal, nasopharyngeal and throat swabs, bronchial washes, and bronchoalveolar lavage specimens
    - Culture: Rhinoviruses A and B (but not C) grow well in cell culture.
        - Good, moderately rapid growth (2 to 6 days) on MRC-5 cells
        - Optimum growth temperature: 33 to 34°C
        - Typical CPE is **rounded, refractile cells**. CPE is difficult to differentiate from other enteroviruses.
        - **Acid lability** assay: may help with differentiation of rhinovirus and enterovirus species. In this test, the unknown virus is added to an acidic buffer (e.g., citrate buffer) and then incubated at several dilutions on a cellular monolayer. Rhinoviruses are inactivated or reduced in titer and show reduced CPE.
    - NAAT: may not differentiate between rhinovirus and enterovirus species, since they are so similar

5. Prevention and treatment:

- Supportive care
- Treatment is generally not needed.
- Combination of antihistamines and nonsteroidal anti-inflammatory drugs may be used.

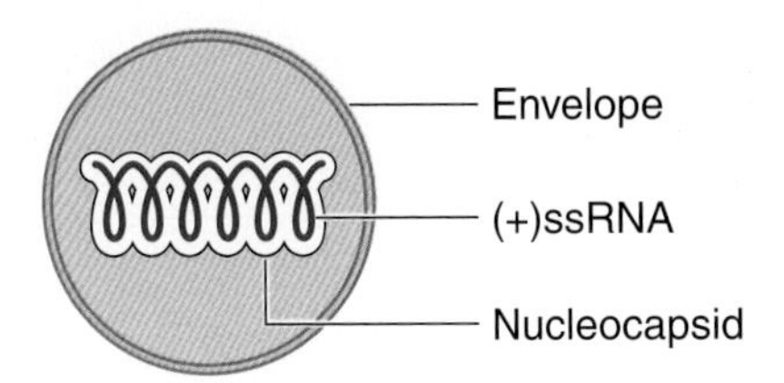

Figure 3.9. Coronavirus.

VII. CORONAVIRUSES. Most human coronavirus (CoV) infections are mild and caused by common species that do not require diagnostic testing. However, two recently discovered species, severe acute respiratory syndrome (**SARS-CoV**) and Middle East respiratory syndrome (**MERS-CoV**) coronaviruses, cause severe disease. These strains are rare and typically restricted to certain geographic areas.

1. **Background:** Members of the *Coronaviridae* family are moderately sized, enveloped, (+) ssRNA viruses (Fig. 3.9). There are 6 known species that infect humans (Table 3.3):
   - Common species: OC43, HKU1, 229E, and NL63
   - Rare species: SARS-CoV and MERS-CoV
2. **Transmission**
   - Respiratory droplets
   - Fomites contaminated with respiratory droplets or other secretions
   - Common coronaviruses are seasonal, with peaks in **winter**
   - Additional risk factors
     - Close, prolonged contact with infected persons
     - MERS-CoV: contact with **camels**, travel to the **Middle East**
     - SARS-CoV: travel to **Asia**, health care workers in a high-risk area or outbreak setting

Table 3.3. Comparison of coronaviruses that infect humans

PARAMETER	OC43, HKU1, 229E, NL63	SARS-COV	MERS-COV
Prevalence	**Common**	**Rare**. It has been detected only in one global outbreak from 2003–2004 and has not been detected since.	**Rare**
Risk factors	Highly prevalent, especially in children	• Travel to **China**, Hong Kong, or Taiwan • Occupational exposure (lab worker)	• Travel (or close contact with someone who traveled) to Saudi Arabia or other **Middle Eastern countries** • Exposure to **camels**
Symptoms	Common cold	Severe acute respiratory syndrome	Severe acute respiratory syndrome
Mortality rate	Very low; disease is generally mild	~10%; disease is mild in children; higher risk of death in older persons	~35%; usually more severe disease in patients with underlying medical conditions

3. **Clinical presentation**
   - OC43, HKU1, 229E, and NL63
     - Common cold-like symptoms (rhinorrhea, pharyngitis, cough, and fever)
     - Usually mild and self-limited
     - May also cause more severe lower respiratory tract infection (bronchitis, croup, bronchiolitis, and pneumonia)
     - Reinfections can occur.
   - SARS and MERS
     - Incubation period: ~5 days
     - High fever, headache, myalgia, shortness of breath, and dry cough
     - Gastrointestinal symptoms (diarrhea, vomiting, and nausea) are common.
     - Rhinorrhea and pharyngitis are uncommon.
     - After 7 days of symptoms: **lymphopenia, leukopenia**, and **pneumonia** (or acute respiratory distress syndrome)
4. **Diagnosis:** OC43, HKU1, 229E, and NL63 are mild viruses and do not need identification. SARS-CoV and MERS-CoV are rare viruses, so clinical suspicion and testing should be based on patient travel to outbreak areas and diagnosis of pneumonia without other potential causes (especially if it occurs in clusters). Testing for these is available at public health laboratories and should be limited to these settings because of low prevalence.
   - Specimen
     - Respiratory specimens are best for common coronaviruses such as nasopharyngeal swab, tracheal aspirate, and bronchial aspirate/lavage samples.
     - Respiratory specimens, other body fluids, and stool can be used for SARS and MERS viruses.
   - NAAT: the preferred test method
     - Samples can remain PCR-positive for weeks to months due to prolonged viral shedding.
     - For SARS, PCR has low sensitivity and negative results do not rule out infection. Positive results should also be confirmed. Because of low prevalence, positive results may be due to false positivity (12).
   - Serology: useful for identifying exposure to SARS and MERS viruses. Enzyme immunoassay (EIA) is performed on paired acute- and convalescent-phase blood/serum samples.
   - Culture: not performed in diagnostic laboratories
     - Most coronaviruses do not grow well in routine cell culture lines.
     - Culture should not be performed on specimens suspicious for SARS and MERS in order to protect lab workers from accidental exposure. These cultures must be performed in a biosafety level 3 (BSL3) lab.
5. **Prevention and treatment:** no vaccines or approved treatments

VIII. MUMPS. Mumps virus is transmitted like other respiratory viruses, but instead of causing a flu- or cold-like syndrome, mumps virus produces parotitis.

1. **Background:** Mumps virus is an enveloped, (–) ssRNA, helical virus in the family *Paramyxoviridae* (Fig. 3.10). There are 12 genotypes (genotype G is the most common in the United States). Sequencing to determine genotype can be useful in outbreak investigations.

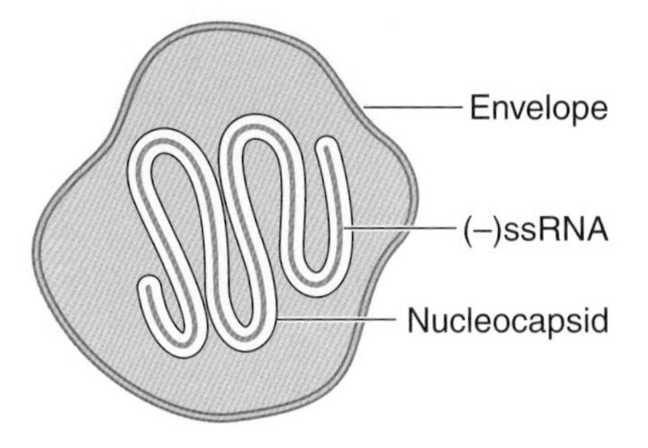

Figure 3.10. Mumps virus.

2. **Transmission**
   - Respiratory droplets (cough or sneeze)
   - Contaminated fomites (shared cups or surfaces)
   - Can be transmitted several days before and after symptoms start to show
3. **Clinical presentation**
   - Often mild or asymptomatic; otherwise fever, headache, and malaise
   - **Parotitis** is a characteristic feature (parotid salivary glands are swollen and painful on one or both sides). Swelling obscures the jawline (Fig 3.11).
   - Complications: permanent **deafness**, **encephalitis**/meningitis, pancreatitis, **orchitis** (swelling of the testicles), myocarditis
   - Death is rare.
4. **Diagnosis**
   - Specimen: oral or **buccal swabs** for PCR or culture. Urine may also be collected but may have reduced sensitivity.
   - Serology: commonly used. Use paired acute- and convalescent-phase sera to identify an IgM response or a 4-fold rise in IgG.
   - NAAT: highly sensitive and specific when specimens are taken early
   - Culture
     - Mumps virus grows moderately fast on RMK cells.
     - Can form syncytia
     - Will show a positive **hemadsorption** reaction, with better binding to guinea pig red blood cells at 20°C

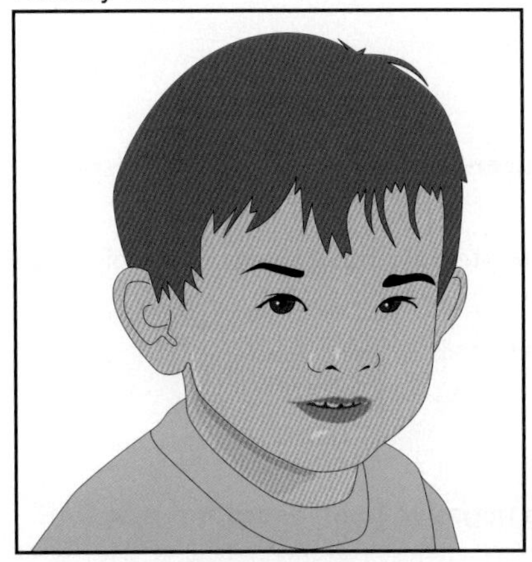

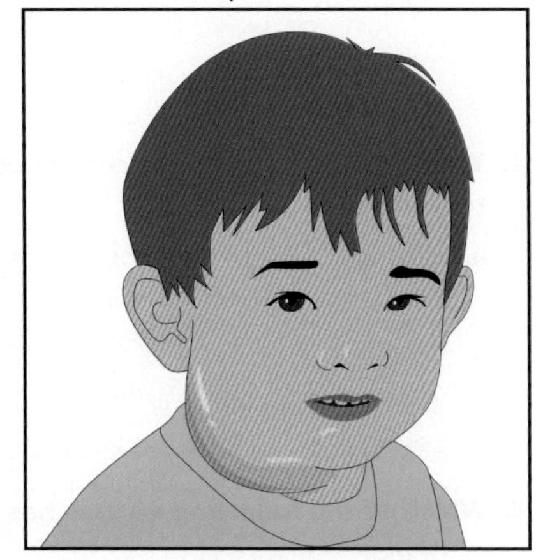

Figure 3.11. Mumps parotitis. Painful swelling occurs in the parotid gland that obscures the jawline.

5. **Prevention and treatment**
   - Prevention:
     - The vaccine is part of the measles-mumps-rubella (**MMR**) vaccine, which is part of the routine childhood vaccination schedule.
     - Mumps vaccine is effective, but even two full doses may not provide complete protection. Vaccine boosters are often administered during outbreaks.
     - Mumps is now **uncommon** in regions with high vaccination rates. Travel to countries that do not vaccinate increases risk of infection.
     - High rates of vaccination in a population provides **herd immunity** and helps prevent outbreaks (*see* chapter 18).
   - Treatment: Supportive; cold or warm compresses for swollen parotid glands or testes

Multiple-Choice Questions

1. **Which of the following viruses is NOT in the *Paramyxoviridae* family?**
   a. Influenza virus
   b. Mumps virus
   c. Parainfluenza virus
   d. RSV

2. **Which of the following viruses is NOT cultured in diagnostic labs?**
   a. Coronaviruses
   b. Influenza virus
   c. Mumps virus
   d. Rhinoviruses

3. **Which of the following respiratory viruses is naked (nonenveloped)?**
   a. Influenza virus
   b. RSV
   c. HPIV
   d. HMPV
   e. Rhinovirus
   f. Coronavirus
   g. Mumps

4. **Which of the following is correct concerning the seasonality of respiratory viruses?**
   a. Most respiratory viruses peak in the winter because they are more infectious when the environment is cold.
   b. Influenza can occur only in the winter.
   c. Viruses can also be seasonal in nontemperate climates.
   d. All respiratory viruses are seasonal.

5. **Which of the following viruses has a negative hemadsorption reaction?**
   a. HMPV
   b. Influenza virus

c. Parainfluenza virus

d. Mumps virus

6. Which group is most commonly at risk for severe RSV infection?

a. HIV patients with CD4 counts of <200 cells/µl

b. Children less than 2 years old

c. Transplant patients

d. Females

7. Despite its rapid turnaround time, rapid antigen testing is not the preferred method of testing for influenza, especially outside the flu season. What is the most important reason for this?

a. It detects only influenza virus and does not identify other causes of infection.

b. It has relatively poor sensitivity.

c. It is expensive because it is rapid and simple to perform.

d. Traditional PCR usually produces even faster results.

8. Which of the following viruses can be specifically treated with antivirals?

a. Influenza virus

b. Parainfluenza virus

c. Coronaviruses

d. Rhinovirus

9. Antigenic shift occurs only

a. In segmented viruses

b. If enough mutations have accumulated over time

c. If neuraminidase is reassorted

d. In winter

10. Which of the following viruses is characteristically associated with parotitis?

a. Rhinovirus

b. Adenovirus

c. Coronavirus

d. Mumps virus

True or False

11. A novel respiratory virus is discovered. It is more likely to contain RNA. **T F**

12. Most respiratory viruses require airborne isolation precautions. *(May require reading outside of this chapter.)* **T F**

13. Respiratory viruses can be transmitted by contact. **T F**

14. Herd immunity is the term used to describe vaccination in animals in order to eliminate the viral reservoir. *(May require reading outside of this chapter.)* **T F**

15. MERS can be spread from person to person. **T F**

4

CHAPTER 4

VIRUSES WITH DERMATOLOGIC MANIFESTATIONS

I. OVERVIEW. Many viral infections result in rashes **(exanthems)**. Viruses covered in this chapter produce characteristic dermatologic manifestations. Often, these rashes can aid in diagnosis of the infection (*see* Fig. 4.11). Viral syndromes that include non-specific rashes (e.g., arboviral fevers) are covered elsewhere.

II. HERPES SIMPLEX VIRUSES 1 AND 2. Herpes simplex viruses (HSV) are highly prevalent in the population, especially HSV-1. They are transmitted through contact with mucosal secretions and cause a lifelong latent infection in neurons. Infections are usually asymptomatic but may present as oral or genital lesions that recur periodically. However, in immunosuppressed individuals or neonates, HSV can cause severe central nervous system or disseminated disease.

1. **Background**
   - HSV-1 and -2 are large, enveloped, dsDNA viruses (Fig. 4.1).
   - They are members of the *Herpesviridae* family and *Alphaherpesvirinae* subfamily (Box 4.1, Table 4.1).
   - HSV-1 and -2 have only 50% sequence homology, but they infect similar cell types and cause similar symptoms (Table 4.2).
2. **Transmission:** HSV-1 and -2 infections typically occur from exposure to mucosal secretions. Contact with these secretions can occur via the following routes.
   - Direct contact (oral-oral, genital-genital, or oral-genital)
   - Vertical transmission. This occurs most commonly during vaginal delivery,

Genital herpes is not transmitted by fomites.

Box 4.1. Features of viruses in the *Herpesviridae* family

- Herpesviruses cause a lifelong, latent infection with asymptomatic or symptomatic reactivation.
- Virions have a tegument (extra space in between the envelope and the capsid that is filled with viral proteins that are used by the virus during replication and growth).
- Symptoms usually involve some dermatologic manifestation. Importantly, these viruses can also cause a wide range of other manifestations.
- Several species are transmitted vertically, from mother to child, and cause congenital disease ("**TORCH**": toxoplasmosis, **o**ther [syphilis, parvovirus, and VZV] rubella, **C**MV infection, and **H**SV infection).

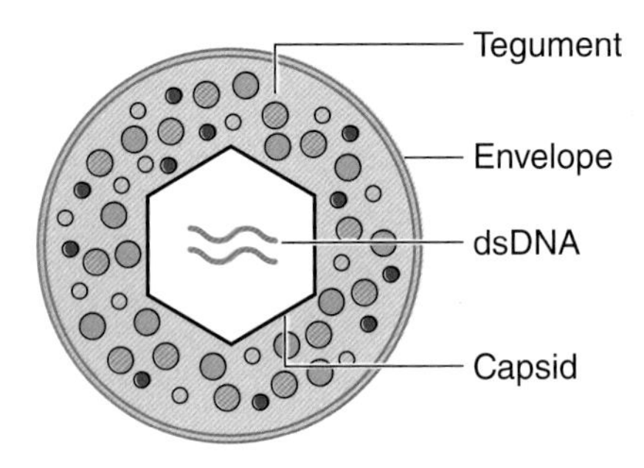

Figure 4.1. **HSV.**

especially when the mother has primary active lesions rather than recurrent lesions or asymptomatic infection. Caesarean delivery may reduce the risk of transmission.

- Contact with secretions: Viruses are shed in large amounts in the **saliva**, active skin/genital/oral **lesions**, and tears. Shedding is highest during symptomatic periods, but HSV can be transmitted during asymptomatic shedding.
- HSV is not acquired from the environment.
  - Genital herpes is NOT acquired through contaminated surfaces (sheets, towels, etc.).
  - The risk of contracting skin or oral herpes is low because HSV does not survive for long in the environment (only a few hours). However, fomites

Table 4.1. **Important characteristics of herpesviruses**

VIRUS ABBREVIATION	VIRUS NAME(S)	OTHER ABBREVIATION	PREVALENCE AMONG HUMANS	NEONATAL DISEASE	VACCINE AVAILABLE	TYPE(S) OF HOST	DERMATOLOGIC PRESENTATION	ONCOGENIC
				ALPHAHERPESVIRINAE				
HHV-1	Human herpesvirus 1, herpes simplex virus 1	HSV-1	High (>60%)	Yes	No	Immuno-competent and -compromised	Yes	No
HHV-2	Human herpesvirus 2, herpes simplex virus 2	HSV-2	Moderate (<30%)	Yes	No	Immuno-competent and -compromised	Yes	No
HHV-3	Human herpesvirus 3, varicella-zoster virus, herpes zoster virus	VZV	Low in vaccinated areas, high globally	Yes	**Yes**	Immuno-competent and -compromised	Yes	No
				BETAHERPESVIRINAE				
HHV-5	Human herpesvirus 5, cytomegalovirus	CMV	High (>60%)	Yes	No	Immuno-compromised	**No** (*see* chapter 7)	No
HHV-6	Human herpesvirus 6		High	No, but can be transmitted through the germline	No	Immuno-compromised	Yes	No
HHV-7	Human herpesvirus 7		High	No	No	Immuno-compromised	Yes	No
				GAMMAHERPESVIRINAE				
HHV-4	Human herpesvirus 4, Epstein-Barr virus	EBV	High (>65%)	No	No	Immuno-competent and -compromised	**No** (*see* chapter 10)	**Yes**
HHV-8	Human herpesvirus 8, Kaposi's sarcoma virus	KSHV	**Rare**, except in certain areas	No	No	Immuno-compromised	Yes (*see* chapter 10)	**Yes**

Table 4.2. **Differences between HSV-1 and HSV-2**

PARAMETER	HSV-1	HSV-2
Latency	Trigeminal ganglia	Sacral nerve root ganglia
Prevalence in the human population (13, 14)	High (>60%)	Moderate (11–30%)
Major site of infection	Oral mucosa Genital mucosa (causes ~40% of genital infections; it has a lower rate of recurrence than HSV-2)	Genital mucosa (causes ~60% of genital infections; it has a higher rate of recurrence than HSV-1)
Main type of CNS infection	Encephalitis	Meningitis

with significant contamination with saliva or skin can rarely transmit oral HSV.

3. **Clinical presentation:** HSV-1 and -2 can infect many different cell types but replicates mainly in epithelial cells of the oropharyngeal or genital mucosa. Most primary infections are asymptomatic, although they can produce severe symptoms in some immunocompetent and immunosuppressed individuals. After primary infection, these viruses are transported into neurons, where they remain latent. They can be reactivated throughout life (unknown triggers) and are transported back through the neurons to the skin and mucosal surfaces.
   - **Primary infection:** typically asymptomatic, but may cause symptoms with a wide range of severity. Several typical infections may occur when HSVs are acquired by the following routes.
     - Oropharynx
       - **Gingivostomatitis** with swollen, red gums, high fever, malaise, myalgia, and cervical lymphadenopathy. Lesions appear on the oral mucosa and can lead to pain and difficulty swallowing. This mostly occurs in children.
       - **Pharyngitis** and tonsillitis: sore throat, fever, malaise, pharyngeal swelling, cervical lymphadenopathy, and ulcerative lesions on the tonsils and pharynx
     - Skin
       - **Herpetic whitlow:** localized infection (usually on the fingers) due to traumatic inoculation of the virus into the skin. This is often seen in people with occupational exposure, like health care workers and dentists.
       - **Herpes dermatitis:** painful cluster of vesicular lesions form on the body from skin-to-skin contact (e.g., wrestlers get "herpes gladiatorum" or "mat herpes")
       - **Eczema herpeticum:** painful, itchy blisters, often on the face and neck. Can occur with underlying conditions that damage the skin (e.g., burns and eczema).
     - Genital mucosa
       - **Lesions:** Multiple blisters or **painful** lesions in the genital area (penis, vagina, vulva, and anus), itching, genital discharge, and inguinal lymphadenopathy
       - Women are more likely to get genital herpes than men.

Both HSV-1 and -2 can cause oral and/or genital herpes. HSV-1 is more commonly associated with oral herpes, and HSV-2 is more commonly associated with genital herpes.

- **Reactivated disease**
  - Oral
    - **Herpes labialis:** red, pruritic, and vesicular lesions form across the margin of the lips and skin (*see* Fig. 4.11). These are also called **cold sores**, or **fever blisters**.
    - Lesions are often preceded by a tingling sensation.
  - Genital (*see* Fig. 4.11)
    - **Lesions** are usually shorter in duration, less severe than the primary episode, and occur less often over time.
    - HSV-2 recurs more often than HSV-1 (~4 times/year without treatment).
- **Complications**
  - Immunocompromised persons present with more severe, frequent, and persistent symptoms.
  - Keratitis/keratoconjunctivitis: eye pain, loss of vision, photophobia, characteristic **dendritic ulcers** on the corneal epithelium. This usually resolves spontaneously in 1 to 3 weeks.
  - Central nervous system infections: can present with headaches, fevers, seizures, and degeneration of the temporal lobe
    - HSVs are the most common cause of viral encephalitis.
    - HSV-1 causes more **encephalitis** than HSV-2. HSV-2 causes more **meningitis** than HSV-1.

Herpes simplex viruses are the most common cause of viral encephalitis. Enteroviruses are the most common cause of viral meningitis.

- **Neonatal herpes** has high morbidity and mortality and occurs when the infant is <30 days old. More severe (i.e., disseminated or CNS) disease can result in long-term sequelae and risk of neurodevelopmental abnormalities. For more, *see* Box 4.2.
  - There are three main categories of disease.
    - **Skin, eyes, and mouth disease**: localized clusters of vesicles
    - **CNS disease:** seizures and brain lesions
    - **Disseminated disease:** fever, rash, multiple-organ involvement (especially lungs and liver), chorioretinitis

4. Diagnosis

- Specimen
  - For lesions: usually vesicular fluid or scrapings of cells from the margin and base of the lesion
  - For neonatal infection: single or combined swab of surfaces, such as the conjunctiva, mouth, nasopharynx, and anus
  - For CNS disease: CSF. PCR should be performed on this specimen since CSF viral culture has very poor sensitivity.
  - Blood is typically a poor specimen for HSV in adults because viremia may not correlate with infection. On the other hand, blood is useful for diagnosis of disseminated neonatal disease.
    - The virus resides mainly in epithelial cells or neurons so blood may be negative during an active infection.
    - Virus is shed in blood transiently, even in asymptomatic people, during reactivation and may not represent the true source of current symptoms.

Viral culture from CSF is not sensitive.

Virus can be shed even when there are no symptoms. Detection or presence of HSV alone (even in blood or CSF) does not necessarily indicate active infection.

  - For eye infection, vitreous fluid and corneal scrapings
- NAAT: the preferred test for diagnosis of HSV-1 and -2 in lesions and CSF. It is the most sensitive and specific method of detection, and the most rapid.
- Cell culture: HSV grows well and rapidly in cells (~2 days).
  - Grows in MRC-5 and A549 cells
  - CPE appears as rounded, refractile cells clumped together. Cells may also fuse together.
  - The enzyme-linked virus-inducible system (ELVIS) cell line turns blue in the presence of HSV (*see* chapter 13).
  - Is highly specific but is less sensitive than PCR
- Tzanck smear: Lesion scrapings or fluid show the presence of **multinucleated giant cells**. This test is rapid but has poor sensitivity and specificity.

The Tzanck smear has poor sensitivity and specificity.

- Histology: HSV-infected cells in tissue biopsies characteristically demonstrate the "three M's": chromatin **margination** (chromatin clusters at the margins of the nucleus which make the nucleus look like it has an outline), cell **molding** (infected cells mold around other cells), and **multinucleation** (*see* Fig. 13.6).
- Serology can be used for screening at-risk pregnant women, in cases with recurrent or atypical presentation, or for screening sexual partners of infected individuals. It can also be used to differentiate between HSV-1 and HSV-2 exposure.
  - Can be performed in serum or CSF (in cases of CNS infection, such as encephalitis)
  - Four-fold rises in IgG titers can indicate recent infection.
  - Antibodies may not increase during recurrent episodes.
  - Antibodies are detected against glycoproteins G1 and G2.

Serology should not be used in newborns because anti-HSV maternal antibodies may be present.

Box 4.2. **Diagnosis and management of neonatal HSV**

HSV is transmitted during delivery, so at-risk pregnant women should be screened. Mothers with recurrent genital herpes or active lesions should be treated prophylactically and offered delivery by cesarean section (C-section). Infected neonates have high morbidity, long-term sequelae, and mortality and are treated with acyclovir. Neonates are tested by culture and/or PCR from an all-surface swab, blood, CSF, and any lesions. Serology is not useful in diagnosing HSV disease in newborns because anti-HSV antibodies may be transferred from the mother.

5. Prevention and treatment

- Prevention
  - There is currently no vaccine against HSV-1 or -2.
  - Minimize close contact, especially when there are active lesions.
  - Barrier protection during intercourse
  - Cesarean delivery of mothers with active genital lesions
- Treatment for genital herpes
  - Prophylaxis: oral acyclovir. Used for uninfected sexual partners, but duration of therapy is not defined.

- Therapeutic: **Acyclovir**, famciclovir, or valacyclovir for active lesions.
  - Antivirals do not eliminate the virus but can reduce transmission, symptoms, and duration of disease.
  - Mild cases do not need treatment. Frequent, severe recurrences indicate the need for chronic therapy. Patients with few recurrences should be treated episodically (just when recurrence occurs).
  - Pregnant women with active lesions anytime during pregnancy should be started on suppressive therapy at 36 weeks to prevent recurrence during delivery.

- Treatment for oral herpes
  - Antivirals are not necessarily needed but can be given. These may include oral (e.g., acyclovir) or topical (e.g., penciclovir) agents.
  - Topical pain relief can also be used.
- Treatment for neonatal infections: intravenous acyclovir.
- Treatment for CNS infections: intravenous acyclovir.

III. VARICELLA-ZOSTER VIRUS. VZV is a highly contagious herpesvirus that causes chickenpox upon primary infection. It causes a lifelong latent infection that may reactivate and present as shingles. VZV is usually not fatal, but it can be lethal in neonates or immunosuppressed individuals.

1. **Background**
   - VZV is an enveloped, dsDNA virus that is part of the *Herpesviridae* family and the *Alphaherpesvirinae* subfamily (Fig. 4.2).
   - It is also known as human herpesvirus 3 (HHV-3), herpes zoster virus, and chickenpox virus (Box 4.1, Table 4.1)
2. **Transmission:** VZV is highly contagious and can be acquired by multiple routes.
   - Contact, such as direct contact with lesions or vesicular fluid, as well as contact with fomites (e.g., bedding)
   - Airborne. Airborne isolation is used when a patient has disseminated infection (localized infection is contact only) (15).
   - Transplacental

Primary infection: chickenpox

Reactivation: shingles

An immunologically naive person will get chickenpox if exposed to shingles.

3. **Clinical presentation:** VZV causes different symptoms in primary and secondary infections. During primary infection, VZV infection can cause chickenpox. Then, the virus causes a lifelong infection and becomes latent in the **dorsal root ganglia.** It may reactivate (unknown triggers), especially in elderly patients, and cause shingles (Fig. 4.3).
   - Primary infection: **Chickenpox.** Prior to vaccination, chickenpox was a common childhood illness; it has low mortality and presents with a characteristic rash.
     - Incubation period: ~2 to 3 weeks
     - Fever for up to 5 days, headache, and malaise
     - Lesions (*see* Fig. 4.10 and Fig. 4.11)
       - Crops of small, red, maculopapular vesicles (like "dew drops on rose petals") form for <5 days.
       - Lesions have **centripetal distribution** (they occur mostly on the face and trunk, with very few lesions on the extremities).

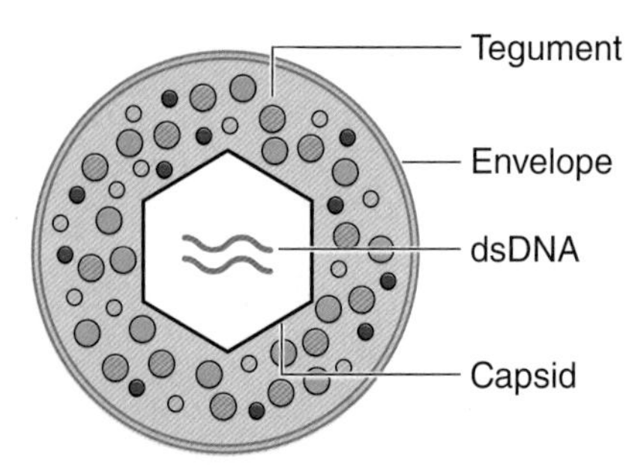

Figure 4.2. VZV.

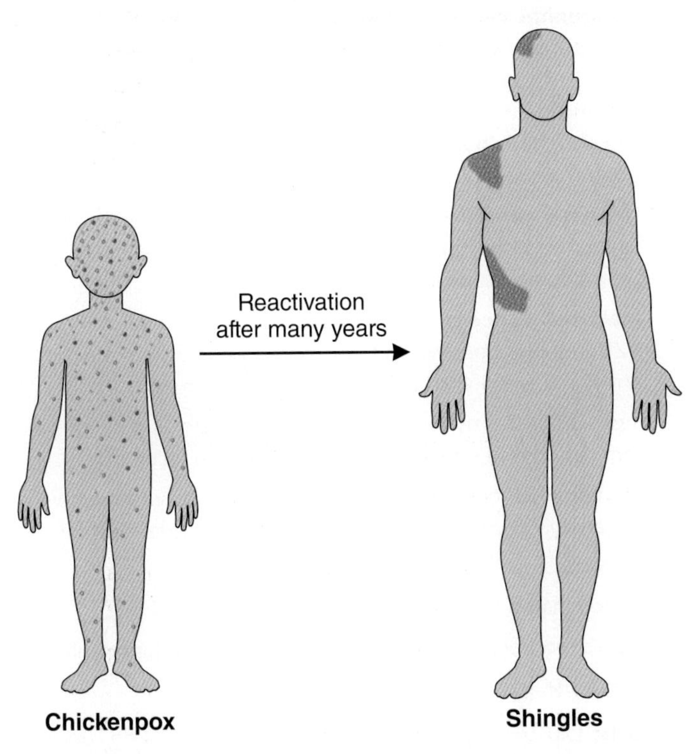

Figure 4.3. VZV presents as chickenpox (crops of vesicles) during childhood. It can reactivate many years later as shingles (unilateral rash with dermatomal distribution).

- Lesions are present at all stages of development (some new, some partially crusted, and some crusted over) on any part of the skin or mucosa. They usually scab and recede in 1 to 2 weeks.
- Patients are infectious until *all* lesions have crusted over.

- Most cases occur in children. Chickenpox acquired in adulthood is more severe than chickenpox acquired in childhood.

- Reactivation: **shingles** (also known as **herpes zoster**) is a painful vesicular rash that characteristically occurs on the trunk. Lesions typically occur unilaterally, with dermatomal distribution (Figure 4.3; see also Fig. 4.11).
  - Only 20% of people get shingles.
  - Pain can be debilitating and can occur before skin symptoms appear Nerve damage from shingles causes pain, sensitivity, and pruritus and may continue even after the lesions disappear. This is called **postherpetic neuralgia**.
  - Most cases occur in adults, especially older adults. Shingles typically resolves in 4 to 6 weeks.
- Reactivation: **herpes zoster ophthalmicus**
  - VZV is reactivated in the **trigeminal** nerve.
  - Symptoms include rash and tingling on the forehead, swollen eyelids, intense ocular pain, and loss of vision. It can cause long-term sequelae and permanent blindness.

Chickenpox: centripetal distribution of lesions

Shingles: dermatomal distribution of lesions

- Infection during pregnancy can cause serious disease in the mother and neonate.
    - **Congenital varicella syndrome:** occurs upon infection of the fetus after primary VZV infection of the mother during early pregnancy (<20 weeks). This is a very rare condition. Affected infants may be asymptomatic at birth or have characteristic birth defects including **scars** on the skin, **limb hypoplasia**, and neurologic, ocular, growth, and other abnormalities. These infants often develop shingles within a year.
    - **Neonatal varicella:** occurs when VZV is transmitted to the infant very late in pregnancy (around delivery) or very soon after birth. It has high mortality, and can cause disseminated infection.
    - **Maternal varicella pneumonia:** a severe complication of primary VZV in a pregnant woman. It may require mechanical ventilation and antiviral therapy and can result in maternal death.

Congenital varicella syndrome = early pregnancy

Neonatal varicella = late pregnancy or post-delivery

- Complications: Immunocompromised hosts are at higher risk of virus dissemination and severe infection, such as the following.
    - **Encephalitis,** meningitis, and cerebellar ataxia. VZV reactivation tends to cause CNS complications more often than primary infection.
    - Hepatitis
    - Disseminated intravascular coagulopathy
    - Reye's syndrome, when aspirin is administered during infection
    - Motor paralysis

4. **Diagnosis:** Lab testing is not recommended for routine, uncomplicated chickenpox and shingles because they can be diagnosed clinically. Testing can be done in some situations, such as in neonates, immunocompromised persons, or in cases with unusual/complicated presentation.
    - Specimen
        - Vesicular fluid, scrapings of cells from unroofed lesions, or biopsy samples for skin lesions. Genital lesions are infrequent but can occur.
        - CSF for CNS presentation
        - Blood is typically a poor source for detection of VZV virions because the presence or absence of viremia does not correlate well with active infection.
    - NAAT: the preferred method. Assays like PCR detect VZV readily from multiple specimen types with high sensitivity and specificity.
    - Culture: is not preferred because VZV grows very slowly (~6 to 14 days) in MRC-5 cells.
    - Tzanck smear: simple to perform but is no longer recommended because of poor sensitivity and specificity (cannot differentiate between HSV and VZV).
    - Serology: used to confirm vaccination or exposure to VZV (in pregnancy or in health care workers)

5. **Prevention and treatment**
    - Prevention: Live, attenuated vaccines are highly effective and are a part of the routine childhood vaccine series.
        - **Varicella vaccine** (Oka strain) can be given alone or in combination with measles, mumps, and rubella (MMR)
        - **Shingles/zoster vaccine** is recommended for people ≥60 years old, whether they have or have not previously had chickenpox or shingles.

- Live attenuated: contains a larger dose of the varicella (chickenpox) vaccine.
- Subunit: contains a recombinant varicella protein and an adjuvant.

- Treatment
  - Dermatologic symptoms are usually self-limited, and treatment is not needed (supportive care only).
  - For individuals with severe disease, immunocompromised patients, and neonates: **acyclovir**, valacyclovir, and famciclovir

IV. MEASLES VIRUS. Measles virus (also known as **rubeola** virus) is a highly contagious virus that presents with a characteristic, morbilliform rash. It can cause severe infection and death, but the number of cases is very low in vaccinated countries.

Like other paramyxoviruses, measles virus also causes a respiratory tract infection. It is described here because of its subsequent, characteristic dermatologic manifestation.

1. **Background:** measles virus is an enveloped, (–) ssRNA virus that is part of the *Paramyxoviridae* family. It only infects humans (Fig. 4.4).
2. **Transmission:** measles virus is transmitted by contact with respiratory secretions or inhalation of large and small respiratory droplets. The patient is infectious before, during, and after symptoms have appeared (~4 days before and after).

Measles is highly contagious and, unlike most viruses, is airborne (can stay suspended in the air of physicians' offices, gyms, etc.).

3. **Clinical presentation**
   - Incubation period: ~14 days
   - Total course of symptoms: 7 to 10 days
   - Initial symptoms: Resembles a severe upper respiratory tract infection, with high fever (up to 105°F), runny nose (coryza), red eyes (conjunctivitis).
   - **Koplik's spots:** Small white lesions on the buccal mucosa with an erythematous base and bluish centers are pathognomonic for measles virus infection ("grains of salt on a red background"). These appear soon (~2 to 3 days) after the initial symptoms begin.
   - **Morbiliform rash:** A blotchy, red, maculopapular rash occurs soon after Koplik's spots; it begins on the face and then spreads downward to the rest of the body (*see* Fig. 4.11).
   - Leukopenia
   - Does not cause congenital abnormalities but may result in spontaneous abortion
   - Complications
     - Common: diarrhea, pneumonia, ear infection caused by measles virus itself, or because of a bacterial superinfection
     - Rare: **subacute sclerosing panencephalitis** (SSPE). The virus can sometimes remain latent after infection at a young age (usually <2 years). After ~6 to 15 years, it can cause SSPE, a chronic (~2 years), usually fatal, neurodegenerative disease. Symptoms include irritability, mental and motor changes, seizures, and difficulty swallowing.

3 C's: cough, coryza, and conjunctivitis

4. **Diagnosis**
   - Serology: the most common method of diagnosis. EIAs are used to detect the appearance of IgM or a 4-fold rise in IgG titer.
   - NAAT: not routinely available in the United States except in some public health labs since measles is uncommon.

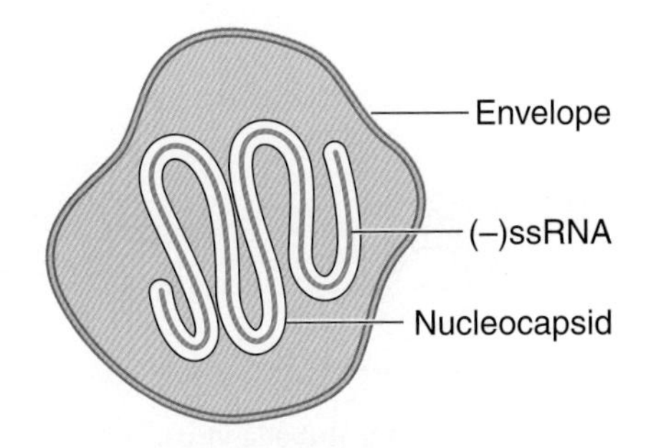

Figure 4.4. Measles virus.

5. **Prevention and treatment**
    - The vaccine contains live attenuated virus and is extremely effective. However, sporadic cases still occur in unvaccinated individuals.
    - The vaccine is formulated with live attenuated mumps and rubella (called the **MMR**) and given as part of the routine childhood series.
    - People at risk of infection are infants too young to be vaccinated and unvaccinated travelers to regions where measles is endemic.
    - Common side effects: temporary fever and rash ~1 week after vaccination

V. RUBELLA VIRUS. Rubella virus causes a disease that is like measles but typically milder. Infections during pregnancy can cause congenital rubella syndrome, which can cause substantial morbidity in the neonate, including hearing loss. Cases of rubella are extremely rare in countries where people are vaccinated.

1. **Background:** Rubella virus is enveloped and has (+) ssRNA. It is a member of the *Togaviridae* family (Fig. 4.5).
2. **Transmission:** inhalation of respiratory droplets and secretions (cough/sneeze)
    - Like respiratory viruses, rubella virus is easily transmitted in communities, families, jails, and schools.
    - It is endemic in countries without good vaccination coverage.
    - It is not endemic in the United States, but imported cases do occur.

Rubella virus produces an illness similar to measles except that it is much less contagious (droplet, instead of airborne, precautions) and is milder and shorter in duration. It is sometimes called **German measles** or **three-day measles**.

3. **Clinical presentation:** Symptoms are not severe except in neonates.
    - **Postnatal infection:** Symptoms occur in >50% of cases.
        - Incubation period: ~2 weeks
        - **Red rash:** A pinpoint maculopapular rash starts on the face and/or oral cavity and lasts for ~3 days (*see* Fig. 4.11). The patient can be infectious before and after the rash appears.
        - Lymphadenopathy
        - Low fever
        - Arthritis in women
        - Others: sore throat, conjunctivitis, and transient joint pain
    - **Congenital rubella syndrome**
        - Occurs when pregnant mothers are infected, especially during the first trimester
        - Affected neonates can have severe birth defects (deafness, heart defects, cataracts, mental retardation, and microcephaly).
        - Seizures
        - Severe rash ("blueberry muffin" appearance, like congenital CMV infection)
        - Progressive rubella panencephalitis
        - Can cause miscarriage or stillbirth
    - Complications in children and adults: encephalitis
4. **Diagnosis**
    - Serology: Detection of IgM from serum or CSF (e.g., ELISA) is the preferred method of diagnosis. False-positive results can occur with parvovirus, EBV, and rheumatoid factor.

Figure 4.5. Rubella virus.

- Cell culture: not routinely used except in public health labs, since rubella is uncommon and CPE is slight and not distinctive.
- NAAT: not routinely available in the United States except in some public health labs since rubella is uncommon. Specimens for isolation or nucleic acid detection include the following:
  - Blood obtained during the viremic phase (~7 days after exposure) or neonatal blood
  - Nasal, nasopharyngeal and throat swabs, aspirates
  - CSF and urine

5. Prevention and treatment

- Supportive treatment
- Vaccine: A highly effective, live attenuated vaccine provides long-term immunity to rubella. It is administered routinely in the childhood vaccination series as part of the **MMR** vaccine.

VI. HUMAN HERPESVIRUS 6 AND HUMAN HERPESVIRUS 7. HHV-6 and HHV-7 are highly prevalent herpesviruses that cause lifelong infection. They are usually asymptomatic or cause mild disease in immunocompetent people but may cause more serious disease in transplant recipients and other immunocompromised patients.

1. **Background:** These viruses are members of the *Herpesviridae* family and the more mild, opportunistic *Betaherpesvirinae* subfamily (Box 4.1, Table 4.1). They are large and enveloped, contain dsDNA, and cause lifelong infection (Fig. 4.6).
2. **Transmission:** Both viruses are highly prevalent in the human population and are acquired during childhood.
   - Probably spread through saliva
   - Unlike other herpesviruses, HHV-6 can integrate into the host cell chromosomes and be transmitted vertically through the germ line. However, it does not cause disease in neonates.

HHV-6A and B **integrate** into host chromosomes (lysogenic). Other herpesviruses persist **episomally** (pseudolysogenic).

3. **Clinical presentation**
   - There are two variants of HHV-6. HHV-6A does not cause disease. HHV-6B causes asymptomatic or self-limited, mild disease (sometimes called "**sixth disease**").
   - HHV-7 causes asymptomatic infection or an even milder disease than HHV-6B.
   - Primary infection: Symptoms occur in ≤20% of children <3 years old and in immunocompromised persons.
     - High fever, lymphadenopathy, and a red, nonitchy rash (**roseola infantum**) on the trunk and extremities can appear when the fever abates (*see* Fig. 4.11).
     - Fever-induced seizures
     - Can cause mononucleosis in some adults
     - Rare: encephalitis
   - Reactivation: Immunocompromised patients, especially transplant patients, are at high risk of reactivation. Symptoms include the following:
     - Viremia
     - Encephalitis, pneumonitis, and liver disease

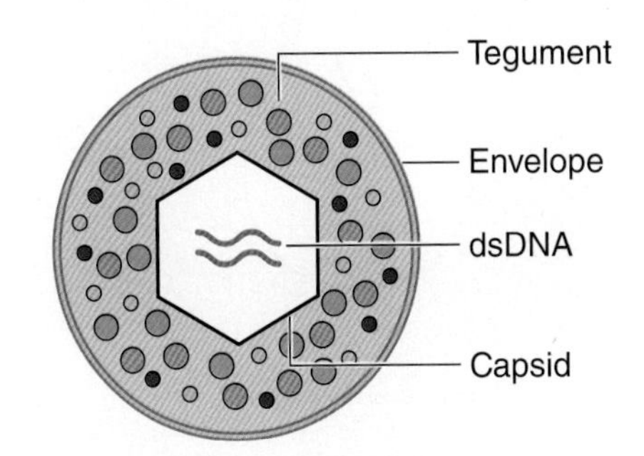

Figure 4.6. HHV-6 and -7.

- ▫ Rash
- ▫ Graft-versus-host disease and organ rejection

4. **Diagnosis:** Testing is used for monitoring in transplant and other immunocompromised patients but is not needed in immunocompetent people.
   - NAAT: the best method to detect active infection. However, like other herpesviruses, detection of HHV-6 and -7 does not necessarily indicate disease.
     - ▫ These viruses can periodically be shed in human specimens with no associated symptoms.
     - ▫ Integrated HHV-6 can be tested in hair clippings/follicles (nonintegrated HHV-6 is not present in these cells) or using quantitative methods (like droplet digital PCR) to show that viral DNA and chromosomal DNA are present in a 1:1 ratio.
     - ▫ If disease is suspected in a patient with the right risk factors, quantitative PCR should be used to monitor a trend of increasing viral loads.
   - Serology: not useful because both viruses are very common

HHV-6 is integrated into the chromosome in ~1% of people. So high levels of virus may be detected even though it may not be responsible for disease symptoms.

5. **Prevention and treatment**
   - There are no approved antivirals or vaccines against HHV-6.
   - No treatment is needed for infections in immunocompetent patients.
   - In immunocompromised patients, reduce immunosuppression when possible. Cytomegalovirus treatments like ganciclovir, cidofovir, and foscarnet may be useful.

VII. MOLLUSCUM CONTAGIOSUM VIRUS. Molluscum contagiosum virus is structurally similar to smallpox virus and also produces lesions. However, molluscum contagiosum is mild, widespread, and causes self-limited disease without significant long-term sequelae.

1. **Background:** Molluscum contagiosum virus is a large, enveloped, dsDNA virus in the *Poxviridae* family (Fig. 4.7).

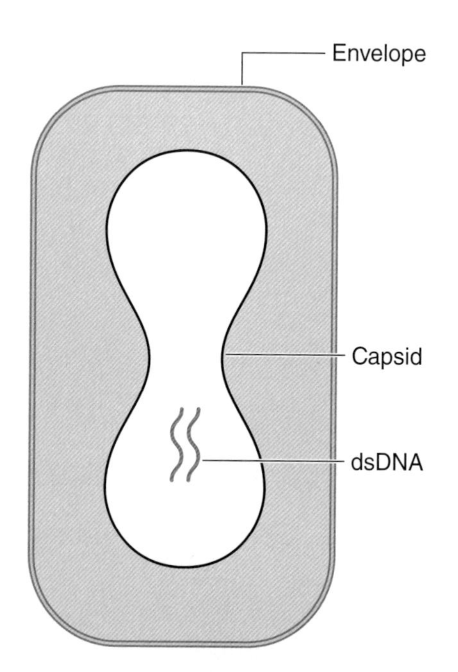

Figure 4.7. Molluscum contagiosum virus.

2. **Transmission:** infects the epithelial layer of the skin and is transmitted through contact with actively infected cells. It does not cause a latent infection.
   - Direct contact with an infected lesion is the most common method of transmission.
   - Autoinoculation onto another region of the body
   - Fomites (such as personal items like clothing and towels). Acquisition of viral infection is associated with warm, humid environments, such as gyms and pools.
   - Sexual contact
3. **Clinical presentation:** Infection with molluscum contagiosum virus may result in small, flesh-colored, smooth, firm papules that have a pearly appearance (sometimes they can become red, irritated, and itchy).
   - Lesions are **umbilicated**, like smallpox lesions.
   - The lesions are benign and resolve spontaneously. Resolution is slow (average of several months; range, 2 weeks to 4 years).
   - Lesions can occur in clusters anywhere on the skin, except palms and soles
   - Symptoms are more common in children
   - Risk factors: atopic dermatitis, immunosuppression, and HIV

4. **Diagnosis:** usually a clinical diagnosis, because of the characteristic appearance of the lesions
   - Skin infection can be confirmed by histology of an excisional biopsy. Epidermal cells have characteristic, large, intracytoplasmic inclusion bodies called **molluscum bodies** (Fig. 4.8).
   - This virus does not grow in cell culture
5. **Prevention and treatment**
   - Prevention: Avoid direct contact with lesions, and do not share clothing/towels.
   - Treatment: Usually not necessary.
     - Lesions can be physically removed.
     - Oral (cimetidine) and topical (podophyllotoxin or salicylic acid) treatments can be used.

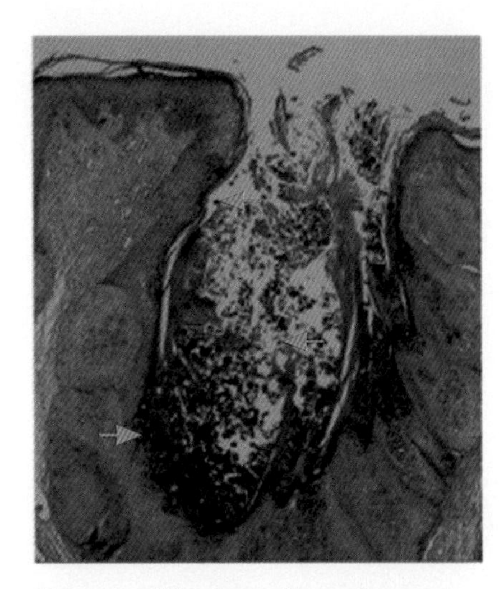

Figure 4.8. Histology of a molluscum contagiosum lesion. The lesion contains many molluscum bodies; arrows point to examples at the surface, middle, and base of the lesion. *See also Fig. 13.6.*

VIII. SMALLPOX VIRUS. Smallpox virus (also known as variola virus) is a highly contagious virus that results in lesions and subsequent scarring over the whole body. It had a high fatality rate, but no infections have been reported worldwide after 1978. In 1980 the WHO declared it to be globally eradicated.

1. **Background:** Smallpox virus is a large (~300-nm), enveloped, brick-shaped virus with a dumbbell-shaped nucleocore. It has dsDNA and is a member of the *Poxviridae* family (Fig. 4.9). There are two strains that cause smallpox.
   - **Variola major:** high mortality rate (up to 50%)
   - **Variola minor:** milder symptoms with low mortality rate (1%)
2. **Transmission**
   - Through respiratory secretions (coughing/sneezing)
   - Can be airborne
   - Contact with vesicular fluid (either directly or from fluid on fomites, like clothes)
   - Contagious until lesions crust over
3. **Clinical presentation**
   - Incubation period: long (2 to 3 weeks)
   - High fever, headache, backache, fatigue, vomiting, diarrhea, and prostration
   - Lesions are characteristic (Fig. 4.10 and Fig. 4.11)
     - They have **centrifugal distribution** (more lesions on face, in the mouth, and on extremities; fewer on the trunk)
     - They begin as a rash and progress to firm, well-circumscribed, fluid-filled pustules. These are typically red or skin colored (smallpox was also known as the "red plague") and are **umbilicated**. After about 10 days, the pustules crust over and form scabs.
     - Lesions occur at the same stage (i.e., all lesions turn from macules to papules to vesicles to nodules and then scabs at the same time. This is unlike chickenpox, which has lesions at all stages simultaneously).
     - Pockmarks: scars of healed lesions. These could appear hyperpigmented on light skin or hypopigmented on dark skin.
   - Risk of severe disease
     - Is higher with more confluent lesions

Smallpox virus was eradicated from the global environment as of 1980. The only remaining vials are stored in 2 laboratories (in the United States and Russia). These carry the potential for use as agents in biological warfare.

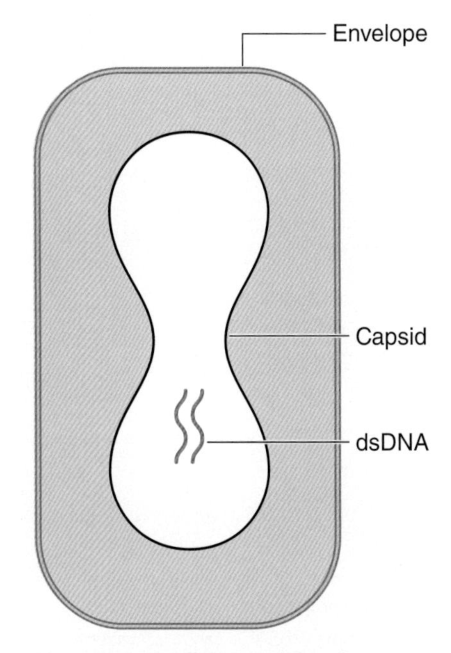

Figure 4.9. Smallpox (variola) virus.

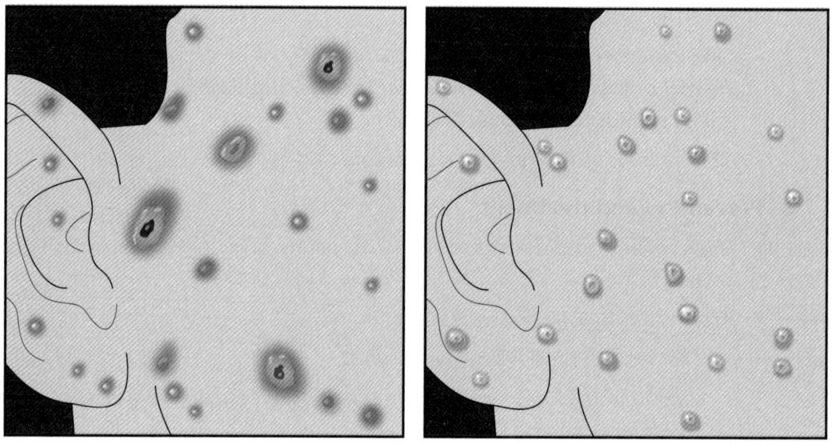

Chickenpox Smallpox

Figure 4.10. Comparison between chickenpox and smallpox lesions. Chickenpox lesions are red and can be at various stages simultaneously (new, mature, scabbed, and resolving); smallpox lesions are umbilicated and flesh colored and are all present at the same stage of resolution.

- Infants and elderly, unvaccinated, and immunosuppressed persons
- Pregnant women. These patients are are at increased risk of hemorrhagic skin lesions and mucous membranes.

4. **Diagnosis:** There are no routine testing methods for smallpox because no new cases have been diagnosed for several decades. Some public health labs can test for smallpox if there is strong clinical suspicion.
   - Specimen: skin lesions or aspirate from vesicles
   - Pock morphology on chicken embryo chorioallantoic membranes
   - **Molluscum bodies** in lesion biopsy or squash preparation
   - Hematoxylin and eosin (H&E) staining: characteristic inclusions called **Guarnieri bodies**
   - Electron microscopy
5. **Prevention and treatment**
   - Prevention:
     - A live vaccine is available containing **vaccinia virus**. This virus is closely related to smallpox virus but is less virulent and does not spread by respiratory secretions.
       - Booster doses are required for optimal protection.
       - Complications: The vaccine can cause vaccinia virus infection.
         - Fever, headache, fatigue, myalgia, vesicular rash, cardiac infection, and encephalitis
         - Progressive vaccinia (extensive tissue necrosis that can cause death) in immunosuppressed persons
         - Vaccinia immunoglobulin can be used to treat vaccine-related infections.
       - The vaccine is no longer administered except in some situations (e.g., emergencies, military, first responders, etc.).
       - Administered by **scarification** (the vaccine dose is injected into the epidermis several times)

Smallpox vaccine = live, attenuated vaccinia virus. This can rarely cause disease.

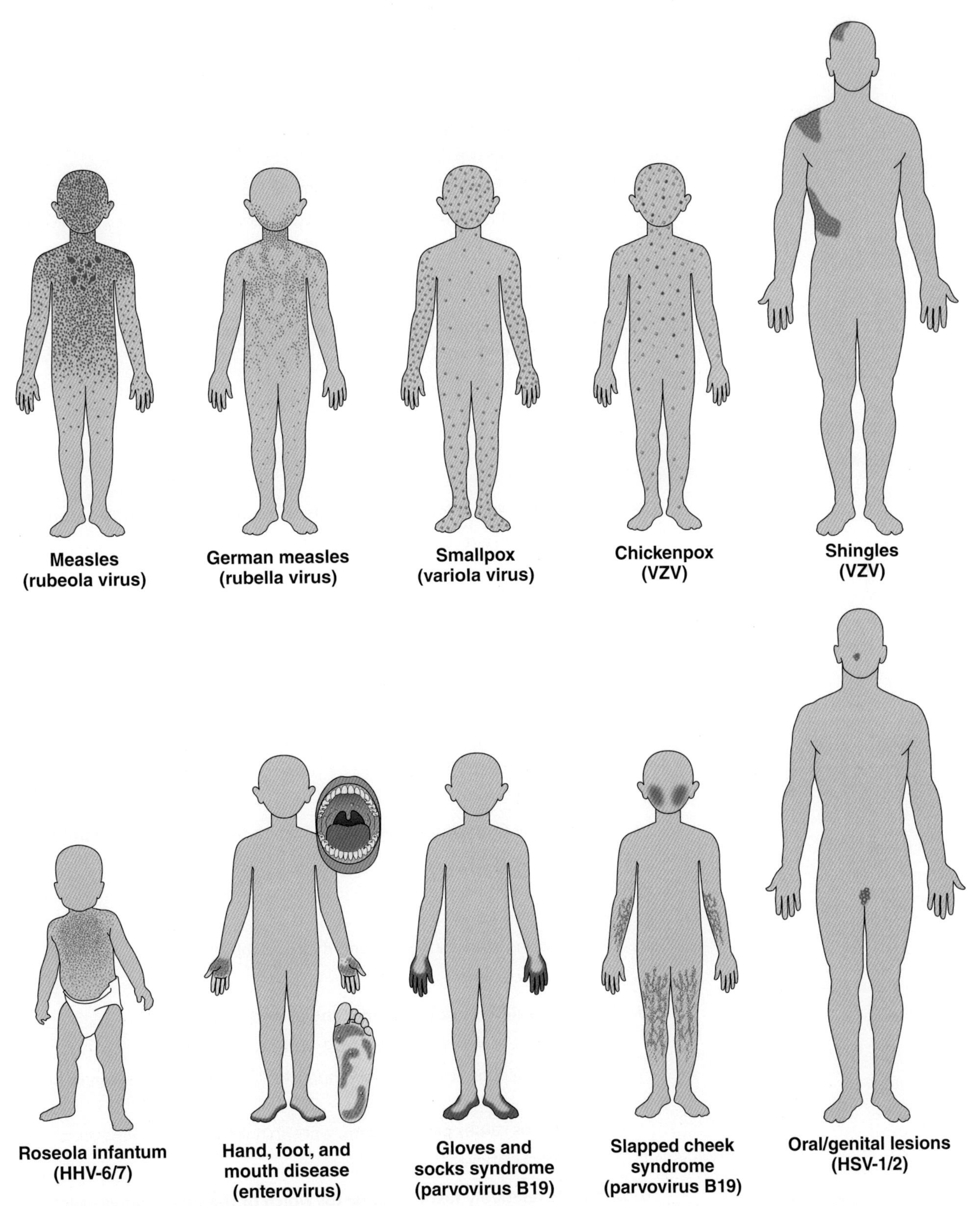

Figure 4.11. Comparison of distinctive viral rashes. Measles (rubeola virus): red, flat, sometimes confluent morbiliform rash that starts on the face and "pours down" to the feet. German measles (rubella virus): pink to red rash that starts on the face and "pours" down. Smallpox (variola virus): umbilicated papules at the same stage of maturation and distributed centrifugally, with more lesions on the extremities than on the trunk. Chickenpox (VZV): red, itchy, vesicular lesions present in multiple stages and distributed centripetally, with more lesions on the trunk and fewer on the extremities. Shingles (VZV): red, painful, vesicular lesions that have dermatomal distribution and generally occur in older individuals. Roseola infantum (HHV-6 and -7): pink to red diffuse rash on the trunk and legs that typically occurs in infants. Hand, foot, and mouth disease (enterovirus): red, painful vesicular rash on the palms, soles, and inside the mouth. Erythema infectiosum or Fifth disease (parvovirus B19): bright red, slapped-cheek appearance that is followed several days later by a lacy, confluent rash on the arms and legs. Oral or genital lesions (HSV-1 and -2): painful, red vesicles around or inside the mouth or genitals.

- Treatment
  - Supportive care for symptoms
  - Antivirals: tecovirimat and cidofovir (16)
  - Vaccinia immunoglobulin: used intravenously to treat complications caused by the vaccinia vaccine

Multiple-Choice Questions

1. Which of the following herpesviruses is generally NOT associated with congenital infections?

a. HSV-2

b. VZV

c. EBV

d. CMV

2. Which of the following is NOT a route of transmission for HSV-1 and -2?

a. Contact with a recurrent lesion

b. Sexual intercourse

c. Human bite

d. Bedsheets

3. Which of the following causes lifelong infection?

a. VZV

b. Rubella virus

c. Molluscum contagiosum virus

d. Smallpox virus

4. Which of the following is the best specimen to test for active genital herpes?

a. A lesion swab

b. Blood

c. Urine, along with testing for gonorrhea and chlamydia

d. Vaginal secretions

5. What is the difference between shingles and chickenpox?

a. Chickenpox is caused by VZV, and shingles is caused by Oka.

b. Primary infection results in chickenpox, while secondary infection results in shingles.

c. Chickenpox causes painful lesions, while shingles causes painless ones.

d. None of the above

6. Culture is relatively fast (≤2 days) for which of the following viruses?

a. Molluscum contagiosum virus

b. HSV

c. VZV

d. None. Viral growth in culture takes a minimum of 7 days.

7. Which of the following viruses are all transmitted by the airborne route and require airborne isolation to prevent transmission?

a. HSV, parvovirus, smallpox virus

b. VZV, measles virus, rubella virus

c. Enterovirus, parvovirus, molluscum contagiosum

d. Measles virus, VZV, smallpox virus

8. Which is the most useful diagnostic assay for molluscum contagiosum virus?

a. Culture

b. PCR

c. Histology

d. Serology

Match the following. Use each answer only once.

9. Types of herpes lesions

Herpes gladiatorum	A. Neonate
Skin, ears, and mouth disease	B. Atopic dermatitis
Herpes whitlow	C. Wrestler
Eczema herpeticum	D. Oral lesion
Herpes labialis	E. Dentist

10. Types of rashes

Morbilliform rash	A. VZV
Dew drop rash	B. Rubella virus
Slapped cheek rash	C. Variola virus
Umbilicated lesions	D. Parvovirus B19
Blueberry muffin lesions	E. Measles virus

True or False

11. HSV-2 causes genital infections and does not cause oral infections. **T F**

12. Measles virus can cause a latent infection. **T F**

13. Smallpox is a severe disease that can cause disfigurement and death. **T F**

14. HHV-6 can be detected in the CSF of asymptomatic patients. **T F**

15. The MMR vaccine should be administered to pregnant women to protect against congenital rubella infection. *(May require reading outside of this chapter.)* **T F**

5

CHAPTER 5

GASTROINTESTINAL AND FECAL-ORAL HEPATITIS VIRUSES

I. OVERVIEW. Viruses are the most common cause of infectious gastroenteritis. They cause an acute, self-limited disease that is usually transmitted by ingestion of contaminated food and water. The viruses covered here can cause gastrointestinal symptoms even in immunocompetent hosts. Other viruses, like CMV (*see* chapter 7) and adenovirus (*see* chapter 6), are usually relatively benign but can cause extended gastrointestinal disease in immunocompromised patients.

Most gastrointestinal viruses are nonenveloped and highly stable in the environment.

1. **Background:** Almost all viruses that cause gastroenteritis are nonenveloped in order to survive the low pH of the gastrointestinal tract.

2. **Transmission**
   - Fecal-oral route. The most common sources of infection are contaminated water and food.
   - Contact with fomites like nonporous surfaces and objects
   - These viruses are highly stable in the environment and relatively resistant to chlorine that is used to disinfect drinking water.
   - Gastrointestinal viruses occur all year long, with some showing seasonal peaks (Fig. 5.1).

3. **Clinical presentation:** Viral gastroenteritis presents as an acute, self-limited, **nonbloody**, watery diarrhea that can last for 1 to 8 days. Reinfections can occur. Other symptoms include the following:
   - Abdominal pain
   - Vomiting, for some viruses
   - Dehydration

4. **Diagnostic testing**
   - Specimen: Stool is the most commonly tested and accepted specimen, but gastrointestinal viruses may also be detectable from vomit. High virus titers are shed in the stool for the first 2 days of symptoms, but shedding can continue for several weeks even after resolution of symptoms. In immunocompromised people, shedding can continue for years.
   - NAATs: usually the method of choice for detecting gastrointestinal viruses. These tests are highly specific and are substantially more sensitive than antigen or electron microscopy methods.

Winter	Spring	Summer	Fall
Norovirus			
Rotavirus			
Enterovirus			
Hepatitis A*			
Hepatitis E*			
Adenovirus			

Figure 5.1. Seasonality of gastrointestinal viruses. *, these viruses can have varied seasonal patterns depending on geographic area.

Gastrointestinal viruses are shed at very high levels. Of concern is contamination of negative samples with very "hot" positives.

- Broad, multiplex PCR for viral, bacterial, and parasitic pathogens can be used to combine detection of multiple etiologies of disease.
- Does not differentiate between active and past infections (i.e., inactive virus that is still being shed)

- Antigen detection: ELISA and latex agglutination are used to detect viral antigens in stool. The presence of antigens can indicate active infection.
- Serology: detecting circulating antibodies against gastrointestinal viruses is not useful because these viruses are common.
- Culture is generally not used.
  - Culture has a slow turnaround time for acute disease.
  - CMV, adenovirus, and some enteroviruses grow in cell culture, but norovirus, sapovirus, and rotavirus do not.
  - Stool specimens are difficult to use in cell culture because of toxicity to the cells and overgrowth of normal flora.

Resolution of viral gastroenteritis is determined clinically. Retesting should not be performed to test for cure because viral shedding may continue for weeks (years, if immunosuppressed) even after the patient is no longer symptomatic.

5. **Prevention and treatment**
   - Handwashing is essential since most transmission occurs through fecal-oral contamination.
   - Supportive therapy with rehydration
   - Antivirals are not typically used because most infections are self-limited.
   - Antibiotics are not appropriate.

II. ROTAVIRUS. Worldwide, rotavirus is one of the most common viral causes of gastroenteritis in children **<5 years old.** It can cause significant morbidity and mortality, especially in developing countries, but this can be effectively mitigated by vaccination.

1. **Background:** Rotavirus is a nonenveloped virus belonging to the family *Reoviridae*. It is one of the few medically relevant dsRNA viruses (Fig. 5.2). It has a wheel-and-spoke appearance on electron microscopy (*rota* = wheel).
   - Its genome is segmented, which means that it can undergo antigenic shift, but unlike reassorted influenza viruses, reassorted rotavirus strains have been mild.
   - Serogroups A to G include human and nonhuman strains (rotavirus is also a major veterinary pathogen).

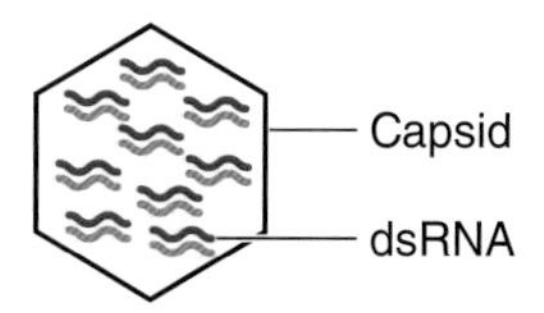

Figure 5.2. Rotavirus.

2. **Transmission**
   - Rotavirus is primarily transmitted by fecal-oral contact. It is highly stable in the environment, and as a result, it is often transmitted directly from person to person or via fomites instead of by contaminated food and water.
   - Associated with day care; there is nosocomial spread in nurseries.
   - Seasonality: peaks in **winter and spring**

3. **Clinical presentation**
   - Incubation: ~2 days
   - Mild to severe diarrhea with no blood or mucus
   - Abdominal pain and fever. Vomiting might occur
   - Dehydration can lead to death from cardiac arrest.
   - Lasts for 3 to 8 days
   - Primary infections in adults are more likely to be mild or asymptomatic, although immunosuppression can result in more severe disease.

4. **Diagnosis**
   - Antigen testing has typically been the most common test for rotavirus.
   - NAATs are the most sensitive and specific type of test. These are becoming more available.

5. **Prevention and treatment**
   - Prevention
     - **Oral**, live attenuated vaccines are part of the routine childhood vaccination series.
     - Vaccine complications: **intussusception** in infants, which usually occurs within 1 to 2 weeks after vaccination. This is extremely rare.
   - Treatment: supportive care

III. NOROVIRUS, SAPOVIRUS, AND ASTROVIRUS. Norovirus and related viruses are common causes of gastrointestinal symptoms. They are highly transmissible and can cause severe vomiting and diarrhea but tend to resolve within 2 or 3 days. These viruses were difficult to detect in the laboratory until NAATs became more common.

1. **Background:** small (~50-nm), nonenveloped (+) ssRNA viruses in Baltimore class IV (Fig. 5.3).
   - **Noroviruses** are categorized within the family *Caliciviridae* (have a capsid with "chalice-like" indentations). There are many different strains of norovirus that are categorized into genogroups I to VI (GI to GVI). Only GI, GII, and GIV infect humans.
   - **Sapovirus** is another calicivirus. It is also a common cause of gastroenteritis, but less is known about it.
   - **Astroviruses** are similar but are categorized within the family *Astroviridae*. They have a star-like appearance on electron microscopy and typically affect children.

2. **Transmission:** water (swimming pools, swimming in lakes, eating shellfish), feces, and vomitus
   - Highly resistant to heat and chlorine inactivation

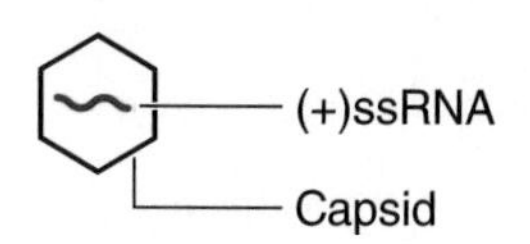

Figure 5.3. Norovirus.

- Outbreaks are often the way these viruses are identified. Outbreaks are usually short (~1 to 2 weeks) and associated with closed or crowded settings (e.g., cruise ships, nursing homes, day care).
- Immunity is not long-lived (lasts only for ~ 6 months), and individuals can be reinfected with the same strain.

Travel on cruise ships is associated with outbreaks of norovirus.

- Seasonality: all year-round, with peaks in winter. Also known as the "**winter vomiting disease**."

3. **Clinical presentation**
   - Incubation: 1 to 2 days for norovirus and 3 to 4 days for astrovirus
   - Abdominal cramps and nonbloody diarrhea (4 to 8 stools/day)
   - Short duration of illness (~2 to 3 days)
   - Nausea and vomiting are common (especially for norovirus).
   - Low-grade fever is often present.
   - No long-term sequelae

Norovirus symptoms are short (~2 days).

4. **Diagnosis**
   - NAATs are the preferred method of detection.
   - Electron microscopy can be used because of the characteristic morphology, but this test is not commonly available in diagnostic labs.
5. **Prevention and treatment:** no vaccines available. Supportive treatment only.

IV. HEPATITIS A VIRUS. Hepatitis A virus (HAV) is acquired fecal-orally, like other gastrointestinal viruses, but it then infects the liver to cause hepatitis (*see* Table 8.1). Disease is characterized by jaundice, fatigue, and abdominal upset, but unlike with blood-borne hepatitis viruses, symptoms are relatively acute and self-limited. HAV is widespread in countries with poor sanitation.

Unlike hepatitis B and C viruses, hepatitis A virus is not latent and does not cause chronic hepatitis.

1. **Background:** HAV is a small virus with (+) ssRNA (Fig. 5.4). It is a member of the *Picornaviridae* family, like enteroviruses (*see* Fig. 1.7).
2. **Transmission**
   - Mainly fecal-oral, especially through ingestion of contaminated food and water
   - Sexual contact (less common)
   - Found worldwide, with higher prevalence in countries with poor sanitation
   - Is highly stable and transmissible, especially because large amounts of virus are excreted in the stool
   - Risk factors for contracting disease
     - Travel to areas where HAV is endemic and vaccination rates are low
     - Close contact with infected persons, especially food preparers
3. **Clinical presentation:** causes an acute and self-limited infection lasting 2 to 6 weeks. Symptoms include the following:
   - Asymptomatic, mild, or fulminant hepatitis
   - **Fatigue**, **jaundice**, pale stools, weight loss, dark urine, and abdominal pain
   - Disease is milder in children.

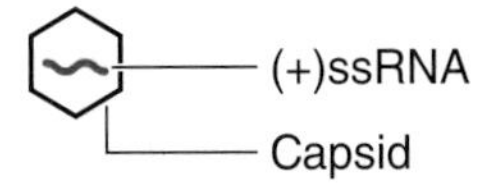

Figure 5.4. HAV.

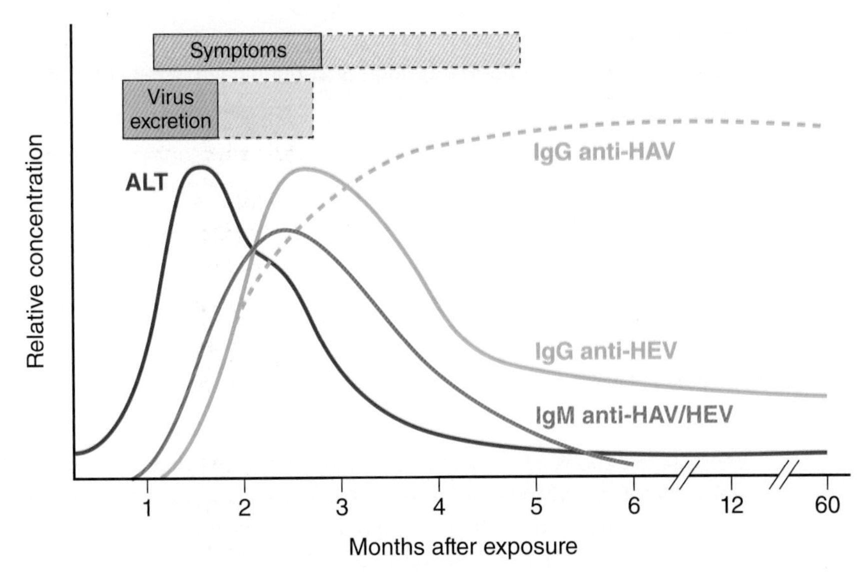

Figure 5.5. Kinetics of HAV and HEV infection. Viral replication, liver damage, and systemic symptoms occur within the first 1 to 3 months of exposure. IgG antibodies provide lifelong protection against HAV but not HEV.

4. Diagnosis

- Serology: the main diagnostic test for HAV
  - Detection of IgM in serum indicates recent or acute infection. Beware of false positives in areas of low prevalence.
  - Detection of IgG is not useful for diagnosis because of vaccination (in areas of low prevalence) or past exposure (in areas of high prevalence).
- NAAT: not generally necessary. Viral load testing may be performed to monitor clearance of the virus.

5. Prevention and treatment

- Prevention: Development of IgG from a vaccine or natural infection provides lifelong protection (Fig. 5.5). An inactivated vaccine against HAV is available and highly effective.
  - It is recommended for people 2 weeks prior to travel into areas of high prevalence.
  - It is part of the routine childhood vaccines in the United States.
- Treatment: supportive care

V. HEPATITIS E VIRUS. Like HAV, HEV is transmitted through ingestion of food and water, causes self-limited hepatitis, and is widespread in developing countries (*see* Table 8.1). Importantly, though, HEV infection during pregnancy can be life threatening for both the mother and the fetus.

1. **Background:** small (+) ssRNA virus. It is similar in structure to caliciviruses, but it belongs to a separate family, *Hepeviridae* ("Hep E viridae"). There is only one serotype but there are 4 genotypes that have different geographic distributions, routes of transmission, and frequencies of outbreaks (17).

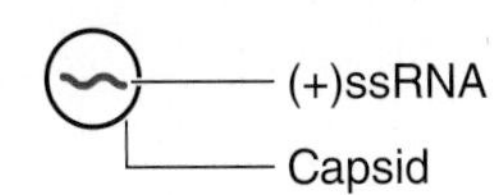

Figure 5.6. HEV.

2. **Transmission**
    - Fecal-oral, especially through ingestion of contaminated food and water
    - Transplacental
    - Found worldwide, but is much more frequent in developing countries, refugee camps, war zones, and other areas of poor sanitation
    - High prevalence in Asia, North Africa, and the Middle East
    - Can be acquired from pigs
    - Risk factors for contracting the disease: travel for a significant amount of time to an area where the virus is endemic
3. **Clinical presentation**
    - Incubation: ~1 to 2 months
    - Almost all infections cause acute, self-limited **hepatitis** that lasts 1 to 2 weeks. Symptoms include jaundice, loss of appetite, hepatomegaly, abdominal pain and tenderness, fever, nausea, and vomiting. Cholestasis (inadequate bile flow) is more apparent than in HAV.
    - In some solid organ transplant cases, HEV genotype 3 infections can progress to chronic infection.
    - Disease is asymptomatic or mild in children. Symptoms are most common in young adults (15 to 40 years).
    - The overall mortality rate is ~1%, which is higher than for HAV.
    - Pregnant women may have severe infection that can progress to **fulminant hepatitis** and death of the mother and the baby.
        - The **maternal mortality** rate is extremely high (~20%).
        - The **fetal mortality** rate is also very high.

Unlike HAV, HEV is transmitted vertically and can cause severe infections and death in pregnant women and fetuses.

4. **Diagnosis**
    - Serology: the main diagnostic test. Identification of IgM, and seroconversion of IgM to IgG, can indicate relatively recent infection.
    - Rapid antigen testing: these tests have been developed but are not currently FDA-approved or commonly available. Data suggests that they have high sensitivity (>90%), and results can quickly identify HEV infection.
5. **Prevention and treatment**
    - Prevention:
        - Minimize exposure to contaminated water and food.
        - One vaccine is licensed in China, but it is not available elsewhere.
    - Treatment: infections are self-limited so no treatment is needed for acute infection. Chronic infection may be treated with ribavirin, but this is not FDA-approved.

Multiple-Choice Questions

1. **Which of the following viruses present with pale stool?**
    a. Enterovirus
    b. Hepatitis A
    c. Norovirus
    d. Rotavirus

2. Which of the following gastrointestinal viruses is a DNA virus?

a. Adenovirus

b. HEV

c. Sapovirus

d. Rotavirus

3. Which of the following gastrointestinal viruses is enveloped?

a. Rotavirus

b. HAV

c. CMV

d. Norovirus

4. Rotavirus vaccine is delivered through which of the following routes?

a. Intramuscular

b. Intranasal

c. Oral

d. Scarification

5. Which of the following is NOT transmitted vertically? *(May require reading outside of this chapter.)*

a. Hepatitis A virus

b. Hepatitis B virus

c. Hepatitis C virus

d. Hepatitis E virus

6. Serology is the main diagnostic test for which of the following pathogens?

a. HEV

b. Astrovirus

c. Rotavirus

d. None of the above; serology should not be used.

7. Which of the following cause peak gastrointestinal illness in winter?

a. Adenovirus

b. HAV

c. Enteroviruses

d. Rotavirus

8. Prolonged shedding can prevent

a. Prolonged transmission

b. Diagnostic tests from being used to measure disease resolution

c. Gastrointestinal symptoms from resolving

d. Resistance from developing

9. Which of the following is the most useful diagnostic test for norovirus?

a. Electron microscopy

b. PCR

c. Serology

d. Culture

Match the following. Use each answer only once.

10. Factors associated with gastrointestinal viruses

Cruise ship	A. Enterovirus
Intussusception	B. HEV
Fetal or maternal death	C. Rotavirus
Peak incidence in autumn	D. Norovirus

True or False

11. Symptoms of norovirus are typically prolonged (>7 days). **T F**

12. Most gastrointestinal viruses are nonenveloped (naked). **T F**

13. Viral agents of gastroenteritis usually cause nonbloody diarrhea. **T F**

14. Rotavirus is typically associated with lakes, streams, hikers, and travelers. **T F**

15. Antibiotics should be used for all cases of gastroenteritis, until a viral etiology is ruled out. **T F**

6

CHAPTER 6

VIRUSES THAT CAN CAUSE MULTIPLE SYNDROMES

I. ENTEROVIRUSES AND PARECHOVIRUSES. Enteroviruses and parechoviruses are very common pathogens that result in significant morbidity but generally low mortality. They can cause an extremely broad spectrum of syndromes, such as respiratory, gastrointestinal, neurologic, dermatologic, and cardiovascular diseases. Pediatric patients are especially at risk of severe disease like meningitis, while mild diseases like colds and gastroenteritis can occur in all age groups.

1. **Background:** small ("pico"), nonenveloped, (+) ssRNA viruses that are part of the *Picornaviridae* family (Fig. 6.1)
   - There are >120 serotypes in the genus *Enterovirus*. Serotypes used to be grouped into 5 subgenera based on similar symptoms. These were polioviruses, echoviruses, coxsackievirus group A (CVA), coxsackievirus group B (CVB), and enteroviruses. Because there is a lot of overlap in symptoms, these groups have since been reshuffled into 4 human species, enteroviruses A through D, based on genetic similarity (Table 6.1).
   - Human rhinoviruses A, B, and C are three more species in the genus *Enterovirus*. However, they cause only respiratory infections ("rhino" = nose), so it can be clinically useful to differentiate them from other enteroviruses. This can be difficult because their growth, structure, nucleic acid sequences, and epitopes are very similar. However, rhinoviruses are destroyed by the low pH in the stomach, while other enteroviruses are resistant to a wide pH range (pH 3 to 10). Rhinoviruses are covered in chapter 3.
   - There are two species in the genus *Parechovirus*. These used to be classified within the *Enterovirus* genus and behave very similarly.

Rhinovirus species = destroyed by low pH and cannot infect the gastrointestinal tract.

Enterovirus species = resistant to wide pH range and can cause gastrointestinal infection.

2. **Transmission:** Enteroviruses are spread easily in crowded locations (e.g., households, schools, and communities) and areas of poor sanitation (e.g., water supply contaminated with sewage).
   - They can be transmitted by several routes.
     - Fecal-oral
     - Contact with respiratory secretions
     - Direct contact with other body fluids (e.g., conjunctival fluid and vesicular fluid)

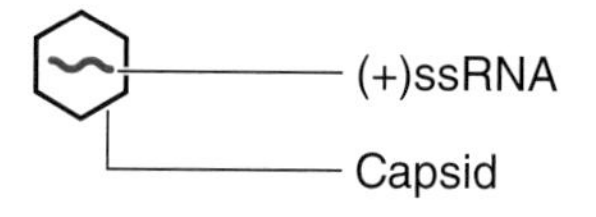

Figure 6.1. Enterovirus.

Table 6.1. Old and new categorization of enteroviruses

GENUS	HUMAN SPECIES	OLDER GROUPING
Enterovirus	*Enterovirus A*	Coxsackie virus group A Enteroviruses
	Enterovirus B	Coxsackie virus group B Echoviruses Enteroviruses
	Enterovirus C	Polioviruses Coxsackie virus group A Enteroviruses
	Enterovirus D	Enteroviruses
	Rhinovirus A *Rhinovirus B* *Rhinovirus C*	
Parechovirus	*Parechovirus A* *Parechovirus B*	Genus *Enterovirus*

- Contaminated fomites
- Vertical transmission during delivery
- Nosocomial spread, especially among newborns

- Seasonality
  - Temperate climates: **summer and fall**
  - Tropical climates: year-round

3. **Clinical presentation:** Most enterovirus and parechovirus species are extremely common causes of infection, although the mortality rate is very low and most infections are asymptomatic or mild. They can cause a wide range of overlapping clinical symptoms, and different syndromes occur depending on the age group (Table 6.2).
   - Incubation period: 7 to 14 days for most strains and syndromes
   - **Common cold:** nasal congestion, rhinorrhea, cough, and pharyngitis
   - **Lower respiratory tract infection:** uncommon but can be severe

Table 6.2. Typical age range and presentation of enterovirus strains

CLASSIC DISEASE	TYPICAL AGE OF INFECTION (YEARS)	IMPORTANT STRAIN(S)[a]
Severe, invasive infection	Neonates	
Poliomyelitis (note: almost eradicated)	<5	PIV1–PIV3
Hand, foot, and mouth disease	<10	CVA16
CNS disease (meningitis, acute flaccid myelitis)	5–15	EV-D68, EV71
Pleurodynia	10–20	CVB3, CVB5
Myocarditis	20–40	
Acute hemorrhagic conjunctivitis	All	EV70, CVA24v
Other (gastroenteritis, common cold)	All	

[a]CVA, coxsackievirus group A; CVB, coxsackievirus group B, EV, enterovirus; PIV, poliovirus

- **Gastroenteritis:** mild, self-limited diarrhea
- **Nonspecific exanthems**: fever with maculopapular rash over face, neck, and trunk
- **Herpangina**: sudden fever with small, vesicular ulcers or blisters that form in the mouth and throat that are often **painful** (Fig. 6.2). It is associated with CVA.
- **Hand, foot, and mouth disease** (*see* Fig. 4.11)
    - Sore throat, low fever, and vesicles that appear in the oral cavity
    - Papules, petechiae, or vesicles also appear on the skin of the hands and feet, including the palms and soles.
    - It is mostly caused by CVA, especially CVA16. EV71 has caused a hand, foot, and mouth disease outbreak that also involved the CNS.
- **Conjunctivitis:** Mild conjunctivitis may occur along with other enteroviral syndromes, with an incubation time of ~5 to 7 days.
- **Acute hemorrhagic conjunctivitis**: a severe infection that has rapid onset (~1 day) and causes epidemics of conjunctivitis (Fig. 6.3)
    - It presents with edema and subconjunctival hemorrhage.
    - It is highly infectious and can be transmitted through direct contact or with fomites contaminated by conjunctival secretions.
    - It typically resolves in 5 to 10 days without any long-term sequelae, except the risk of bacterial superinfection.
    - Associated with EV70 and CVA24v
- **Neonatal infection:** Virions are likely transmitted during delivery from a mother that has been recently infected. Infection can cause newborn myocarditis or fulminant hepatitis in the first week of life, which can lead to death.
- **Meningitis and encephalitis:** Enteroviruses are the most common causes of viral, or "aseptic", meningitis. Other features of enteroviral infection, such as rash and upper respiratory and abdominal symptoms, are usually also present.
    - Symptoms resolve within ~7 to 10 days, usually with no long-term sequelae.
    - More frequent in children <5 years old
- **Cardiac inflammation:** Enteroviruses (especially coxsackievirus) are frequent causes of viral myocarditis and pericarditis in children or young adults.
- **Poliomyelitis:** a biphasic illness with initial nonspecific symptoms followed by severe CNS infection that can result in paralysis
    - Minor illness: fever, headache, abdominal pain, lethargy, and pharyngitis. Symptoms last ~1 to 3 days.
    - Major illness: in about 1% of infections patients appear to recover for a few (2 to 5) days but then progress to severe disease, including the following.
        - **Meningitis**, fever, and malaise
        - Severe **myalgia**, paresthesias, and muscle spasms
        - Lesions in anterior horn of the spinal cord that affect motor neurons over the course of 2 to 3 days

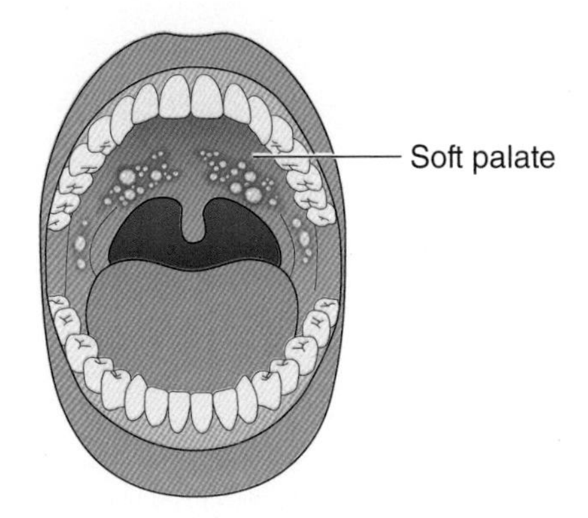

Figure 6.2. Herpangina. Painful oral lesions.

Enteroviruses are the most common cause of viral meningitis.

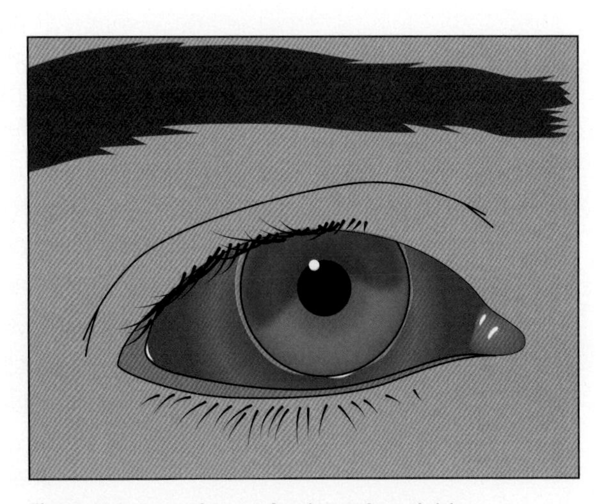

Figure 6.3. Acute hemorrhagic conjunctivitis.

- **Acute flaccid paralysis:** paralysis occurs over the course of hours to days with loss of muscle strength that is usually asymmetric (paralysis begins in an arm or leg on only one side of the body). Paralysis is irreversible. Respiratory paralysis can cause death in 5 to 10% of people.

- Poliomyelitis is caused by polioviruses 1 to 3 **(PIV1 to PIV3)**. These wild-type viruses have been globally eradicated, except in Pakistan, Afghanistan, and possibly Nigeria. However, cases of vaccine-derived poliovirus have continued to occur in some countries of Africa and the Middle East.

Wild (indigenous) polioviruses have been eradicated except in Pakistan, Afghanistan, and possibly Nigeria.

- Polioviruses are acquired fecal-orally, like other enteroviruses, through contaminated water (e.g., swimming pools) and food. After exposure, viruses can be detected for 2 weeks in the throat and are shed for up to 4 months in the stool.
- Risk factors
  - Travel to areas of endemicity by unvaccinated individuals
  - Paralysis is more likely in males.
  - Pregnancy and exercise increase severity of the disease.

Nonpolio enteroviruses can also cause a polio-like illness called acute flaccid myelitis.

- **Acute flaccid myelitis:** similar to polio but is caused by enteroviral strains other than PIV1 to PIV3 (e.g., EV-71 and **EV-D68**) and some arboviruses (e.g., West Nile virus). Infection begins with fever and upper respiratory symptoms and progresses to severe muscle weakness. Symptoms include the following:
  - Drooping face
  - Weakness, paralysis
  - Slurred speech
- **Pleurodynia:** fever and sharply painful, unilateral chest spasms that make it hard to breathe. It is also known as "the devil's grip" or Bornholm disease and is associated with CVB.

4. **Diagnosis:** Most enteroviral infections are mild, self-limited, and do not need laboratory confirmation.
   - Detection of enteroviruses does not necessarily indicate infection.
     - Virions and virus DNA can be detected in multiple specimen types, even during localized infections. So it is very important to sample the precise location of infection (e.g., CSF for meningitis, not stool or nasopharynx).
     - They are highly prevalent in humans and may be present as a coinfecting pathogen.
     - They may be shed for a long time after infection.
   - Specimen
     - Rectal swab or stool
     - Respiratory specimens (including swab of nose or throat)
     - Vesicular fluid
     - CSF, if CNS disease is suspected
     - Blood and urine in neonates
   - NAAT: highly specific and sensitive and is very useful for detection and differentiation of enteroviruses and parechoviruses

- Culture: Most, but not all, *Enterovirus* and *Rhinovirus* species can be grown in cell lines.
  - They grow in primary cells (e.g., RMK cells, like other RNA viruses) but also in diploid cells (e.g., MRC-5 cells, unlike other RNA viruses).
  - They are lytic viruses that cause CPE within several days.
  - CPE: refractile small, rounded cells; may be tear shaped
- Identification of enteroviruses from stool is not specific and is difficult to correlate with the clinical picture. Virus can be shed in the stool for many months after a previous infection, or patients may have subacute infection unrelated to the symptoms being investigated. Enterovirus can also be detected in environmental sources.

There are many enteroviruses and related species; culture and molecular amplification assays may not detect all of them.

5. Prevention and treatment

- **Prevention:** Vaccines are available for poliovirus but not for other enteroviruses. These are highly efficacious and are part of routine childhood vaccinations. There are two formulations.
  - **Oral poliovirus:** the most commonly used formulation in areas of endemicity. It is sometimes called the **Sabin vaccine**.
    - Live attenuated vaccine
    - Provides excellent protection against the three poliovirus strains. However, it very rarely causes cases of paralytic polio.
  - **Inactivated poliovirus:** primarily used in developed countries where the risk of polio is very low. This is also known as the **Salk vaccine**.
    - An inactivated polio vaccine that does not cause disease but produces a slightly lower level of protection against the three strains of poliovirus
    - Hard to make safely, because a large amount of virulent virus has to be produced before it can be inactivated
- Treatment
  - None, only supportive care for symptoms
  - For polio, ventilators may need to be used to help with breathing. Positive pressure ventilators have replaced now-obsolete negative pressure ventilators ("iron lungs").

Sal**k** vaccine: inactivated ("**k**illed") virus

Sabin vaccine: live attenuated poliovirus. It may cause vaccine-derived polio.

II. ADENOVIRUSES. Adenoviruses are common causes of respiratory, ocular, and gastrointestinal infections. They can also cause urinary tract and CNS infections. They cause severe and prolonged infections in immunocompromised patients, although some strains can also cause severe infections in immunocompetent persons.

1. Background: medium-sized (~90 nm), nonenveloped, dsDNA virus (Fig. 6.4)

- There are 7 species that infect humans. These are classified into >60 genotypes/serotypes (serotyping was used for classification of earlier strains, while genotyping is used for newer strains) that cause distinct or overlapping clinical syndromes (Table 6.3).
- There are many animal-specific species as well.
- Adenovirus is a lytic virus. It produces many thousands of virions in each host cell before bursting it open. Occasionally, adenoviral DNA remains in the cell episomally and causes a long-term infection (months to years).

Adenovirus can be latent in humans for many months.

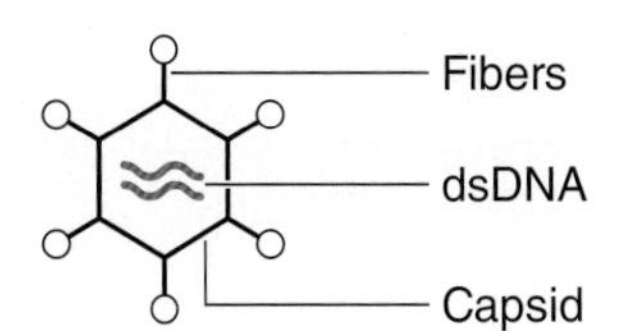

Figure 6.4. Adenovirus.

2. Transmission

- Primarily transmitted through respiratory droplets (i.e., it is found in the adenoids)
- Fecal-oral (can be waterborne, e.g., transmitted through swimming pools)
- Body fluids and secretions like tears (direct contact or contaminated fomites)
- Associated with crowded environments (e.g., schools and barracks)
- Adenoviruses are highly stable and can survive for several weeks on environmental surfaces. They are relatively resistant to alcohol disinfectants. Contact precautions should be used for incontinent inpatients.
- They are not seasonal and occur year-round.

3. Clinical presentation: Adenoviruses are common causes of respiratory, gastrointestinal, and ocular infections. Most infections are mild, but some can be severe and prolonged, especially in immunocompromised patients. Different serotypes/genotypes are often associated with specific infections (Table 6.3).

- Respiratory tract infection: Most serotypes cause an asymptomatic or mild respiratory infection. They are often acquired during childhood.
  - Incubation period: ~1 week
  - **Common cold:** self-limited, cough, fever, pharyngitis, and rhinorrhea
  - **Acute respiratory distress syndrome** (ARDS): difficulty breathing, pleural effusions, and respiratory failure. ARDS can be fatal. Epidemics occur in adults.
  - **Pneumonia:** Children <2 years old have high morbidity and mortality rates.
- Ocular infection
  - **Conjunctivitis:** Adenovirus is one of the most common causes of follicular conjunctivitis ("pinkeye").
    - Infection is usually self-limited, but patients are highly contagious when they have watery discharge (e.g., household contacts have a high risk).
    - Redness, inflammation, itchiness, and blurry vision. Conjunctivitis is first unilateral and then bilateral.

Military recruits are at risk of severe respiratory disease by Ad4 and 7

Table 6.3. Adenoviral serotypes associated with particular diseases

DISEASE	COMMONLY ASSOCIATED ADENOVIRAL GENOTYPE(S)/SEROTYPE(S)
Epidemic keratoconjunctivitis	8, 19, 37
Mild upper respiratory tract infection	1, 2, 5, 6
Acute respiratory distress syndrome	**4, 7**, 14
Gastroenteritis	**40, 41**
Acute hemorrhagic cystitis	11, 21
Pharyngoconjunctival fever	3
Obesity	36

  - **Pharyngoconjunctival fever:** upper respiratory tract symptoms with low fever and conjunctivitis. Children in summer camps are at high risk.
  - **Epidemic keratoconjunctivitis:** conjunctivitis for 1 to 4 weeks which then progresses to keratitis.
    - Affects adults but not children
    - Causes ocular pseudomembranes
    - Can cause loss of visual acuity for weeks to months
    - Can cause epidemics
- **Hemorrhagic cystitis:** occurs in children. It is more common in boys (unlike bacterial cystitis, which is more common in girls).
- **Gastroenteritis:** watery, nonbloody diarrhea
- Other diseases: hepatitis, encephalitis, myocarditis

4. **Diagnosis:** Testing is not necessary in most cases since infections are mild and/or self-limited. Also, adenovirus can be shed in asymptomatic people for many months after a primary infection, so detection does not necessarily indicate an active infection. In serious infections, adenovirus can be diagnosed in several ways.
   - Specimen: should be taken from the infected area only. Specimens that are often used are as follows.
     - Respiratory (throat/nasopharyngeal swabs, tissue, sputum, and bronchial aspirate/lavage samples)
     - Stool
     - Conjunctival swab
     - Urine
     - Plasma: for disseminated disease, especially in transplant patients
     - Tissue biopsy
   - NAAT: a commonly used platform for all specimen types. However, note that there are many strains of adenovirus and NAATs may not detect all of them.
     - It is more sensitive, specific, and rapid than culture.
     - Quantitative assays can be used for monitoring prolonged infection or treatment (e.g., for immunocompromised patients).
   - Culture and shell vials: can be used, especially for respiratory adenoviral infections, because most serotypes grow well in culture. It is not used for gastrointestinal infections because Ad40 and Ad41 do not grow in culture.
     - Most adenoviruses show good growth in A549 (immortalized cells) and human diploid cells (e.g., MRC-5 cells).
     - They have a moderate rate of growth (~3 to 7 days) in culture. They can be detected in 1 to 2 days from shell vials
     - Adenoviral CPE is typically seen as rounded cells in clumps ("**grape-like clusters**")
     - To increase diagnostic sensitivity, the specimen can be freeze-thawed to help break open cellular membranes and release more viral particles.
   - Histology: Infected cells have enlarged nuclei that look smudged (**smudge cells**) (*see* Fig. 13.6). Identifying the virus in tissue is specific for active infection, rather than viral shedding.
   - Antigen detection

Detection of adenovirus does not necessarily indicate active infection. Virions can be shed in respiratory and stool specimens for weeks to months after infection. Many latent respiratory adenoviruses can also be swallowed and then detected in the stool.

- EIA can indicate active infection in patients with diarrhea. These assays have low specificity.
- Point-of-care lateral-flow assays for conjunctivitis are rapid and have good sensitivity.

- DFA: used to rapidly screen respiratory specimens if NAATs are not available. However, culture should continue to be performed because DFA has poor sensitivity.
- Serology: A 4-fold increase in IgG titer between acute- and convalescent-phase sera can indicate recent infection. However, serology is not typically very useful because adenoviruses are common in humans.
- Strain typing is not commonly available or performed but helps to identify serotypes that are associated with certain diseases (Table 6.3) and for outbreak investigations.

5. **Prevention and treatment**
   - Prevention: There is an oral, live vaccine against Ad4 and Ad7. It is not part of the routine vaccination series but it is administered to military recruits.
   - Treatment
     - Most infections are self-limited. Immunocompromised patients with nonresolving infections may need treatment.
     - There are no FDA-approved drugs, but **cidofovir** has been used and may improve outcomes.
     - Ocular infections: supportive care, such as cold compresses and drops containing antihistamines.

III. PARVOVIRUS B19. Parvovirus B19 is usually acquired during childhood. It causes an asymptomatic or mild respiratory infection with a characteristic "slapped cheek" rash. If acquired during adulthood it can cause painful, swollen joints and anemia. Infection during pregnancy can cause severe fetal edema, anemia, and even fetal death. Rarely, parvovirus B19 can cause a constellation of other symptoms, including meningoencephalitis and myocarditis.

1. **Background:** Parvovirus B19 is a small, naked ssDNA (either positive or negative sense) virus in the family *Parvoviridae* ("parvus" means small) (Fig. 6.5).
2. **Transmission:** Parvovirus B19 is transmitted primarily through respiratory secretions. It then infects red blood cell precursors, so it may also be transmitted by blood products.
   - Respiratory droplets (cough or sneeze)
   - Transplants, blood transfusion
   - Transplacentally
   - Seroprevalence is high (~60%).
   - Persons at risk of severe disease: immunosuppressed people, fetuses (pregnant women should avoid primary exposure)

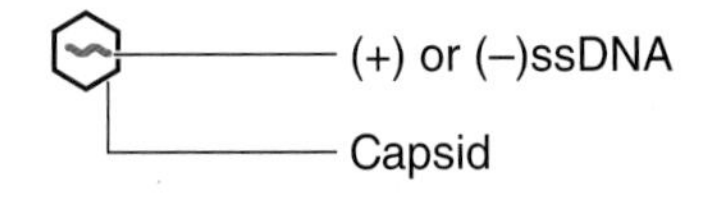

Figure 6.5. Parvovirus B19.

3. **Clinical presentation:** Parvovirus B19 infection is most common in children and adolescents. Adult infections are less common and usually present with severe polyarthralgia. Previous exposure results in immunity.
   - Incubation period: 1 to 2 weeks
   - **Fifth disease** (or **erythema infectiosum**): In children, infection usually presents with characteristic rashes (*see* Fig. 4.11).

- First phase: The patient may be asymptomatic or have headache or rhinitis, with or without fever. A characteristic red "slapped-cheek" rash may appear on the face that lasts for 2–3 days. This stage is infectious.
- Second phase: A second, often itchy, rash can appear on the trunk, arms, and/or legs. This has a **lacy** pattern as it fades.

- **Polyarthralgia** and painful joint swelling is a common presentation. Symptoms usually improve in 2 weeks, but can recur or persist for months to years. These symptoms are more severe in adults.
- **Papular purpuric gloves and socks syndrome** (*see* Fig. 4.11):
  - Painful red rash on palms and soles, with a sharp demarcation at the ankles and wrists
  - This is a rare presentation that occurs mostly in young adults.
  - Lasts 1 to 2 weeks
- **Fetal infection:** Parvovirus B19 can cause serious complications in a fetus, such as anemia, **hydrops fetalis** (abnormal accumulation of fluid in the fetus), preterm birth, or fetal loss.
- **Transient aplastic crisis:** production of red blood cells stops temporarily. People with anemias (e.g., sickle cell disease, thalassemia, HIV) are at risk of severely low hemoglobin levels.
- Other: may also cause cardiovascular, liver, neurologic, renal, and respiratory disease. Immunosuppressed persons may present with pure red blood cell aplasia.

Infection in children results in slapped-cheek rash. Infection in adults results in joint pain.

4. **Diagnosis:** typically needed only for pregnant women or in cases of serious or congenital infections
   - Serology: a common method of diagnosis for immunocompetent individuals. Immunocompromised patients may not mount an adequate antibody response and can have false-negative results.
   - Immunohistochemistry: Parvovirus B19 antigens are detected from placental or fetal tissue. This test has low sensitivity but high specificity.
   - NAAT: the preferred method for testing immunosuppressed patients because they may not mount a robust antibody response. The ideal specimens for PCR are as follows.
     - Blood and bone marrow because the virus infects red blood cells. Viremia during an acute infection is extremely high.
     - The virus can also persist in tissues and sterile body fluids. However, this may or may not represent active disease.
   - Culture: Parvovirus B19 does not grow in commonly used cell culture lines.

Serology should be used for pregnant women only if they are exposed or develop symptoms. Routine testing is not recommended because it increases the chance of false-positive results.

Parvovirus B19 DNA persists in tissues and synovial fluid (detection by PCR may not represent active disease).

5. **Prevention and treatment:** supportive. No treatment or vaccine is available.

IV. HUMAN BOCAVIRUS. Human bocavirus (HBoV) is a potential human pathogen. It may cause respiratory or gastrointestinal symptoms, but it is present in asymptomatic individuals and has also been detected in the presence of other likely pathogens.

1. **Background:** small, nonenveloped, ssDNA virus (positive or negative sense) (Fig. 6.6). HBoV is a member of the *Parvoviridae* family, like parvovirus B19.
   - There are four human genotypes, 1 to 4.

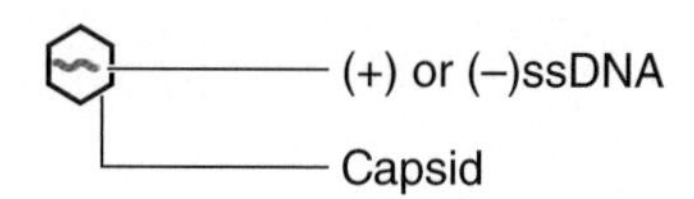

Figure 6.6. Human bocavirus.

- The genus ***Boca**virus* also contains several animal species, including *bo*vine and *ca*nine viruses.

2. **Transmission:** spread by respiratory and gastrointestinal routes. It is highly prevalent (>70% of the population) and is usually acquired during childhood.

3. **Clinical presentation:** The role of HBoV as a pathogen is debated because it is frequently detected in asymptomatic individuals. However, it is also linked with the following symptoms:
   - Upper and lower respiratory infections: associated with HBoV1
     - Cough, rhinorrhea, croup, fever, and bronchiolitis
     - Pneumonia in immunocompromised patients
   - Acute flaccid paralysis
   - Gastroenteritis: associated with HBoV2, -3, and -4

HBoV may or may not be a pathogen.

4. **Diagnosis:** Detection of HBoV does not necessarily indicate disease.
   - NAATs are the primary method of detection but are not commonly available.
   - Specimens
     - Respiratory infection: respiratory samples. Virus may be detected in the stool after being swallowed.
     - Gastroenteritis: stool

5. **Prevention and treatment:** none

Multiple-Choice Questions

1. **Which of the following is NOT a picornavirus?** *(May require reading outside of this chapter.)*
   a. Enterovirus
   b. Rhinovirus
   c. HAV
   d. HEV

2. **Which of the following is NOT a Baltimore class II virus?** *(May require reading outside of this chapter.)*
   a. HPV
   b. Parvovirus B19
   c. HBoV
   d. Torque teno virus

3. **Canine parvovirus is highly contagious and can cause fatal disease in unvaccinated dogs. Which of the following is true?** *(May require reading outside of this chapter.)*
   a. This is the same species of parvovirus that causes disease in humans.
   b. This virus is not transmitted to humans.
   c. This virus is airborne.
   d. This virus has been eradicated.

4. **Which characteristic feature can be used to differentiate between rhinoviruses and other enteroviruses?**
   a. Rhinoviruses cause respiratory infection but enteroviruses do not.
   b. Enteroviruses are transmitted by the fecal-oral route but rhinoviruses are not.

c. Enteroviruses are ~10% bigger than rhinoviruses.

d. All of the above

5. Which of the following causes a rash on the palms and soles? *(May require reading outside of this chapter.)*

a. *Treponema pallidum*, *Rickettsia rickettsii*, and enterovirus

b. *Rickettsia typhi*, *Treponema pallidum*, and VZV

c. Enterovirus, adenovirus, and parvovirus B19

d. All of the above

6. Which of the following adenoviral serotypes causes gastrointestinal disease?

a. Ad5

b. Ad7

c. Ad36

d. Ad41

Match the following. Use each answer only once.

7. Which viruses are most commonly associated with these important clinical presentations?

Acute respiratory disease syndrome	A. CMV
Hydrops fetalis	B. Parvovirus B19
Pleurodynia	C. Adenovirus
Mononucleosis	D. Enterovirus

True or False

8. Poliovirus has been eradicated. **T F**

9. Acute flaccid myelitis or poliomyelitis is transmitted by the fecal-oral route. **T F**

10. Detection of enterovirus in stool samples by PCR does not necessarily indicate current, active infection. **T F**

11. Pregnant women should receive the parvovirus B19 vaccine. **T F**

7

CHAPTER 7

OPPORTUNISTIC VIRUSES ASSOCIATED WITH IMMUNOSUPPRESSION

I. OVERVIEW. People that are significantly immunosuppressed, like solid-organ transplant, hematopoietic stem cell transplant, or AIDS patients, are susceptible to severe, opportunistic viral infections.

1. **Background:** The most frequent opportunistic infections are caused by common latent viruses. Many of these are dsDNA viruses.
    - Important **transplant-associated** viruses: CMV, EBV, and BK, JC, and respiratory viruses
    - Important **AIDS-associated** viruses: HHV-8, CMV, HPV, VZV, HSV, and EBV (*see* Table 9.1)
2. **Transmission:** Opportunistic viruses are highly prevalent in the population and are usually acquired in childhood. Occasionally, they are transmitted within donor-derived material.
3. **Clinical presentation:** Primary infection with opportunistic viruses in immunocompetent individuals is usually asymptomatic or mild. Primary or reactivated disease in immunocompromised patients can have high morbidity and mortality.
4. **Diagnosis:** Opportunistic infections are difficult to diagnose because detection may not represent true disease.
    - NAATs: can be used to quantitate the presence of virus and to monitor viral loads over time
    - Serology: may not be helpful. Immunosuppressed patients may not produce a robust immune response and may have falsely negative viral serology.
5. **Prevention and treatment:** The approach to treatment of opportunistic viruses is usually to reduce immunosuppression (e.g., reduce immunosuppressive drug dosage or administer antiretrovirals for HIV^+ patients), although some effective antiviral agents are available (e.g., ganciclovir for CMV infection).

II. CYTOMEGALOVIRUS. CMV is highly prevalent in the population (>60%). Like the other herpesviruses, it typically causes asymptomatic or mild primary symptoms that progress to latent, lifelong infection. However, congenital infection or reactivation in immunosuppressed individuals can cause severe disease in virtually any organ in the body.

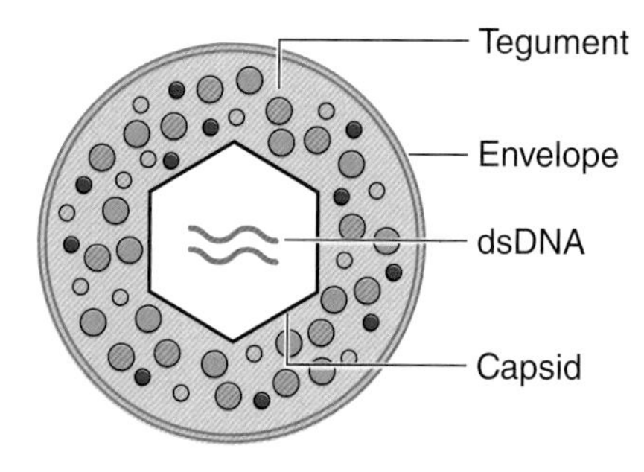

Figure 7.1. CMV.

The heterophile antibody test will be negative in CMV mononucleosis.

1. **Background:** CMV is an enveloped dsDNA virus that is part of the *Herpesviridae* family (Fig. 7.1). It is also known as HHV-5 (*see* Box 4.1, Table 4.1).
2. **Transmission:** mostly occurs via virus shedding in body fluids (saliva, blood, tears, semen, breast milk, and urine). Other routes of transmission include the following.
   - Transfusions
   - Organ transplants
   - *In utero*, when primary infection or reactivation occurs during pregnancy
   - Groups at risk of severe disease: HIV^+ persons, people receiving immunosuppression, and transplant patients. Frequency of CMV infections is especially high for the following.
     - Solid-organ transplant patients: seronegative recipients with seropositive donors (primary disease)
     - Hematopoietic stem cell transplant patients: seropositive recipients (reactivated disease)
3. **Clinical presentation:** CMV can infect many different cell types and cause a wide range of symptoms. Infections in immunocompetent hosts are asymptomatic or show nonspecific symptoms, like fever, pharyngitis, malaise, and lymphadenopathy. In immunocompromised hosts, disease can occur in almost any part of the body. Common disease presentations include the following.
   - **Mononucleosis** in immunocompetent hosts
     - Fever, lymphadenopathy, and lymphocytosis in peripheral blood. Mild hepatitis and/or a scarlatiniform rash may also occur.
     - Unlike with EBV, the heterophile antibody test will be negative.
   - **Gastrointestinal** disease in immunocompromised patients: fever, abdominal pain, and bloody diarrhea
   - **Esophagitis in HIV+ and transplant patients:** nausea, vomiting, painful swallowing, diarrhea, and chest pain. HSV is another common cause of infectious esophagitis.
   - **Retinitis in AIDS patients with a CD4 count of <50 cells/µl:** (*see* Table 9.1). Without treatment, it can progress to blindness.
   - **Congenital disease in neonates:** disease within the first 2 to 3 weeks of life. After this time, the infant may have acquired CMV postnatally. Most infected infants are asymptomatic, but some have neurologic birth defects, such as microcephaly, anemia, jaundice, hearing loss, seizures, or death.
   - **Transplant associated**
     - Fever, leukopenia, thrombocytopenia, organ inflammation (hepatitis, pancreatitis, or pneumonitis), graft rejection, and death
     - Disease occurs early (within ~3 months) after transplant
     - Interstitial pneumonia in bone marrow transplant patients. This has a high mortality rate.
4. **Diagnosis:** Testing is not performed for mild, self-limited infections but is used to assess disease and risk of disease in immunocompromised persons.
   - Specimen: The virus can be present even in asymptomatic infections, so disease should be confirmed in specimens collected only from the site of infection.

- Body fluids: ocular fluid, blood fractions, CSF, amniotic fluid, urine, and saliva
- Tissue biopsy: from any organ, especially intestinal, retinal, skin, and esophageal
- Respiratory: throat swab/wash, bronchoalveolar lavage fluid

- Serology
    - Used to identify the CMV status of transplant donors and recipients or to identify primary infection in pregnant women
    - Difficult to interpret because CMV is highly prevalent in the population. Also, immunosuppressed patients (the main population for CMV disease) may not produce adequate antibody levels.
- NAATs
    - Highly sensitive and specific
    - PCR from blood is used to quantitate viremia in patients with disseminated disease and measure resolution of infection.
    - There is no standardized viral load threshold to begin treatment (it varies, usually between 1,000 and 3,000 IU/ml).
- Culture: CMV grows on MRC-5 cells but is a very **slow-growing** virus (7 to 14 days). Shell culture assays improve turnaround time.
- Histology: Actively infected cells have large nuclear inclusions surrounded by a clearing, so the cells look like "**owl's eyes**" (Fig. 7.2; *see also* Fig. 13.6).
- Antigenemia: determined by an antigen-based assay that is more sensitive than culture-based methods for the detection of CMV in blood. It is used to quantitate viremia and monitor progression of the virus in patients with disseminated disease (e.g., transplant or HIV^+ patients). It is not performed often anymore because it is labor-intensive and PCR is more rapid and sensitive. The method is as follows.

Culture of CMV in blood has poor sensitivity. PCR or antigenemia assays should be used on blood instead.

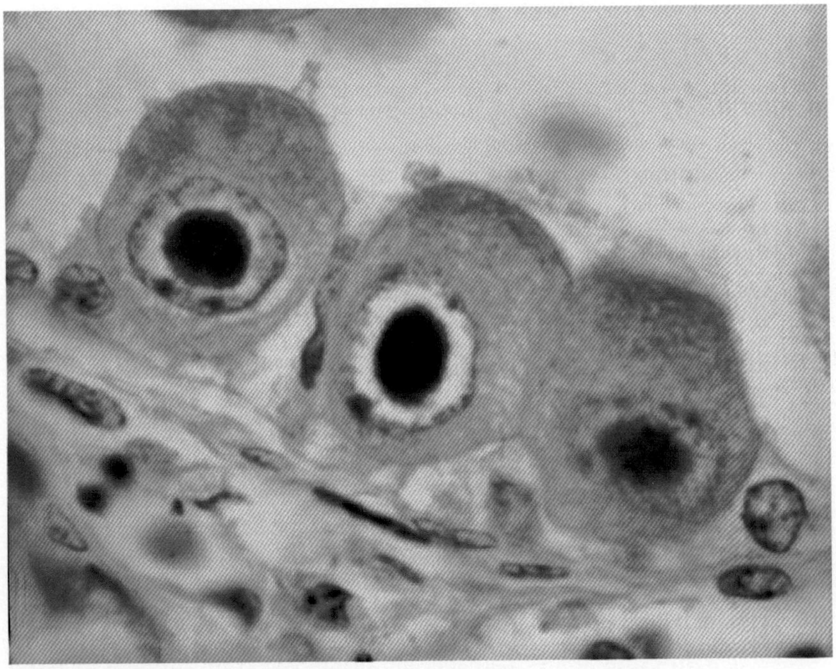

Figure 7.2. Owl's eye inclusions caused by CMV infection. Image courtesy of CDC/ Rosalie B. Haraszti, M.D. (CDC PHIL ID#1155).

- Dextran is added to patient blood. The mixture is centrifuged to separate polymorphonuclear leukocytes (PMLs) out into the supernatant. Ammonium chloride is added to the PMLs to lyse any remaining red blood cells.
- The PMLs are counted with a hemacytometer. Approximately 2×10^5 leukocytes are centrifuged onto a slide (more blood should be collected from leukopenic patients in order to get an adequate number of cells).
- The cells are fixed with formaldehyde and then permeabilized.
- Fluorescently tagged monoclonal antibodies are added that bind to **pp65**, which is a CMV protein that collects in the nuclei of leukocytes.
- The amount of CMV is quantified by determining the number of stained cells per total number of leukocytes present.

- Resistance testing: Resistance to antivirals can occur in immunosuppressed patients treated for a long time.
  - PCR amplification and then sequencing of the UL97 and UL54 genes can identify mutations associated with resistance to ganciclovir, cidofovir, and foscarnet.
  - Sequencing may not work if viral load is too low.

5. **Prevention and treatment**
   - Prevention: no vaccine is available. Letermovir is a recently approved antiviral that can be used for prophylaxis of hematopoietic stem cell transplant patients.
   - Treatment: not needed in immunocompetent people, but mortality is high without treatment in patients with immunosuppressive conditions.
     - Reduce immunosuppression.
     - Antivirals: mainly **ganciclovir** and valganciclovir
     - CMV intravenous immunoglobulin is a pool of human IgG that can be given in addition to antivirals.

Unlike the other herpesviruses, CMV is intrinsically resistant to acyclovir.

III. BK VIRUS. BK virus is a highly prevalent virus that is named after the initials of a patient. It causes asymptomatic, lifelong infection in the kidneys and is shed in urine. The virus can reactivate in transplant and other immunosuppressed patients and cause severe kidney damage.

1. **Background:** nonenveloped, small, and contains dsDNA. It is a member of the *Polyomaviridae* family (Fig. 7.3).
2. **Transmission:** acquired in childhood, and the prevalence is very high (up to 100%). The route is unknown but is probably respiratory or transplacental.
3. **Clinical presentation**
   - Primary infection: almost always asymptomatic but remains latent for life in many tissues, especially uroepithelial cells
   - Reactivation: almost exclusively occurs when patients are immunosuppressed, especially renal and bone marrow transplant patients. The virus reactivates and replicates in the kidneys. It is excreted in the urine at high titers. The patient can also become viremic.
     - Can cause kidney damage and failure called **BK virus-associated nephropathy (BKVAN)**
     - Can also cause nonhemorrhagic or hemorrhagic **cystitis**

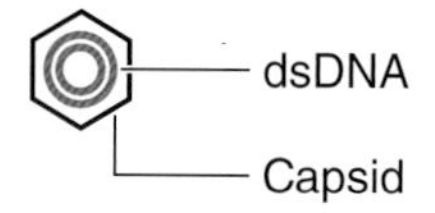

Figure 7.3. **BK virus.**

4. **Diagnosis:** Virus is shed periodically and asymptomatically in healthy individuals and transplant patients. So detection does not necessarily indicate infection.
   - NAATs
     - Testing is performed in cases of severe disease or to monitor levels in renal transplant patients.
     - Specimens: primarily urine (highest sensitivity), but blood has greater correlation with active infection (highest specificity)
   - Immunohistochemistry
     - Is the gold standard for diagnosis of BKVAN because it is highly specific for active infection. The number of infected tubular cross-sections should be determined.
     - Immunostaining is actually performed using an **anti-SV40 antibody**, which cross-reacts with BK virus.
     - Specimen: renal biopsy
   - Urine cytology: Identification of **decoy cells** (cells that look like cancer cells) may indicate BK virus (or CMV) infection, but this test has very poor sensitivity and poor specificity compared to NAATs.
   - Serology: not useful for diagnosis because of high prevalence
     - Can be used for screening transplant donors and recipients. However, there are no formal guidelines for screening patients pre- and post-transplant.
     - BK virus-seropositive donors increase the risk of BK virus-associated nephropathy in recipients (even if a recipient is seropositive too).
5. **Prevention and treatment**
   - Treatment is not needed for immunocompetent individuals.
   - For other symptomatic patients, reduce immunosuppression.
   - Can use cidofovir in low doses

IV. JC VIRUS. Like BK virus, JC virus is highly prevalent and causes an asymptomatic lifelong infection. In immunosuppressed patients it can reactivate to cause severe disease in the brain called progressive multifocal leukoencephalopathy.

1. **Background:** is a member of the *Polyomaviridae* family, like BK virus. It is nonenveloped and small and contains dsDNA (Fig. 7.4). It is also named after the initials of a patient.
2. **Transmission:** acquired in adolescence. The route is unknown. Prevalence is very high.
3. **Clinical presentation**
   - Primary infection: almost always asymptomatic, and infection remains latent for life in many tissues
   - Reactivation: almost exclusively occurs when patients are immunosuppressed. It is associated with malignancy, AIDS, transplantation, and immunosuppressive therapy.
     - **Progressive multifocal leukoencephalopathy (PML)**: reactivation that causes severe CNS disease with seizures and cognitive, motor, and visual defects
       - PML has a high fatality rate.
       - Asymmetric lesions can be seen with neuronal imaging.

Patients taking natalizumab are at higher risk for PML.

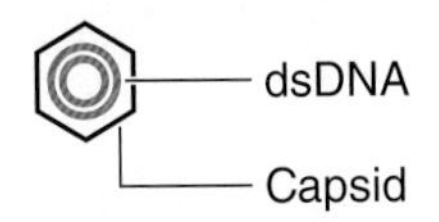

Figure 7.4. JC virus.

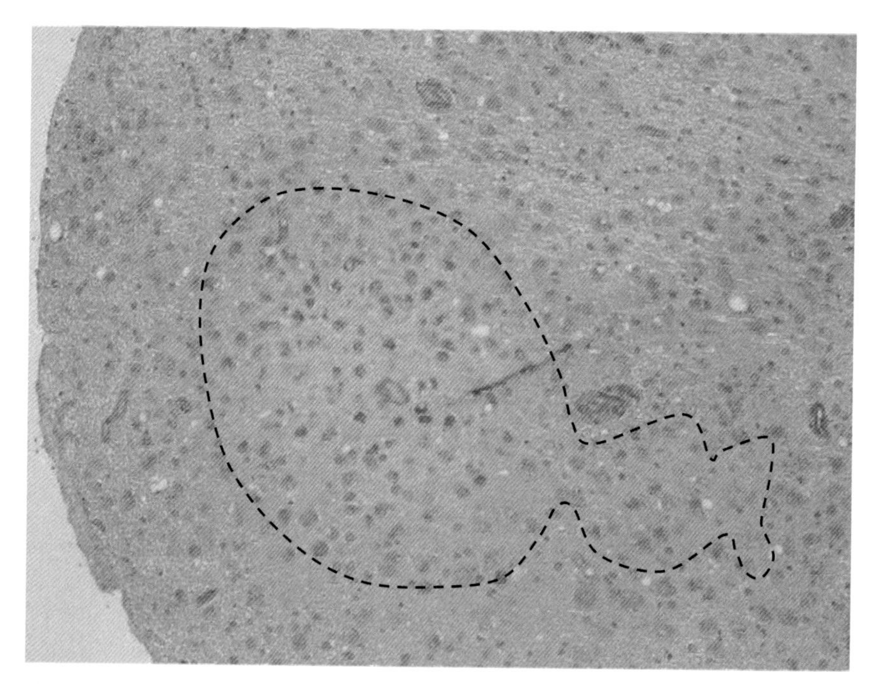

Figure 7.5. **JC virus-infected brain tissue stained with luxol fast blue.** Blue, staining of myelin; pink (dotted area), demyelination. Courtesy of Dennis W. Dickson, M.D., Mayo Clinic.

- Virus infects oligodendrocytes and astrocytes. Neurons are demyelinated.
- PML is insidious and progresses over the course of weeks to months
- Reactivated JC virus can also cause nephropathy, but this is milder and occurs less often than with BK virus

4. **Diagnosis:** Like BK virus, JC virus can be detected in many types of specimens (such as blood, CNS, urine, urine of pregnant women, and tonsils) even in asymptomatic, healthy people. So detection may not indicate disease.
   - Specimen: Tissue, such as renal or brain tissue, is the best sample for identifying active infection. Other sample types include blood and CSF.
   - NAATs: The clinical picture should be correlated with multiple PCRs over time (i.e., trends).
   - Tissue staining: Immunohistochemistry (for proteins) or *in situ* hybridization (for DNA) is the gold standard for detecting JC virus because it shows active infection. Other histologic features include demyelination (Fig. 7.5), bizarre astrocytes, and oligodendrocytes with enlarged nuclei (*see* Fig. 13.6).
   - Serology: not useful for diagnosis because of high prevalence. It may be used for screening transplant patients before starting immunosuppressive therapy.
5. **Prevention and treatment:** Reduce immunosuppression.

Multiple-Choice Questions

1. **Transplant-associated viruses refer to viral infections**
   a. Acquired from a donor specimen
   b. From an endogenous virus that has reactivated
   c. In a transplant patient that was acquired from the community
   d. All of the above

2. A 66-year-old male with a history of a liver transplant complained of left-sided weakness. Which of the following results would be most indicative of progressive multifocal leukoencephalopathy? *(May require reading outside of this chapter.)*

a. A magnetic resonance image showing multiple white matter lesions

b. A single PCR positive result for JC virus in a urine specimen

c. A negative SV40 stain on a brain biopsy

d. A culture for JC virus on brain biopsy

3. Like BK virus, which of the following has been associated with hemorrhagic cystitis? *(May require reading outside of this chapter)*

a. Adenovirus

b. JC virus

c. Ebola virus

d. Enterovirus

4. Which of the following can be used to treat CMV?

a. Acyclovir

b. Ganciclovir

c. Zanamivir

d. All of the above

5. Which of the following viruses is the most common cause of retinitis in AIDS patients?

a. Adenovirus

b. HSV

c. CMV

d. EBV

6. Which of the following diagnostic methods can result in a definitive diagnosis of BK virus nephropathy?

a. Clinical symptoms

b. A single PCR positive result from urine

c. Culture of BK virus from a renal biopsy

d. *In situ* hybridization from a renal biopsy

7. Which of the following is a member of the family *Herpesviridae*?

a. CMV

b. BK virus

c. JC virus

d. SV40

True or False

8. An effective strategy to treat transplant-associated infections is to reduce immunosuppression. **T F**

9. CMV can cause pneumonia. **T F**

10. Primary CMV infection during pregnancy can cause congenital infection. **T F**

8

CHAPTER 8

BLOOD-BORNE HEPATITIS VIRUSES

I. OVERVIEW. Hepatitis is inflammation of the liver. Most often it is caused by viruses, but it can also be caused by drugs, alcohol, or autoimmune diseases. Viral agents of hepatitis infect **hepatocytes** (liver cells), which damages the liver and disrupts its normal functions. The most important viral causes of hepatitis are hepatitis A, B, C, D, and E viruses (HAV to HEV). See Table 8.1 for a comparison of these five viruses. HBV, HCV, and HDV are blood borne and are covered in greater depth here, while HAV and HEV are transmitted by the fecal-oral route and are covered in chapter 6. Other viruses associated with hepatitis are CMV, EBV, and yellow fever virus.

HAV and HEV = nonenveloped = gastrointestinal
HBV, HCV, and HDV = enveloped = blood borne

1. **Transmission:** HBV, HCV, and HDV are transmitted primarily through body fluids such as blood, semen, and vaginal fluid. They can also be transmitted via the following:
   - Microabrasions. Tiny tears in skin and mucosal membranes can lead to exposure to infected body fluids or objects contaminated with infected blood, such as shared razors and toothbrushes.
   - Contaminated needles
   - Blood transfusions and organ transplants
   - Sexual intercourse
   - Vertical transmission during delivery, due to exposure of the newborn to maternal blood (transmission typically does not occur during gestation)
   - High-risk groups
     - Household contacts
     - Sexual contacts, or people with multiple partners
     - Men who have sex with men
     - Users of intravenous drugs
     - Health care workers with needlestick injuries
     - Those who have received dialysis or blood transfusions, especially prior to routine blood and organ donor screening
     - Incarcerated persons
     - Excessive alcohol consumption will worsen liver damage.

Rule of thumb: The risk of a susceptible person being infected by a needlestick exposure from an infected patient is ~30%, 3%, and 0.3% for HBV, HCV, and HIV, respectively.

Table 8.1. **Comparison of viral agents of hepatitis**

PARAMETER	HAV	HEV	HCV	HBV	HDV
Transmission	Ingestion	Ingestion	Blood	Blood	Blood
Envelope	No	No	Yes	Yes	Yes
Virus grouping	Pico**rna**virus	Hep**e**virus	Flavivirus	Hepa**dna**virus	**D**eltavirus
Nucleic acid	(+) ssRNA	(+) ssRNA	(+) ssRNA	ds**DNA**	(–) ssRNA
Average incubation before acute phase (weeks)	4	6	8	12	12
Chronic phase	No	No	**Frequent** (~75% of individuals)	Infrequent (~5% of adults, **95% of infants**)	Infrequent
Risk factors or groups	Travel to areas of high prevalence	High risk of fulminant hepatitis during **pregnancy**	Intravenous drug use, transfusions (prior to HBV/HCV blood product screening), hemodialysis, incarceration, men who have sex with men, babies born to positive mothers		
High-prevalence areas	Areas with poor sanitation	Areas with poor sanitation	Central Asia, North Africa (especially **Egypt**)	Asia and Pacific Islands, Africa, the Mediterranean, Southeast Asia	The Mediterranean, Middle East, Pakistan, Central Asia
Frequency of occurrence	Most common cause of hepatitis in the world	Common in areas of endemicity	Most common blood-borne hepatitis in the USA	Most common blood-borne hepatitis in the world	Infrequent (5% of HBV)
Vaccine	Yes	No	No	Yes	No, but HBV vaccine effectively protects against infection

2. **Clinical presentation:** hepatitis viruses can cause acute, fulminant, or chronic symptoms.
   - **Acute hepatitis**
     - Low-grade fever and flu-like symptoms
     - Fatigue
     - **Jaundice.** The liver is not able to excrete a pigment called bilirubin adequately. This results in yellowing of the skin and sclera of the eyes, pale stool, and dark urine.
     - Nausea and loss of appetite
     - Hepatomegaly (enlarged liver) and abdominal pain
   - **Fulminant hepatic failure:** a rare condition of severe and sudden liver failure. It has a high mortality rate. Management is limited to supportive treatment and liver transplantation.
   - **Chronic hepatitis:** an unresolved liver infection that has occurred for at least 6 months. Patients may
     - Not have had preceding symptoms of acute hepatitis
     - Still be asymptomatic or have nonspecific symptoms (such as low-grade fever, malaise, and abdominal pain). This is called the carrier state.

Acute hepatitis: infection for <6 months
Chronic hepatitis: infection for ≥6 months

- Progress to scarring of liver tissue (**fibrosis and cirrhosis**), liver cancer (**hepatocellular carcinoma**), and liver failure over many years.

3. **Diagnosis of hepatitis**
   - Serology: the primary screening method for blood borne and gastrointestinal hepatitis viruses. Symptoms between hepatitis viruses overlap, so they are often tested as a panel.
   - NAAT: routinely used for confirmation of diagnosis and for monitoring of HBV and HCV viral load during treatment. NAATs are generally not necessary for HAV and HEV.
   - Liver biopsy: used to determine inflammation and grades of fibrosis. It cannot differentiate between different viral etiologies.
   - Transient elastography: measures the stiffness of the liver with ultrasound in order to grade fibrosis noninvasively.
   - Other laboratory markers of hepatitis
     - Elevated liver injury enzymes, like aspartate transaminase (AST) and alanine aminotransferase (ALT)
     - Alkaline phosphatase, gamma glutamyltransferase (GGT), blood urea nitrogen (BUN), and prothrombin time (PT)
     - Total serum bilirubin, elevated conjugated bilirubin, and presence of conjugated bilirubin in the urine
   - Regular screening (e.g., annual) is recommended for people in high-risk groups.

II. HEPATITIS B VIRUS. HBV can cause significant morbidity and mortality. It is the most common cause of blood-borne hepatitis (worldwide), and approximately 2 billion people have been infected (Table 8.1).

***Hepadna**viridae*: Hepa + dna = hepatitis-causing DNA virus

1. **Background:** HBV is a member of *Hepadnaviridae* and is the only medically relevant **Baltimore VII** virus (Fig. 8.1). It has several important features.
   - It is enveloped.
   - It has a partial dsDNA genome (some parts are single stranded). When it infects a cell, DNA polymerases in the host cell nucleus complete the genome to form **covalently closed circular DNA** (**cccDNA**).
   - It replicates its genome through an RNA intermediate. Because of this intermediate step, the following apply.
     - HBV has a much higher error rate during replication than most DNA viruses. This allows it to mutate rapidly.
     - It produces reverse transcriptase.
     - It can integrate into the host cellular genome.
   - Important proteins used as markers of infection
     - Surface antigen (HBsAg): a protein embedded in the envelope. The virus produces huge amounts of HBsAg, which makes it an excellent marker of HBV infection. HBsAg can even assemble into noninfectious spherical or filamentous/tubular particles.
     - Core antigen (HBcAg): the capsid protein
     - Early antigen (HBeAg): associated with the nucleocore

HBV produces reverse transcriptase and can integrate into the host genome. In this way it is similar to HIV (an RNA virus) and different from other DNA viruses (which generally persist episomally and do not integrate).

2. **Clinical presentation:** HBV infection can be acute or chronic.
   - Incubation period: ~1 to 3 months after exposure

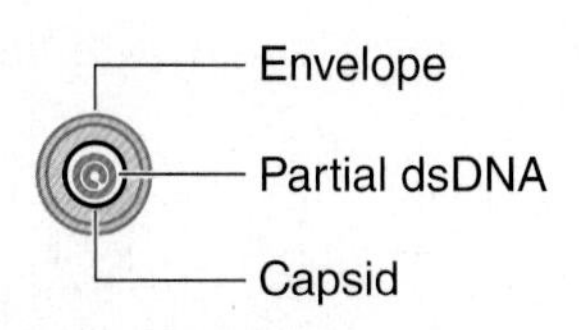

Figure 8.1. HBV.

- **Acute, self-limited infection**
  - Is typically asymptomatic (50–80% of infections) but may present with fatigue, hepatitis, jaundice, low fever, pale stools, dark urine
  - Most adult infections are acute and self-limited (>90%) and resolve within ~6 months of exposure (i.e., HBsAg and DNA become undetectable after ~3 to 6 months).

Most adults get acute, self-limited infection. Most neonates get chronic infection.

- **Chronic infection** (old term: "HBV carriers")
  - Persistent infection (i.e., HBsAg is still present while anti-HBs is still absent) for more than 6 months. Chronic infections can last for years or decades.
  - More likely to occur when infection occurs at an early age. Unlike adult infections, most (95%) neonatal infections are chronic.
  - **Inactive carrier state**: most chronically-infected patients do not have active infection and are asymptomatic
    - Mild to no liver inflammation
    - HBeAg is negative and anti-HBe is positive.
    - DNA levels are low or undetectable.
  - **Active infection:** some patients have chronic infection that is symptomatic, with anywhere from mild to severe disease
    - HBeAg and DNA are detectable.
    - Patients are highly infectious.
    - Liver necrosis and inflammation continue to occur and can be seen on biopsy. This can progress to liver cirrhosis and hepatocellular carcinoma.

There are two main types of chronically infected people: inactive carriers (low viral load) and active carriers (high viral load).

Most HBV infections are acute and self-limited. Most HCV infections are chronic.

3. **Diagnosis**: Done through a combination of serology and PCR (Table 8.2, Fig. 8.2, Fig. 8.3). A general diagnostic algorithm is described in Fig. 8.4.
   - Serology: the primary method of diagnosis and monitoring of infection. An HBV immunoassay panel will detect the presence of a variety of antibodies against HBV antigens.

Table 8.2. Serologic and nucleic acid markers of HBV infection

STAGE OF INFECTION	DNA	HBsAg	HBeAg	ANTI-HBc IgM	ANTI-HBc TOTAL	ANTI-HBe	ANTI-HBs
None (naive)	–	–	–	–	–	–	–
Immunized	–	–	–	–	–	–	+
Early acute	+	+	+	–	–	–	–
Acute	+	+	±	+	+	±	–
Chronic	+	+	±	–	+	+	–
Resolved	–	–	–	–	+	±	+
Unclear (occult infection, resolved/resolving infection, false positive, HBsAg mutant)	±	–	–	–	+	–	–

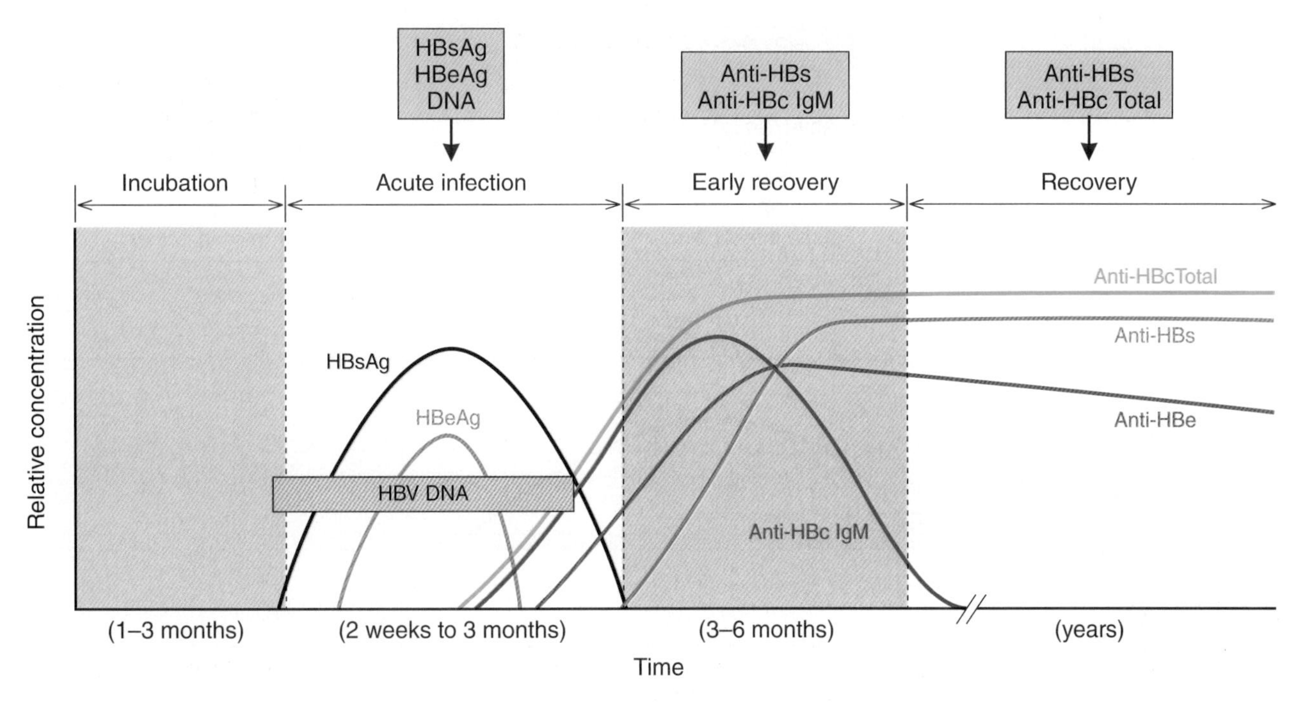

Figure 8.2. Diagnosis of acute HBV infection using serologic and virologic markers.

- **HBsAg**
  - If present, the patient has current infection. Infection may be acute, chronic inactive, or chronic active.
  - Production of HBsAg means that the person is infectious. Chronically infected people can be infectious for years.
  - Disappearance of HBsAg and appearance of anti-HBsAg indicate that the disease is resolving.

The presence of surface antigen (HBsAg) indicates current infection (either acute or chronic).

- **HBeAg**
  - If present, the patient has current infection. It is usually a marker of early acute infection, but can frequently be present in chronic infection.
  - It is a marker of active replication, so it often correlates with high viral load.
  - Disappearance of HBeAg and appearance of anti-HBeAg indicate that the disease is starting to resolve.

The presence of **E** antigen (HBeAg) usually indicates active replication, **e**arly infection, and/or high infectivity.

- **Anti-HBe**
  - If present, the patient is either starting to resolve the infection or the patient has chronic inactive infection.
  - It only appears when HBeAg disappears.

The presence of anti-HBe indicates that either the acute infection is resolving or that the patient is progressing to chronic inactive infection.

- **Anti-HBc IgM**
  - If present, the patient has acute infection.
  - IgM will usually be undetectable after 6 months.

The presence of IgM against core protein indicates acute infection.

- **Anti-HBs**
  - If present, the patient is immune. Antibodies against surface antigen are neutralizing.
  - Both vaccination and naturally resolved infections will result in production of anti-HBs.

The presence of anti-HBs indicates immunity to HBV

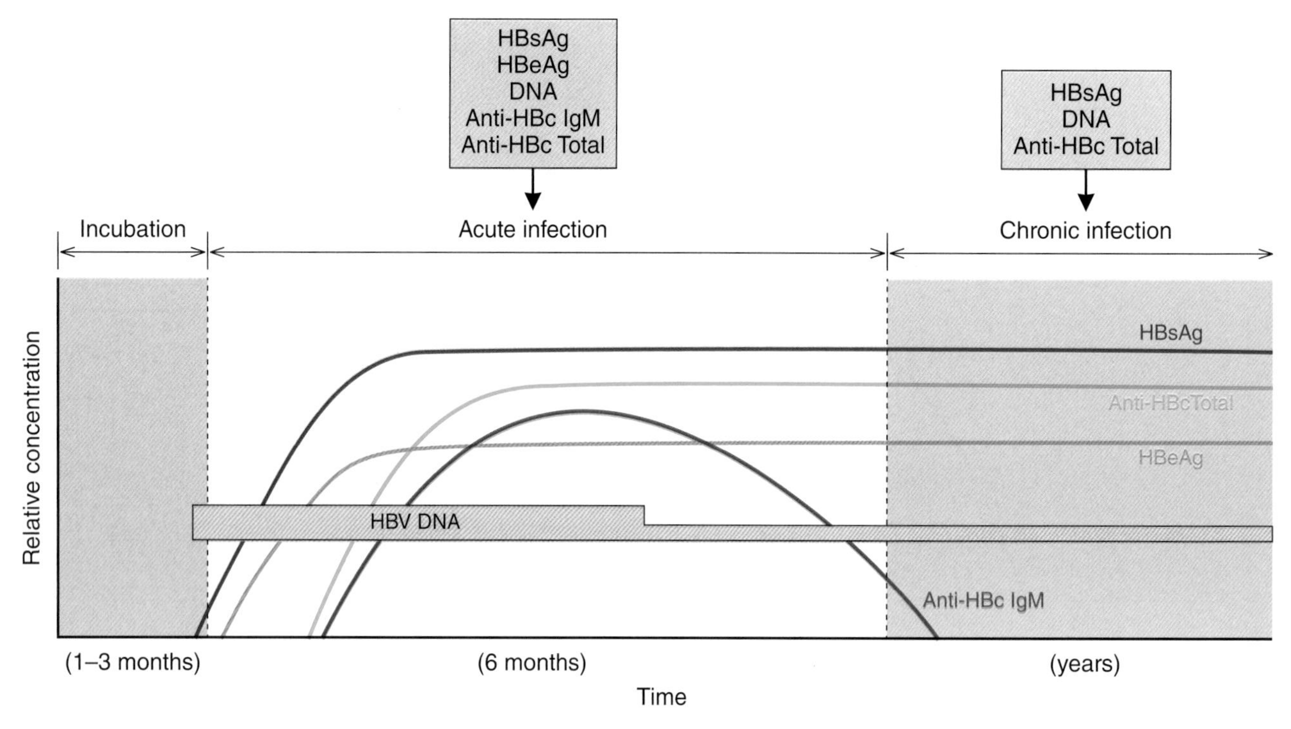

Figure 8.3. Diagnosis of chronic HBV infection using serologic and virologic markers.

Antibodies against core protein or HBeAg indicate natural infection instead of exposure through vaccination.

- **Anti-HBc total** (pronounced "core total")
  - Is the sum of both IgM and IgG antibodies against the core protein
  - If present, the patient has had a natural infection instead of immunity through vaccination.
  - The infection may be either acute, chronic, or resolved.
- Serology results can be falsely positive or negative. This may occur due to the performance of the assay but also if the mutants do not produce some HBV proteins.
  - Consider a false positive if a positive result does not fit the typical profile or if there is low clinical suspicion of viral infection.
  - **Precore mutant:** may be negative for anti-HBe but other serologic and DNA markers will be positive. Precore mutants are variants of HBV that cause chronic infection but do not produce HBeAg. This virus is still infectious.
  - **HBsAg (escape) mutant:** may show a false negative serology for HBsAg but the patient will still have a positive viral load. These variants produce a slightly different sAg from the wild-type virus and can evade vaccine-induced anti-HBs antibodies.

- NAAT: a highly sensitive tool that can measure the presence of HBV DNA. Serology is preferred as an initial screening test, but NAATs are usually used for confirmation.
  - NAATs cannot differentiate between acute and chronic infections, but can be used to help identify chronic inactive carrier states (low viral load) versus chronic active (high viral load) infections.
  - NAATs may be used for diagnosis in very early infection, even before detection of HBsAg.

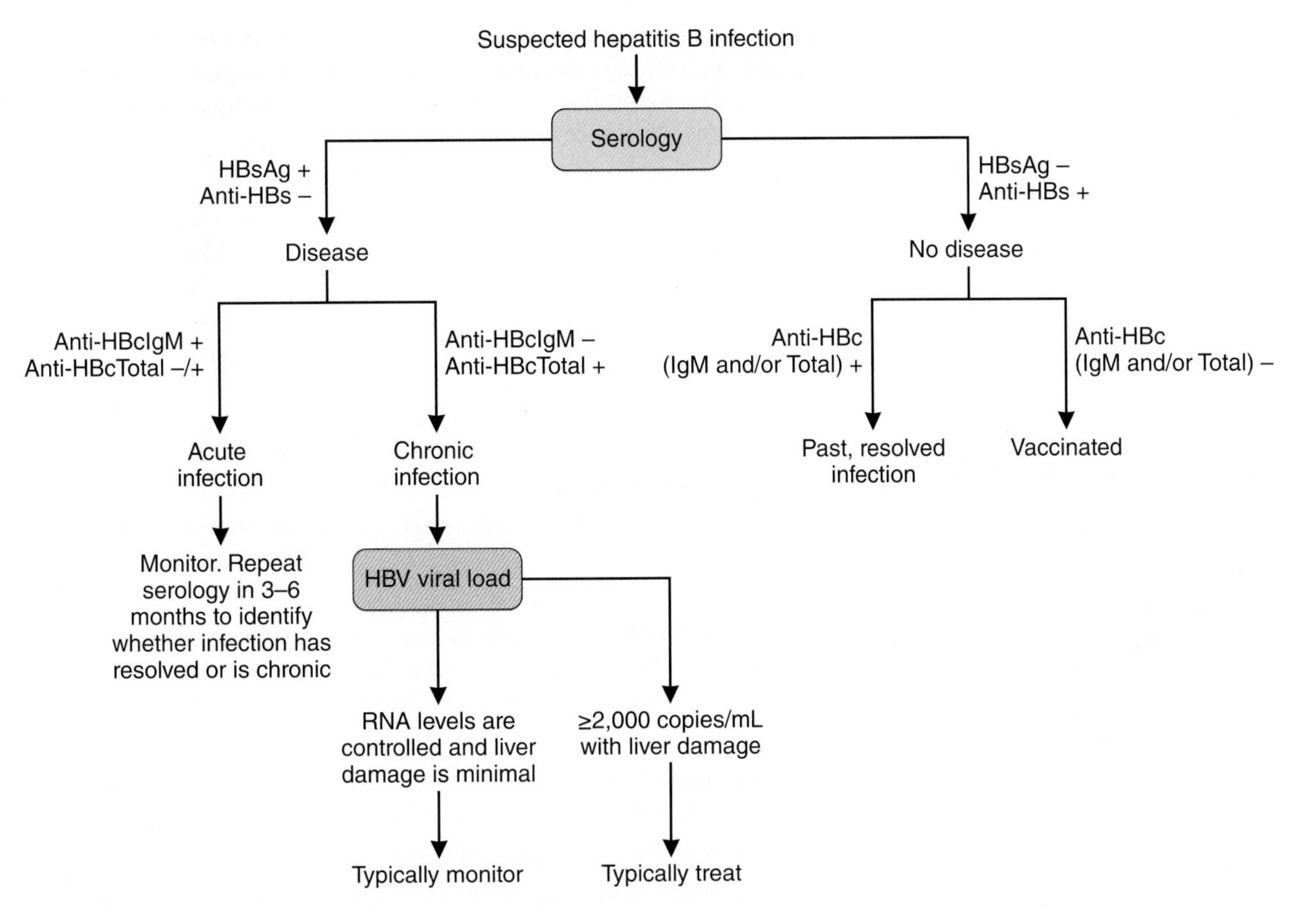

Figure 8.4. General algorithm for diagnosis and management of HBV.

- Quantitative PCR is used for monitoring disease and response to therapy. It is performed before treatment to determine baseline viral load and then every 3 to 6 months.
- **Occult infection:** HBV DNA is present and detectable in the blood or liver but HBsAg is absent.
- NAAT results can be incorrect.
  - False positives can occur. Be suspicious of positive viral load and negative serology (or vice versa). Viral loads are not recommended as the primarily screening method and may cause confusion if false positives are detected.
  - False negatives can occur in cases of an inactive carrier state.
- HBV genotyping: Like HCV genotypes, HBV genotypes occur in different geographic areas and may have clinical significance. However, testing is not commonly available or used.

4. Prevention

- Vaccination is highly effective and is part of the routine vaccination schedule. There are 3 doses within the first year to prevent the risk of chronic infection, which occurs with higher frequency in infants (*see* Fig. 18.4).
- Postexposure prophylaxis: hepatitis B immunoglobulin within 12 hours of exposure

Most adult infections resolve on their own.

5. **Treatment:** Not all HBV infections are treated because many infections resolve on their own. However, HBV cannot be considered "cured" yet because the virus integrates, and neither resolution of symptoms nor treatment eliminates all traces of the virus.
   - Supportive therapy for most infections
   - Treatment recommendations are complex, and guidelines are updated continually. Antivirals are given based on viral load, amount of liver damage, and ALT levels (18).
   - Therapeutics (*see* Table 19.8)
     - Nucleos(t)ide reverse transcriptase inhibitors **(NRTIs)**. First-line treatments: entecavir and tenofovir. Other treatment: lamivudine. Oral NRTIs are usually continued for years to prevent reactivation of HBV disease. NRTIs are also used to treat HIV because both viruses contain reverse transcriptase.
     - **PEGylated interferon** alpha-2a or -2b. Treatment is usually for 48 weeks.
   - Liver transplant

HBV cannot be totally cured. Once a person has been infected, cccDNA can remain in cells and pose a lifelong risk of reactivation (even in patients that have an immunologically resolved infection or have sustained viral DNA suppression).

III. HEPATITIS D VIRUS. HDV (or delta virus) is a less frequent cause of hepatitis because it can cause infection only in the presence of hepatitis B (Table 8.1). Vaccination and therapy against HBV are effective in preventing and treating HDV infection.

1. **Background:** HDV is a small, enveloped virus with circular (–) ssRNA (Fig. 8.5). It is not yet classified in a family but is in the genus *Deltavirus*.
   - It makes only one protein, D antigen.
   - It replicates only in the presence of HBV, since HBV surface antigens are used in the HDV envelope.
   - High prevalence: the Mediterranean, Middle East, Pakistan, and Central Asia

HDV can replicate and cause infection only in people with HBV. About 5% of HBV-positive individuals are also infected with HDV.

2. **Clinical presentation:** HDV increases the mortality rate of HBV-infected individuals 10-fold.
   - Incubation time similar to that for HBV (~1 to 3 months)
   - Acute infection
     - Occurs simultaneously with HBV infection. This is called **coinfection.**
     - Acute HDV and HBV coinfection is more severe than HBV infection alone, but >95% of cases are still acute and self-limited (similar to HBV infection alone).
   - Chronic infection
     - A small number of HDV-HBV coinfections may progress to chronic infection. However, HDV is usually acquired by an individual that already has chronic HBV. This is called **superinfection**. Superinfection is the primary cause of chronic HDV infection.
     - Like acute infection, chronic HDV and HBV infection is also more severe than chronic HBV alone.

HDV and HBV together cause a more severe infection than HBV alone.

3. **Diagnosis**
   - Serology: the most common method of diagnosis
     - Anti-HDV: total IgM and IgG antibodies against HDV. IgM clears within a few months (self-limited infection) or remains persistent for several years (chronic infection).
     - HDV infection may cause a higher titer of anti-HBc IgM

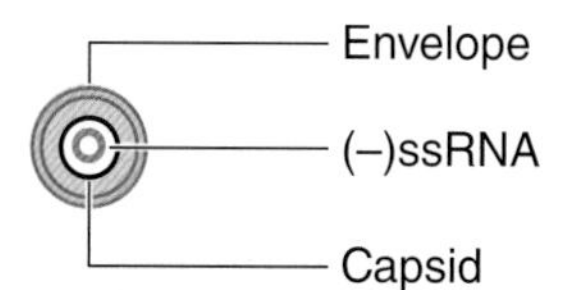

Figure 8.5. HDV.

- Serologic detection of HDAg and immunohistochemistry techniques are not sensitive.
- NAAT: highly sensitive and specific, and is available for serum samples only at reference laboratories. Primers and probes need to be designed carefully to capture all 8 genotypes.

4. **Prevention and treatment**
   - Vaccination against HBV prevents infection with HDV.
   - Treat with PEGylated interferon.

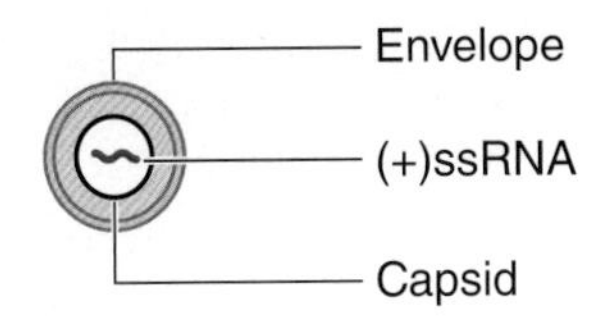

Figure 8.6. HCV.

IV. HEPATITIS C VIRUS. HCV (old term: non-A, non-B hepatitis virus) is a cause of significant morbidity and mortality worldwide. Unlike HBV, most HCV infections progress to chronic disease (Table 8.1).

1. **Background:** HCV has (+) ssRNA and is part of the *Flaviviridae* family (Fig. 8.6).
   - It is enveloped, like other blood-borne viruses.
   - Is a Baltimore class V virus, so it uses a specialized RNA-dependent RNA polymerase for replication. This can be used as a drug target.
   - Is categorized into genotypes 1 to 7
   - Its highest prevalence is in Egypt due to use of contaminated needles during public health efforts to treat schistosomiasis.
2. **Clinical presentation:** HCV infection is often less symptomatic than other hepatitis viruses. So patients are more likely to be unaware of their infection.
   - Incubation: ~1 to 3 months
   - Slight malaise, often without jaundice
   - May/may not have mild fatigue, nausea, or abdominal pain
   - Usually does NOT cause fulminant hepatic failure
   - Associated with **cryoglobulinemia**
   - Causes mostly chronic infections

HCV is the most common cause of posttransfusion hepatitis because there were no FDA-approved blood product screening tests for the virus prior to 1992.

Rule of thumb: 75% of HCV infections are chronic, but only 5% of adult HBV infections are chronic.

3. **Diagnosis:** Testing is done for people with symptoms or risk factors or who are born between 1945 and 1965 (Fig. 8.7 and Table 8.3).
   - Serology
     - Detection of total IgM and IgG against HCV. This is the preferred method for screening.
       - Serology has very high sensitivity and specificity
       - EIAs (i.e., ELISAs) on serum/plasma are the most common platforms.
       - Point-of-care assays are becoming more available. These can be done on fingerstick whole blood and oral fluid.
     - Detection of HCV core antigen can be done but is not currently used or available in the United States.
     - Unlike HBV serology, HCV serology does not differentiate between recent, chronic, and resolved infections.
   - NAAT: done on serum and plasma
     - Used to confirm a case of positive serologic results
     - Quantitative PCR is used to monitor infection (i.e., viral load testing).
       - Used prior to treatment, either to establish a baseline or to monitor if the patient is clearing the infection without therapy
       - Used for monitoring the patient's response to treatment

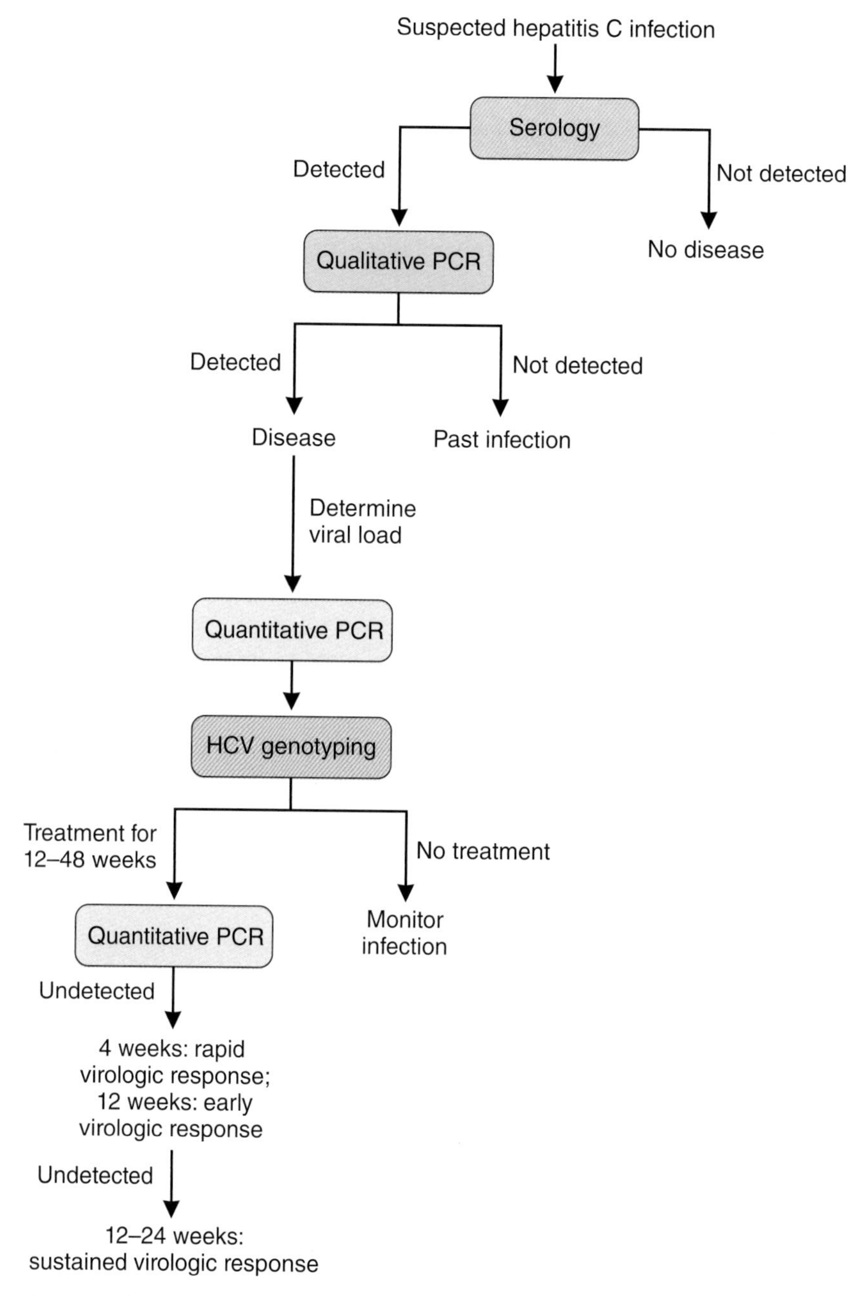

Figure 8.7. A general algorithm for HCV testing.

- **Genotyping:** The 7 HCV genotypes are concentrated in different geographic areas (Table 8.4). They do not correlate with severity of infection but affect choice of treatment regimen and likelihood of treatment success.
  - Genotype is typically identified from the NS5b gene, 5′ noncoding region, core, or other components of the HCV genome.
  - Diagnostic assays for identifying the genotype
    - RT-PCR
    - Sequencing
    - Line probe assays
  - Genotyping needs to be performed only once (genotypes do not switch during infection)

Table 8.3. Results of HCV testing with serology and PCR

INFECTION	SEROLOGY	PCR
None (naive)	–	–
Current (active/chronic) infection	+	+
Past (resolved) infection	+	–

- Need to have a minimal viral load (usually around 500 to 1,000 IU/ml) to perform genotyping.

4. **Prevention:** Antibodies against HCV are nonneutralizing, so there is no vaccine available and postexposure prophylaxis with immunoglobulins does not work. Even patients that have cleared the virus can be reinfected.

Unlike for HBV, antibodies against HCV do not provide immunity. As a result there is **no vaccine.**

5. **Treatment**
   - The goal of therapy is to reduce end-stage liver disease. Like with HBV, not all infections are treated because some resolve spontaneously. But, unlike HBV, HCV can be cured (Fig. 8.7).
   - **Response-guided therapy** is the main approach to HCV treatment. It means that the drugs and duration of therapy are tailored to patients depending on how they respond to a treatment regimen.
     - Generally, treatment regimens last 12 to 48 weeks but can be shorter depending on the patient's response and infecting genotype.
     - Quantitative PCR is used to monitor response to treatment at several time points.
       - Week 0: Baseline viral load is established.
       - Week 4: Undetectable viral load indicates rapid virologic response **(RVR)**.
       - Week 12: Undetectable or 100-fold reduction of viral load indicates early virologic response **(EVR)**.
       - Week 24 or 48 (depending on drug regimen): Viral loads are measured at the end of treatment (EOT).
       - 12 to 24 weeks after completion of treatment: Undetectable viral load indicates long-term suppression of the virus, or sustained virologic response **(SVR).**

Unlike HBV, HCV can now be **cured.**

Table 8.4. Genotypes of HCV

GENOTYPES AND SUBTYPES	GEOGRAPHIC LOCATION	COMMENTS
1a, 1b	Worldwide, especially Europe and North America	Very **hard to treat** with previous interferon + ribavirin therapy (better response if patient has the **IL28** gene containing the "CC" genotype)
2a–2d	Worldwide	
3a–3f	Worldwide	
4a–4j	Middle East, Africa	Infrequent
5	South Africa	Rare
6	Southeast Asia	Rare
7	Central Africa	Rare

- Treatment regimens (*see* Table 19.9, Table 19.10, and Table 19.11)
  - **PEGylated interferon alpha-2a or -2b** with ribavirin has been the most widely used treatment but is rapidly being replaced with newer, curative antivirals.
    - Used for chronic infection
    - Has high failure rates for genotype 1
    - Has a large number of side effects
    - Is noncurative
  - **Direct-acting antivirals (DAAs):** These directly interfere with HCV and can be curative. There are 4 main classes.
    - NS3/4A protease inhibitors
    - NS5A inhibitors
    - Nucleos(t)ide polymerase inhibitors (NPI)
    - Nonnucleoside polymerase inhibitors (NNPI)

Sofosbuvir (NPI) can cure ~90% of cases.

Multiple-Choice Questions

1. Which of the following viruses is nonenveloped?

a. HAV

b. HBV

c. HCV

d. HDV

2. Which of the following viruses is NOT associated with inflammation of the liver (hepatitis)?

a. HBV

b. VZV

c. CMV

d. Yellow fever virus

3. Which hepatitis virus produces reverse transcriptase?

a. HAV

b. HBV

c. HCV

d. HDV

4. Genotyping of HCVs

a. Affects choice of antivirals

b. Should be performed annually

c. Is used to determine whether resistance to antivirals has developed

d. All of the above

5. Which of the following viruses is most likely to cause chronic infection in adults?

a. HAV

b. HBV

c. HCV

d. HEV

6. **Which of the following viruses is the least likely to be oncogenic?**
   (May require reading outside of this chapter.)
   a. HBV
   b. HCV
   c. EBV
   d. CMV

7. **Serologic testing of a 64-year-old male showed only core total and HBsAb as reactive. Which of the following is the most likely interpretation of the results?**
   a. The patient had a positive vaccine response.
   b. The patient is immune after natural infection.
   c. The patient has acute HBV infection.
   d. The patient has chronic HBV infection.

8. **What is the best test to diagnose HBV infection in neonates?**
   a. Serology
   b. PCR
   c. Genotyping
   d. None of the above

9. **Neonates with exposure to HBV are at highest risk of**
   a. Chronic hepatitis
   b. Birth defects
   c. Hepatocellular carcinoma
   d. Death

10. **A 39-year-old female with a history of intravenous drug use has no symptoms of hepatitis. She is tested for hepatitis C for the first time, and the serology is positive. Which of the following is the most likely interpretation of the results?**
    a. The patient has chronic hepatitis C.
    b. The patient does not have hepatitis C.
    c. The patient has cross-reactivity with other flaviviruses.
    d. Molecular testing is required for a final diagnosis of hepatitis C.

True or False

11. Hepatitis C vaccine is highly effective. **T F**
12. Hepatitis B virus can integrate into host cells. **T F**
13. Most adult-acquired cases of hepatitis B are chronic. **T F**
14. Treatment can be used to cure HCV infection. **T F**
15. HCV genotyping is used to identify drug resistance mutations prior to therapy **T F**

CHAPTER 9

HUMAN RETROVIRUSES

I. OVERVIEW. Retroviruses are Baltimore class VI viruses. They have (+) ssRNA as their nucleic acid, but instead of using it directly as mRNA (like Baltimore class IV viruses), they first convert it into a DNA intermediate. Importantly, their RNA makes them highly mutable, while the DNA intermediate allows them to integrate into the host genome, evade the immune system, cause a lifelong infection, and passively produce many more copies of (+) ssRNA.

II. HUMAN IMMUNODEFICIENCY VIRUS.

1. **Background:** HIV consists of a conical capsid surrounded by an envelope. The capsid contains a diploid genome (i.e., two copies of (+) ssRNA) (Fig. 9.1).
   - There are three main genes that encode structural proteins.
     - ***gag*** encodes the Gag polyprotein, which is cleaved into proteins that form the nucleocore, including **p24**.
     - ***pol*** encodes the Pol polyprotein, which is cleaved into 3 enzymes: **reverse transcriptase, integrase, and protease.**
     - ***env*** encodes envelope proteins, including **gp160**. gp160 is cleaved into two smaller proteins, called **gp120** and **gp41**, which are used for binding to cells.
   - There are two main **types** of HIV, HIV-1 and HIV-2 (Fig. 9.2). HIV-1 is present worldwide and accounts for the vast majority (~95%) of infections.
     - HIV-1 has four **groups**. Group M is the most prevalent and is the primary cause of infections. Group O is the next most common but accounts for only ~2% of infections. Groups N and P are exceedingly rare and may not be detected by commonly used diagnostic tests. Group N is sometimes called the "not-M not-O group."
     - Group M has multiple **subtypes**. Subtype C is the most common worldwide, but subtype B is the most common in Western Europe, Australia, and North and South America.

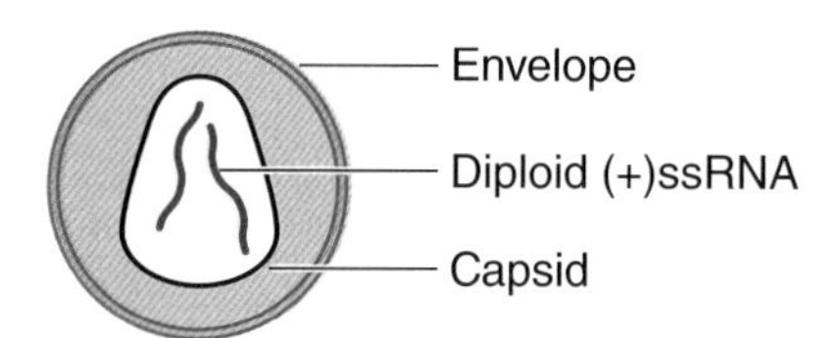

Figure 9.1. HIV.

Primary receptor: **CD4+**
Initial coreceptor: **CCR5**
Late stage coreceptor: **CXCR4**

CCR5 antagonists are a class of drugs that bind to CCR5 and block binding of R5 tropic viruses. CCR5 antagonists are not effective against X4 tropic or dually tropic viruses.

Fusion inhibitors bind to gp41, which prevents the virus from fusing with the host cell.

  - Subtypes can recombine to cause hybrid subtypes, known as **circulating recombinant forms**.
- HIV-2 is not transmitted as efficiently as HIV-1 and is predominantly confined to West Africa. Group A is the most prevalent.

2. **HIV life cycle** (Fig. 9.3)
   - **Cell binding:** gp120 is complexed with gp41 on the viral membrane and must bind to 2 receptors (a primary receptor and a coreceptor) on the host cell.
     - The primary receptor for HIV is CD4, which is present on helper T cells (also known as CD4+ T cells), macrophages, and monocytes.
     - The two most common coreceptors are CXCR4 and CCR5, which normally bind to chemokines. HIV tropism depends on which coreceptor is utilized. Tropism can change during the course of infection.
       - **R5 viruses** use CCR5. They are also known as **M** tropic viruses because they primarily infect macrophages and monocytes. HIV strains that cause initial infection are almost always R5 tropic viruses.
       - **X4 viruses** use CXCR4. They are also known as **T** tropic viruses because they primarily infect T cells. X4 tropic viruses tend to form during later stages of infection.
       - Dually tropic strains can use either receptor.
       - Mixed tropic strains are a combination of viruses with different tropisms.
   - **Fusion and entry:** Once bound to the cell, gp41 fuses the viral and host cell membranes together so that the virus can enter the cell.
   - **Reverse transcription:** uses a viral enzyme (an RNA-dependent DNA polymerase) called **reverse transcriptase** to convert single-stranded RNA into double-stranded DNA.
     - Reverse transcriptase binds to the ssRNA genome and makes the corresponding DNA strand, forming an **RNA-DNA hybrid.**
     - It then degrades the original RNA template using its RNase activity and creates the complementary DNA strand to form a dsDNA molecule. This dsDNA contains a complementary strand to the original RNA template and is called complementary DNA, or **cDNA**.

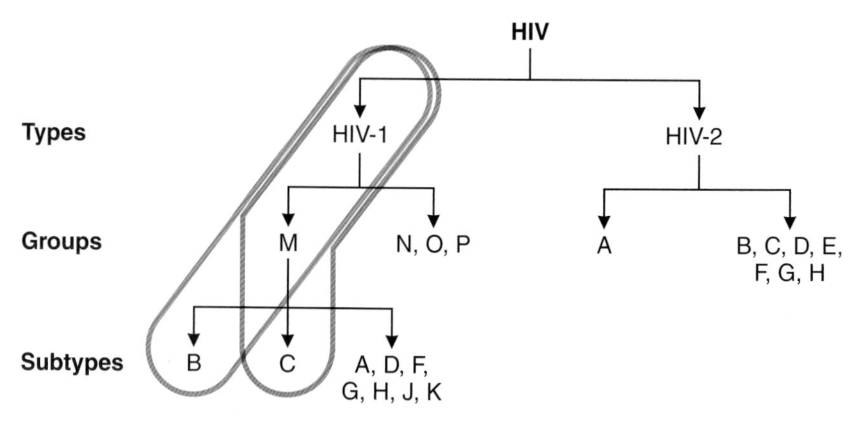

Figure 9.2. HIV genotypes. For HIV-1, Group M subtype C is most common worldwide (orange) and subtype B is most common in Western Europe, Australia, and North and South America (blue). Group A is most common for HIV-2.

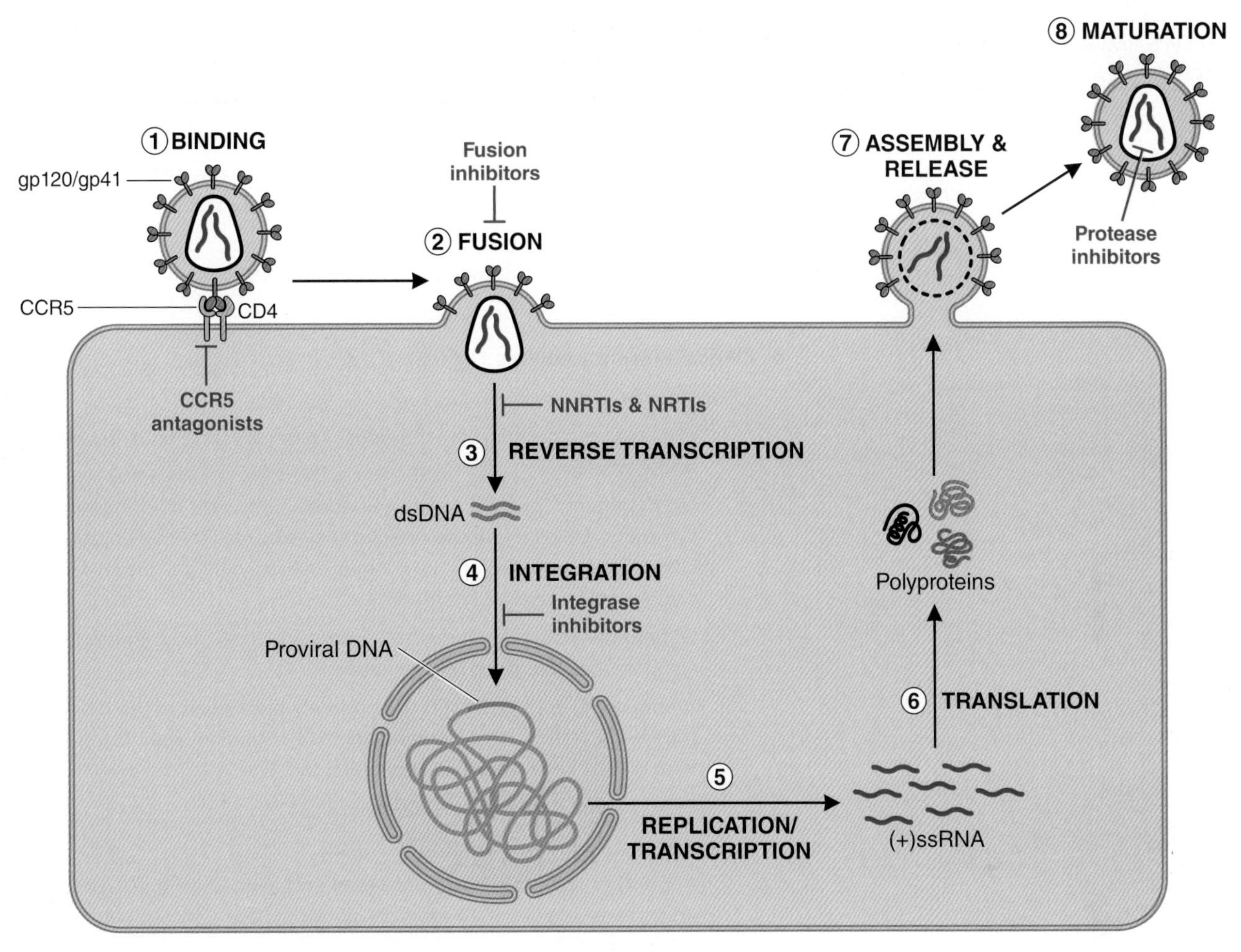

Figure 9.3. Life cycle and drug targets (shown in red) for HIV.

- Reverse transcription has a high error rate, which translates to a high mutation rate for HIV.

- **Integration:** a required part of the HIV life cycle. **Integrase** integrates the viral cDNA into the host cell DNA (i.e., lysogenic life cycle).
  - The integrated virus is called the **provirus** (*see* Fig. 19.5).
  - Integration tends to occur more often in regions of active transcription.
- **Virus replication and release**
  - New (+) ssRNA genomes are transcribed from the proviral DNA.
  - Proteins are translated from *gag*, *pol*, and *env* as long polyproteins.
  - Viral genomes are packaged within nucleocore proteins.
  - Immature virions gain an envelope as they exit the cell.
  - **Protease** cleaves viral polyproteins into different individual enzymes and the virus matures into a functional virion (*see* Fig. 19.6).

3. **Transmission:** HIV is transmitted through direct contact with blood, semen, vaginal and rectal secretions, breast milk, and genital mucosa.
   - Can occur during acute, latent, or late disease stages of infection but happens most efficiently during acute infection

Nonnucleoside reverse transcriptase inhibitors (NNRTIs) bind directly to reverse transcriptase and prevent it from forming cDNA.

Nucleoside reverse transcriptase inhibitors (NRTIs) are nucleoside analogs. If incorporated by reverse transcriptase, they will terminate any further DNA elongation.

Integrase inhibitors bind the active sites in integrase and block integration of viral DNA.

Protease inhibitors bind protease and prevent cleavage of polyproteins. This prevents maturation of new infectious virions.

HIV can be transmitted through breast milk.

- Risk factors/groups
  - Sexual intercourse without barrier protection, partners of infected individuals, sex workers, and men who have sex with men
  - Intravenous drug use and sharing of needles
  - Maternal transmission (infected mothers may pass HIV *in utero*, by exchange of blood during labor and delivery, or through breast milk)
  - Contact with contaminated blood products (e.g., blood transfusions) and associated risk groups (e.g., hemophiliacs)
  - African Americans and Latinos

4. **Clinical Presentation** (*see* Fig. 9.5)

- Acute infection
  - The incubation period for the initial phase of infection is between 1 and 4 weeks.
  - Most patients present asymptomatically or with mild nonspecific symptoms called the **acute retroviral syndrome**. These symptoms include fever, rash, weight loss, night sweats, malaise, myalgias, sore throat, and/or swollen lymph nodes.
  - Differential diagnoses: influenza, tuberculosis, infectious mononucleosis, *Streptococcus pyogenes* infection
  - Viremia: The virus replicates to extremely high levels in the blood (>1 million copies/ml). This increases the risk of transmission from people who do not know they have an infection.
  - CD4 cells are depleted to ~500 cells/µl.
- **Latency**
  - Antibodies against HIV are produced and keep the virus in check for an extended period (~10 years). Most people are asymptomatic during this phase.
  - CD4 cell numbers initially recover, but they slowly deplete over time in the absence of treatment.
  - **Viral set point:** the point at which the viral load stabilizes because cytotoxic T cells hold the virus in check
  - Antibodies are nonneutralizing and do not provide complete protection.
- Late-stage disease: As the levels of CD4 cells go down, increasing numbers of **opportunistic infections** occur (Table 9.1).
  - **Acquired immunodeficiency syndrome (AIDS)** occurs when HIV-infected individuals develop AIDS-defining illnesses, or if their CD4 cell counts are ≤200 cells/µl (regardless of whether illness has occurred).

Table 9.1. Pathogens associated with HIV infection (AIDS-defining illnesses)[a]

<50 CELLS/µL	50–100 CELLS/µL	100–200 CELLS/µL	200–500 CELLS/µL	ANY COUNT
Mycobacterium avium complex	*Cryptosporidium*, *Toxoplasma*, CMV, *Cryptococcus*	*Pneumocystis jirovecii*, JC virus (PML), dimorphic fungi (especially *Histoplasma* and *Coccidioides*), HSV	*Candida* (thrush), HHV-8 (KSHV), VZV	*Mycobacterium tuberculosis*

[a] PML, progressive multifocal leukoencephalopathy.

- Virus titers increase from 10^4 to 10^6 copies/ml.
- AIDS-defining illnesses: infections or conditions that are associated with or occur almost exclusively in AIDS patients

AIDS is diagnosed when an HIV+ individual either has 1) a CD4 count of ≤200 cells/μl or 2) develops an AIDS-defining illness.

- **Nonprogressors** (or **controllers**) are individuals who do not succumb to the typical course of HIV infection. Even in the absence of treatment they maintain normal levels of CD4 cells and have low or undetectable viral titers (19).
  - Controllers are defined as maintaining low viral titers, usually for at least 1 year. Elite controllers have <50 RNA copies/ml, while viremic controllers have 50 to 2,000 RNA copies/ml.
  - Long-term nonprogressors is an older term used for individuals that have a controlled infection of long duration (>7 to 10 years). Their viral loads may not have been measured because they predate current viral load testing.
  - The **CCR5-Δ32 mutation** inhibits CCR5 from acting as a cofactor for HIV entry into cells. This mutation is present at a higher frequency in Caucasian individuals than in individuals of African descent or East Asians.

In 2008, the **Berlin patient** was transplanted with CD4 T cells containing a CCR5-Δ32 deletion. The patient was able to stop taking antiretroviral therapy, his viral titers became undetectable, and his CD4 counts recovered. This was the first patient considered to be cured for HIV.

5. **Diagnosis (Fig. 9.4 and Fig. 9.5):** Serology is the recommended screening method for HIV. Qualitative NAATs can be used for confirmatory testing and screening in patients suspected of having acute retroviral syndrome. Quantitative viral load assays are used to measure viral load in order to monitor response to treatment. All adults and adolescents should be informed and tested at least once, unless they explicitly opt out. (An opt-in testing model is when patients must explicitly elect to get tested.) High-risk individuals should be tested more frequently.

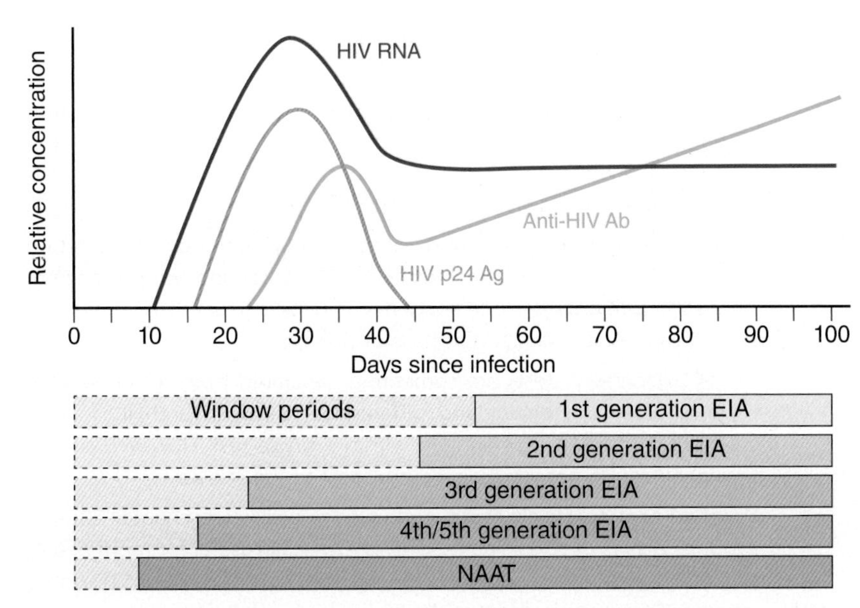

Figure 9.4. Improvement in early detection of HIV over 5 generations of assays. Markers such as viral RNA, antigens, and antiviral antibodies are used to detect the virus in plasma or serum.

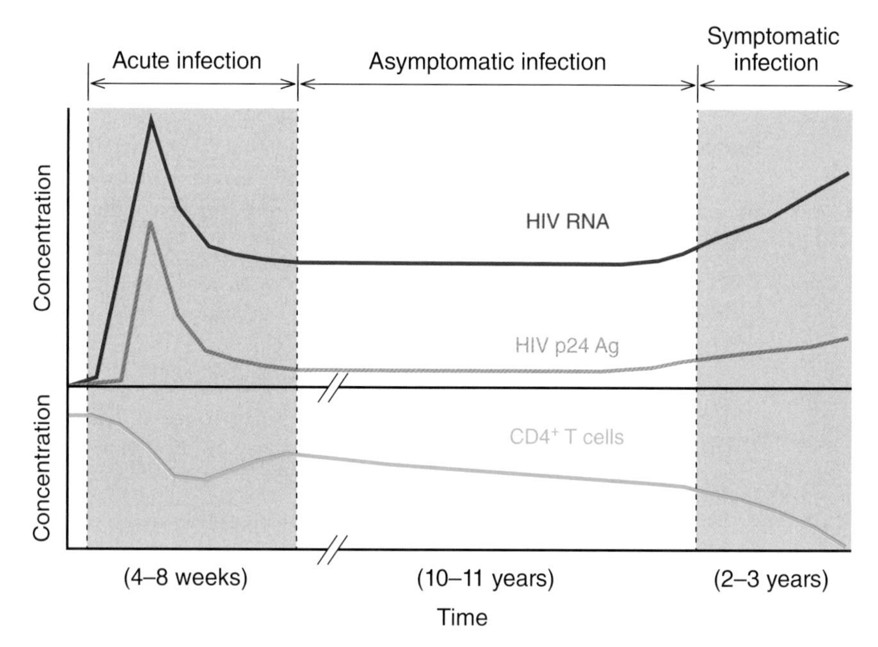

Figure 9.5. Kinetics of HIV infection.

Note: Serology should not be used on infants <18 months for diagnosis of HIV because of the presence of maternal antibodies. Instead, NAATs for HIV DNA or RNA should be performed and then repeated for confirmation.

- **Laboratory serologic tests** are moderate or high complexity (*see* chapter 20), are often automated, and typically take several hours to perform. There are several "generations" of serologic tests that use different formulations of HIV markers to detect the presence of HIV or antibodies against HIV (Table 9.2).
    - The **window period** is the time between infection with HIV and the time the test can actually detect an infection. Early-generation tests had longer window periods, while newer-generation tests have shorter window periods. This means that newer tests can detect infection earlier.
    - 1st, 2nd, and 3rd generation tests used viral lysates or specific HIV proteins to detect antibodies the patient had formed against the virus. These assays are no longer recommended because they are relatively insensitive and may not detect early infection.
    - 4th-generation tests are also called antigen-antibody combination assays because they detect anti-HIV antibodies and the presence of the HIV **p24 antigen**. p24 antigen is an early marker of HIV-1 infection and can result in earlier detection of disease.
    - "5th"-generation tests are sometimes separated from 4th generation ones. They can detect and differentiate between anti-HIV-1, anti-HIV-2, and HIV-1 p24 antigen within a single test. However, current algorithms do not yet describe how best to use this assay.
- **Point-of-care serologic tests** are low complexity (*see* chapter 20) and produce results in ≤30 minutes. This is especially useful for rapid detection of HIV in women and infants during labor and delivery, or STD clinics when patients may be lost to follow-up.
    - Some are designed to detect antibodies only while others detect antibodies and p24 antigen.

Table 9.2. Comparison of HIV tests (20)

TEST	HIV ANTIGENS USED FOR DETECTION	COMPONENT BEING DETECTED	EXAMPLE(S)[a]	TIME FROM INFECTION TO DETECTION (WINDOW PERIOD)
1st-generation serology	Viral lysate	Any antibodies, including cross-reactive ones	Western blot, IFA	~7 weeks
2nd-generation serology	Synthetic/recombinant peptides	IgG	EIAs, rapid tests	~5 weeks
3rd-generation serology	Synthetic/recombinant peptides	IgG and IgM (HIV-1 and HIV-2)	EIAs	~3 weeks
4th-generation combination serology	Synthetic/recombinant peptides	**p24**, IgG, and IgM (HIV-1 and HIV-2)	EIAs, rapid tests	~2 weeks
"5th-generation" combination serology	Synthetic/recombinant peptides	Detects **and differentiates** positivity for HIV-1 p24, HIV-1 IgG/IgM, and HIV-2 IgG/IgM	EIAs, rapid tests	~2 weeks
NAAT	Nucleic acid	RNA	RT-PCR, bDNA, TMA	~10 days
	Nucleic acid	Proviral DNA	PCR	~10 days

[a]IFA, immunofluorescence assays; bDNA, branched DNA assays; TMA, transcription-mediated amplification assays.

- These methods may have generally high sensitivity and specificity but laboratory serologic methods are generally preferred.
- These tests are useful when patients are at risk of being lost to follow-up.
- Some home collection assays are available.

- **Confirmatory testing** should follow all currently recommended serologic testing. These tests include the following:
  - HIV-1/2 differentiation tests, which indicate whether the individual has HIV-1 or HIV-2 antibodies
  - NAATs, which detect RNA down to just a few (e.g., 20 to 50) copies/ml in blood, serum, or plasma.
  - Other tests, if necessary. **Western blot** or indirect immunofluorescence assays were needed to confirm the results of early-generation tests but are not recommended for 4th- and 5th-generation ones. These are less sensitive and specific methods of confirmation and may take several months of infection to become positive.
- **NAATs**
  - Are highly sensitive and have a very low limit of detection
  - Have a shorter window period (can detect disease earlier)
  - Most assays detect only HIV-1. Some assays are available for HIV-2.
  - Used for confirmation or monitoring of adult HIV$^+$ patients (Fig. 9.4)
  - Used to diagnose newborns because serology is not reliable (maternal HIV antibodies can persist in newborns and result in positive serologic tests)

- **RT-PCR** is the most common NAAT.
  - Quantitative RT-PCR is used to quantitate the amount of viral RNA in the blood in order to monitor response to therapy.
  - Viral loads are expressed in two ways: as total copies per milliliter of blood and as log values (*see* chapter 15).
  - Monitoring recommendations vary. Typically viral loads are measured before treatment (to establish a baseline), when treatment is modified, and every 3 to 4 months (monitoring frequency can be reduced to every 6 months if the patient is suppressed or stable).
- Other molecular assays: transcription-mediated amplification, nucleic acid sequence-based amplification, and proviral DNA detection (used to detect integrated HIV DNA)

A change in viral load that is greater than 0.5 log is considered significant.

- **Specimen**
  - Laboratory-based serology: blood, plasma, or serum
  - Rapid assays: fingersticks or oral swabs
  - NAATs: blood, plasma, or serum without heparin (it is a PCR inhibitor). The preferred anticoagulant is sodium citrate. CSF can be tested for neurologic disease.
- **Genotyping** is analysis of HIV's viral genome to assess antiviral resistance.
  - The genes for reverse transcriptase, protease, and sometimes integrase are sequenced by Sanger sequencing directly from a plasma sample.
  - The sequences are analyzed for mutations that are known to confer resistance. Known mutation sequences are maintained in curated databases that are continually updated for real-time resistance information (for example, the Stanford HIV RT and Protease Sequence Database).
  - Genotyping is done before therapy is initiated and in cases of suspected therapeutic failure.
  - Genotyping can fail or be unreliable when
    - Viral loads are <500 copies/ml.
    - There are mutations within the sequencing primer binding region.
    - The relevance of the detected mutations is not understood.

HIV "genotyping" is not used to identify genotypes but is actually a genetic analysis of resistance mutations. This is unlike genotyping for other viruses, which *is* used for typing strains (HCV genotyping, for example, identifies genotypes 1 to 6).

- **Phenotyping** is an analysis of how HIV actually grows and replicates in the presence of antivirals.
  - Viral RNA is purified from the specimen.
  - RT-PCR is used to amplify critical regions (such as protease and reverse transcriptase genes).
  - Amplified genes are inserted, or cloned, into a vector containing the rest of the genes to make viable HIV virions, as well as a reporter gene called the luciferase gene.
  - The vector DNA is introduced into a cell line so that new HIV virions containing the genes of interest are produced.
  - These virions are then used to infect cells in the presence of different antivirals. As the virions replicate, they produce luciferase, which allows

them to produce light. This light output can be measured and reflects the ability of the virus to replicate in the presence of certain drugs.

- The levels of replication are compared to that of a reference strain, and the fold changes are correlated to drug susceptibility.
- A patient's viral load must be >500 copies/ml to have enough starting material for accurate phenotyping.
- Phenotyping is a highly manual, labor-intensive, time-consuming, and costly test. It is performed only rarely and in cases of multidrug resistance.

- **Tropism testing** is used for identifying whether CXCR4 or CCR5 coreceptors are being used by the infecting viral strains and whether CCR5 antagonists will be useful for treatment purposes.
  - Phenotypic assays: The RNA for tropism-related HIV genes (e.g., the gp160 gene) is amplified from a patient specimen. It is cloned into viral constructs in order to form lab-derived virions expressing the patient's binding proteins. These virions are used to infect cells expressing CCR5 and CXCR4 in order to determine which coreceptor those binding proteins are able to use. These assays are labor-intensive, with a slow turnaround time.
  - Genotypic assays: The portion of the *env* gene that encodes gp160 is sequenced, and algorithms are used to predict which coreceptor the patient's HIV strain is able to use.

6. **Prevention and treatment:** All HIV-positive individuals should be treated regardless of CD4 count and viral load level (Fig. 9.6).

The United Nations has a 90-90-90 treatment goal by 2020:

90% of HIV+ people will be diagnosed.
90% of diagnosed cases will be receiving treatment.
90% of people receiving treatment will have viral suppression.

- There is no vaccine.
- Goal of therapy: to achieve undetectable viral loads in the blood
- Antiretroviral therapy (ART) or antiretrovirals (ARV) consist of 6 main groups of drugs (*see* chapter 19)
  - CCR5 antagonists
  - NRTIs
  - NNRTIs
  - Integrase inhibitors
  - Protease inhibitors
  - Fusion inhibitors
- A typical regimen consists of two NRTIs and one integrase inhibitor, protease inhibitor, or NNRTI. CCR5 antagonists and fusion inhibitors are not considered first-line treatments and are reserved for patients who failed or are failing treatment.
- Highly active antiretroviral therapy (**HAART**) is also known as combination antiretroviral therapy (**cART**). It is the standard of treatment for HIV-infected individuals. It is a treatment approach that uses 3 or more drugs in combination to reduce the risk of resistance (*see* Table 19.7).
- Boosters: drugs that can enhance or prolong the activity of other HIV drugs so that they can be given in less toxic doses.
- **Preexposure prophylaxis** (**PrEP**)
  - Once-daily tenofovir and emtricitabine
  - Used by people at high risk of HIV infection, such as partners of HIV-infected individuals

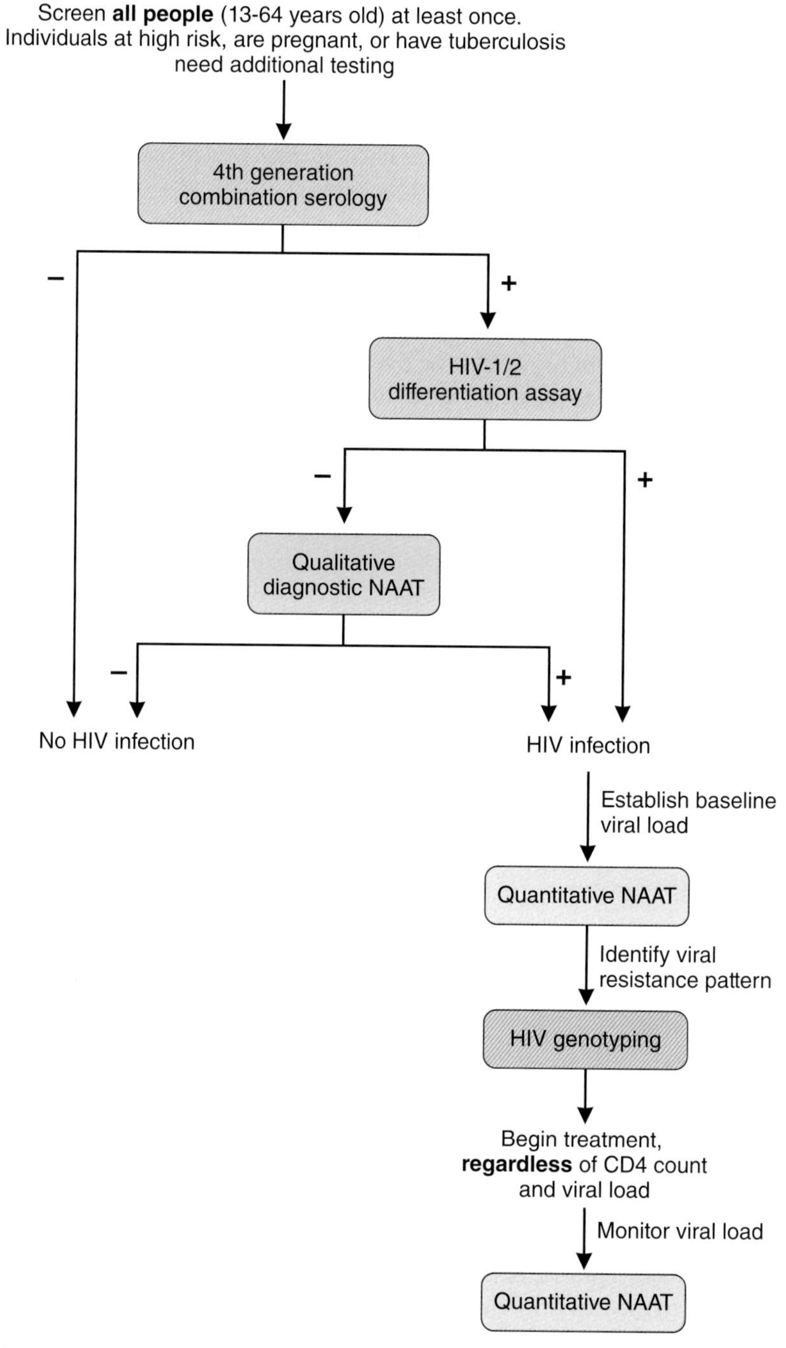

Figure 9.6. General overview of HIV testing.

- **Postexposure prophylaxis (PEP)**
  - For adults or newborns with exposure
  - Initiated within 72 hours of exposure
  - Is a 2- or 3-drug cocktail administered for 28 days
- **Immune reconstitution inflammatory syndrome (IRIS)**: when HIV-related symptoms get worse after treatment because the immune system recovers. This occurs because the immune system is able to produce a

robust inflammatory response to the preexisting infection. Factors associated with IRIS include the following:

- Very low CD4 cell count
- Occurs within 2 weeks to 3 months of beginning therapy

III. HUMAN T-CELL LYMPHOTROPIC VIRUS. HTLV has a biology and structure similar to that of HIV (in fact, HIV used to be part of the HTLV group). However, HTLVs can cause leukemia and myelopathy.

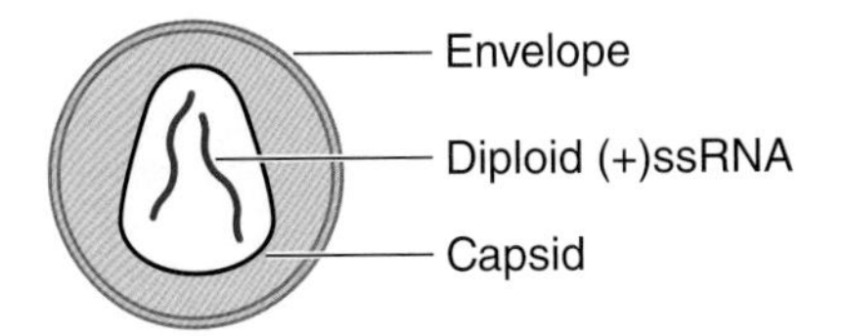

Figure 9.7. HTLV.

1. **Background:** They are ~100-nm, enveloped (+) ssRNA retroviruses (Baltimore class VI). There are 4 distinct types (Fig. 9.7).
   - HTLV-1 is prevalent in Japan, Central Africa, the Caribbean, Melanesia, and pockets of South America (e.g., Brazil).
   - HTLV-2 is prevalent in Native Americans.
   - Little is known about HTLV-3 and HTLV-4.
2. **Transmission**
   - Sexual intercourse is the most common mode.
   - Vertically; mainly through **breast milk** but also transplacentally
   - Infected blood, transfusions, and intravenous drug use
   - Naturally infects humans and primates
   - Integrates into the genome and causes a lifelong infection
3. **Clinical presentation:** Only 1 to 5% of HTLV-1-infected persons get clinical disease.
   - **Adult T-cell leukemia:** an aggressive T-cell cancer
     - Is caused by HTLV-1 only.
     - Lymphadenopathy, hypercalcemia, dermatitis, and bone lesions
     - There are 4 main forms: smoldering, chronic, lymphoma/leukemia, and acute.
     - Poor prognosis (death in ~6 to 24 months), depending on the form of infection
     - Acquisition during infancy is the largest risk factor.
     - The incubation period is ~30 to 50 years, so disease often occurs during middle age.
     - Opportunistic pathogens (e.g., *Pneumocystis jirovecii* and *Strongyloides*) are associated with adult T-cell leukemia.
   - **HTLV-associated myelopathy (HAM)**, also known as tropical spastic paraparesis, is caused by both HTLV-1 and -2.
     - Causes debilitating neurogenerative disease because neurons in the CNS are demyelinated
     - Paralysis
     - Less likely to occur in people with HLA-A 02 haplotypes
     - Chronic

Adult T-cell leukemia: HTLV-1

HTLV-associated myelopathy: HTLV-1 and -2

4. **Diagnosis:** Serology is the primary method of HTLV detection. Antibodies form 30 to 60 days after exposure.
   - Serology
     - Primary screening is done by ELISA. This has high sensitivity but cannot differentiate HTLV-1 and HTLV-2.

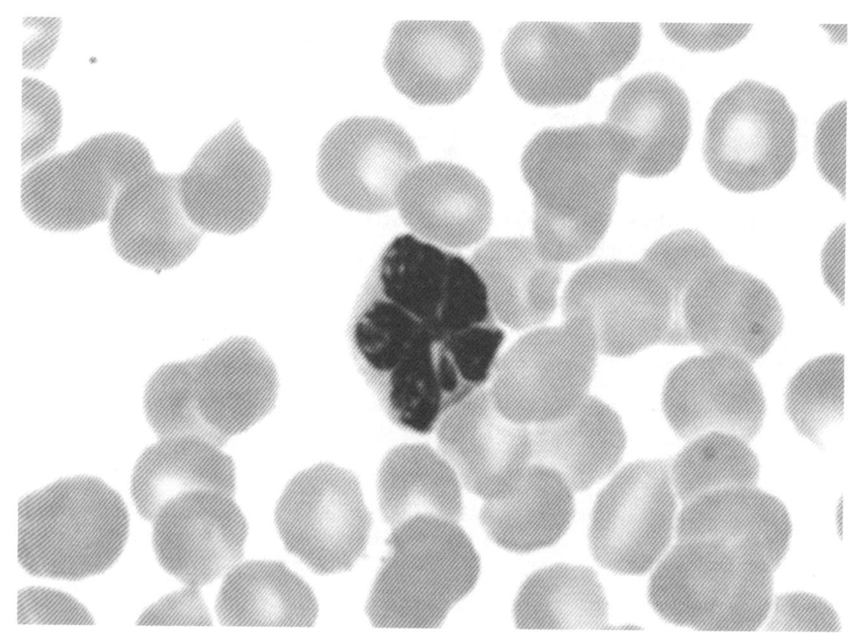

Figure 9.8. **Characteristic "flower" or "cloverleaf" cells in HTLV-1 infection.** Reprinted from reference 56 [Luca DC, August CZ, Weisenberg E. Adult T-cell leukemia/lymphoma in a peripheral blood smear. *Arch Pathol Lab Med.* 2003; 127(5):636] with permission from *Archives of Pathology & Laboratory Medicine*. Copyright 2003 College of American Pathologists.

    - Western blot is done for confirmation and is available at some reference labs.
        - It is able to differentiate between HTLV-1 and -2.
        - It can be indeterminate even upon repeat testing (especially cases from Africa and Melanesia).
- Histology: Peripheral blood smear shows T cells with abnormal, multilobular, or deeply indented nuclei ("flower" or "cloverleaf" cells) (Fig. 9.8; *see also* Fig. 13.7).
- NAAT: not commonly available. It can be used if Western blot is indeterminate or to follow an infection over time.
- Culture: not used in clinical labs

5. Prevention and treatment

- There is no defined treatment regimen, although it can be treated with NNRTIs and some protease inhibitors.
- There is no vaccine.
- Blood products in the United States are screened for HTLV.

Multiple-Choice Questions

1. A 25-year-old female presents to her family care physician for an annual physical. She is sexually active with a single partner, has no symptoms of illness, has a body mass index of 21, does not smoke or use drugs, and has not been tested for HIV. Which of the following is true regarding follow-up testing for sexually transmitted diseases?

a. No testing is recommended because she is at very low risk.

b. An HIV screen should be performed.

c. An HIV screen should be performed, followed by mandatory annual screening.

d. None of the above

2. Which of the following is an AIDS-defining illness?

a. BK virus-associated nephropathy

b. Adenovirus-induced acute respiratory disease syndrome

c. Kaposi's sarcoma

d. Congenital rubella

3. What is the purpose of HIV protease?

a. To cleave the HIV polyprotein

b. To cleave enzymes released by macrophages

c. To excise the provirus from the host DNA

d. To kill the host cell

4. Which of the following is the preferred method of screening for HIV in adults?

a. Nucleic acid testing

b. Genotyping

c. Serology

d. Culture

5. Which of the following is the preferred method for diagnosis of HIV in newborns?

a. Nucleic acid testing

b. Genotyping

c. Serology

d. All of the above

6. Which of the following is the preferred method for monitoring HIV infection in adults?

a. Nucleic acid testing

b. Genotyping

c. Serology

d. Culture

7. Retroviruses have reverse transcriptase. This makes them similar to

a. Baltimore class I viruses

b. Baltimore class IV viruses

c. Baltimore class V viruses

d. Baltimore class VII viruses

8. Retroviruses have (+) ssRNA. This makes them similar to

a. Baltimore class I viruses

b. Baltimore class IV viruses

c. Baltimore class V viruses

d. Baltimore class VII viruses

9. Preexposure prophylaxis should be administered

a. Within 72 hours of an exposure

b. To HIV^+ patients, prior to sexual intercourse

c. To individuals with high risk of exposure to HIV

d. As a three-drug regimen

10. An HIV⁺ patient with pneumocystis pneumonia and a CD4 count of 50 cells/μl is treated with HAART for the first time. After 4 weeks, his fever recurs, he has shortness of breath, and X-rays show substantial pulmonary infiltrates. Diagnostic testing shows that virus titers have decreased and the CD4 count has increased. Which of the following is most likely?

a. This is a highly resistant strain of HIV.

b. This is a highly resistant strain of *Pneumocystic jirovecii*.

c. This is immune reconstitution imflammatory syndrome.

d. This is a secondary opportunistic infection.

11. Which of the following is true regarding HIV resistance testing?

a. Genotyping should be performed for every patient.

b. Phenotyping should be performed for every patient.

c. Genotyping and phenotyping should be performed for every patient.

d. Genotyping or phenotyping should be performed only if patients begin failing therapy.

True or False

12. HIV can be transmitted through breast milk. **T F**

13. Blood products are screened for HIV and HTLV. **T F**

14. HTLV-2 causes adult T-cell leukemia and HTLV-associated myelopathy. **T F**

15. All HIV infected individuals will progress to AIDS. **T F**

10

CHAPTER 10

ONCOGENIC VIRUSES

I. OVERVIEW. Some viruses are **oncogenic** (or **transformative**), meaning that they can transform normal cells into cancerous cells. Transformation usually occurs when the normal host cell replication cycle is disrupted. For instance, genes that stimulate replication (**oncogenes**) can be turned on. Or, proteins like **tumor suppressors** that normally prevent uncontrolled replication are inhibited. Oncogenic viruses that can do this are usually ones that can persist and/or integrate into the host genome (such as Baltimore class I, VI, and VII viruses).

II. HUMAN PAPILLOMAVIRUSES. Human papillomaviruses (HPVs) are widely prevalent viruses that are transmitted by direct skin or sexual contact. They typically cause asymptomatic infection or benign growths called **papillomas** but can also cause malignancies, like cervical cancer.

1. **Background:** HPVs are small, nonenveloped viruses in the *Papillomaviridae* family (Fig. 10.1). They contain circular dsDNA (Baltimore class I).
   - HPVs only infect humans.
   - The virus can integrate into the host genome as well as persist episomally to cause a long-lived infection.
   - There are >200 genotypes of HPV categorized into 5 genera.
   - HPV E6 and E7 proteins inhibit cellular tumor suppressors p53 and Rb. Without tumor suppressors to regulate cell division, cells can proliferate uncontrollably and cause cancer.

HPV DNA can persist for years in the host.

2. **Transmission**
   - HPVs are primarily transmitted through direct contact with infected epithelial cells. This includes direct cutaneous contact and oral, anal, or vaginal intercourse.
   - They are not transmitted by body fluids like blood, sperm, and saliva.
   - They are common in the population (among adults the presence of oral HPV is ~7% and the presence of genital HPV is ~45%) (21).

Oncogenic HPVs are called **high risk**. Nononcogenic HPVs are called **low risk**.

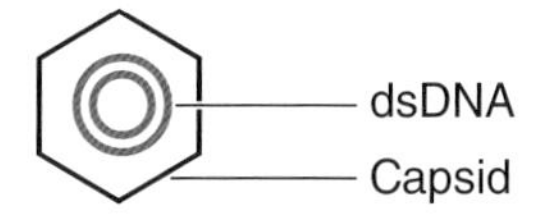

Figure 10.1. HPV.

Important: the majority of HPV infections will resolve on their own. A small proportion of infections will become cancerous, but only after many years/decades.

HPV6 and -11 are low risk, and the most common causes of papillomas.

- Risk factors for the following types of infection
  - Genital infection: multiple sex partners
  - Common warts: young age
  - Genital warts: adolescents, young adults

3. **Clinical presentation:** HPV infects the epithelial basal cells in the lower layers of skin and mucous membranes. Most infections are asymptomatic, but some initiate growth and replication of host cells. Some genotypes cause malignant cellular changes that can lead to cancer, while other genotypes cause more benign papillomas. Papillomas are outgrowths or warts that originate from epithelial cells (Table 10.1).
   - **Cutaneous papillomas**
     - Include common, flat, and plantar warts
     - May appear several months after initial infection
     - Diagnosis can be made clinically or histologically from a biopsy in order to rule out malignancy.
     - Warts can be removed immediately in a doctor's office.
     - **Epidermodysplasia verruciformis** ("tree man disease"): patients with a rare, autosomal recessive defect in cell-mediated immunity are hypersusceptible to HPV. They present with a lifelong, uncontrolled formation of papillomas and lesions.
   - **Anogenital papillomas (condylomata)**
     - Include condyloma acuminatum and giant condyloma
     - More common in women
     - Men who have sex with men have increased rates of rectal warts.
     - Should be differentiated from condyloma lata, which is caused by syphilis, and condyloma subcutaneum, which is caused by molluscum contagiosum virus
   - **Oral papillomas**
     - **Recurrent respiratory papillomatosis:** benign warts on the oral mucosa and larynx that cause airway obstruction, stridor, shortness of breath, and vocal changes. It primarily affects children and can be transmitted during vaginal delivery from mothers with active condylomata.
     - **Focal epithelial hyperplasia:** warts in the mouth and soft palate
   - **Malignancies:** Most infections are asymptomatic, but high-risk genotypes (especially HPV16 and -18) are oncogenic. Some infections produce

Table 10.1. Symptoms caused by HPV genotypes

MANIFESTATION	COMMON CAUSATIVE HPV GENOTYPES
Cutaneous papillomas	1, 2, 3, 4, 10
Epidermodysplasia verruciformis	5, 8
Anogenital papillomas	**6, 11**
Oral papillomas	**6, 11**
Focal epithelial hyperplasia	13, 32
Head and neck cancers	**16, 18**, 31, 33
Anogenital cancers	**16, 18**
Cervical cancer	**16, 18** (cause ~70% of cases), 31, 45

abnormal cells that are precancerous. Most of these abnormalities (as much as 90%) resolve on their own within 2 years. A small proportion of abnormal cells progress to cancerous cells over the course of years to decades. The following malignancies are associated with HPV.

- Anal cancer
- Cancer on external genitalia
- Skin neoplasm: red, scaly, nonpainful patches on skin surface (Bowen's disease)
- Head and neck cancers occur in the oropharynx and symptoms include sore throat, painful swallowing, earache, swollen lymph nodes, and unexplained weight loss. They are most common on the tonsils and on the back of the tongue.
- **Cervical cancer.** This may be asymptomatic, or else symptoms can include bleeding, discharge, and pain during intercourse.

HPV16 and -18 are the most common causes of HPV-related malignancies.

HPV causes almost all cases of cervical cancer.

4. Diagnosis

- Cytology: a technique used to evaluate the structure of individual cells that are in a suspension. Cervical cells are brushed or scraped from the cervix and stained with Papanicolaou, or **Pap**, stain.
  - There are two types of Pap smear. In the **direct smear**, collected material is smeared directly onto a slide. In **liquid-based cytology**, collected material is placed into a solution containing fixatives in order to preserve morphology and separate out debris and nonrelevant cells. The two most commonly used in solutions used are ThinPrep (contains alcohol) and SurePath (contains alcohol and formaldehyde).
  - The cells are read by a pathologist for characteristic morphologic changes according to the Bethesda system, a cytology classification system.
    - **ASC-US:** atypical squamous cells of undetermined significance
    - Low-grade squamous intraepithelial lesions (**LSILs**): Mild abnormalities are present, such as **koilocytic atypia**. Koilocytes are enlarged cells with enlarged nuclei and a large clear space around the nucleus called a perinuclear halo (*see* Fig. 13.7).
    - High-grade squamous intraepithelial lesions (**HSILs**): moderate to severe abnormalities are present.
    - **ASC-H:** atypical squamous cells are present; cannot exclude HSIL
- Colposcopy: a technique used to evaluate reproductive structures and tissues macroscopically for abnormalities.
  - Acetic acid test: vinegar is applied to the cervix. Normal tissue is unaffected; neoplasms turn white.
  - Lugol's iodine: used to stain normal tissue brown
  - These tests have low sensitivity and specificity but are inexpensive and rapid.
  - A biopsy is taken if abnormalities are seen during examination of a patient.
- Histology: A biopsy is taken from the cervix. Abnormal cell growth in the basal epithelial layer is called **cervical intraepithelial neoplasia (CIN)**.
  - The extent of abnormal cells is described in 3 levels: CIN 1 usually correlates with a cytology result of LSILs, while CIN 2 and CIN 3 correlate

CIN typically appears as disordered, undifferentiated, and dividing cells.

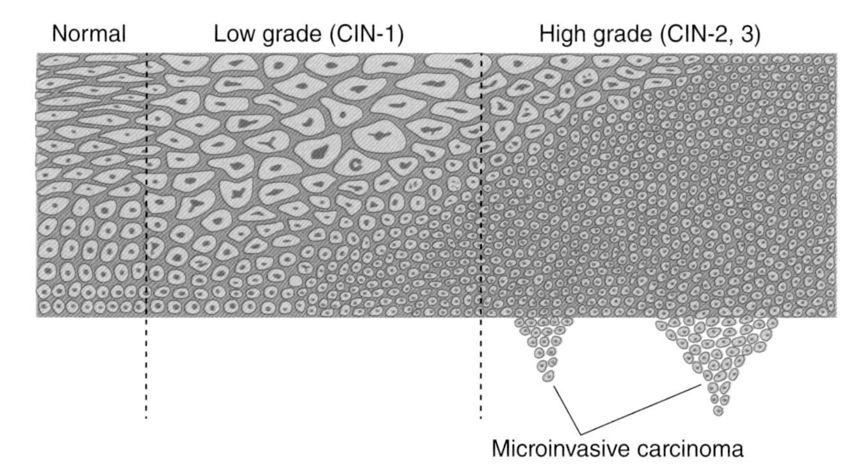

Figure 10.2. Diagram of CIN. Adapted from reference 57, with permission.

with HSILs (Fig. 10.2). High-grade neoplasia progresses to cancer more frequently than low-grade neoplasia.

- **CIN 1:** mild nuclear abnormalities, usually restricted to the lower one-third of the epithelial layer
- **CIN 2:** moderate numbers of cells with nuclear abnormalities, usually restricted to the lower two-thirds of the epithelial layer
- **CIN 3:** Nuclear abnormalities span the entire depth of the epithelial layer.

- Immunohistochemistry: p16 is a tumor suppressor produced by human cells. It can be used as a marker for high-risk HPV because its expression is upregulated during HPV-induced cancer.
- NAAT: highly sensitive and specific. The target detected is usually E6/E7 mRNA.
  - Specimen: usually cervical; sometimes vaginal or oral. Sample must contain epithelial cells, such as ThinPrep/SurePath cytology specimens, mucosal scrapings/swabs, or biopsy tissue.
  - Detection of high-risk versus low-risk HPV genotypes can increase specificity for the disease and escalates the need for more invasive procedures. For example, HPV16 and -18 are the most common high-risk genotypes. If detected in cervical specimens, a colposcopy should be performed.
- Inappropriate for diagnosis
  - Culture is not used. HPV does not grow in cell monolayers that are typically used.
  - Serology is not used since HPV is highly prevalent.
  - Blood and other body fluids do not contain adequate levels of HPV and should not be tested.

5. Prevention and treatment

- Vaccines are available and highly effective at preventing cervical cancer caused by the most common high-risk HPV. Vaccines may protect against non-cervical cancers too.
  - Vaccines mainly target high-risk HPV16 and -18 (some formulations include additional high- and low-risk genotypes)

  - The vaccines target L1, which is the antigenic major capsid protein of HPV.
  - Recommended routinely for preteen (before sexual activity) girls and boys
  - 2 or 3 doses, depending on the formulation
- **Cervical screening**
  - Should be performed only on women ≥21 years old
  - Pap smear screening is performed every 3 years.
  - Alternative: For women between 30 and 65, screening can be done every 5 years if a Pap smear screen is also performed with HPV nucleic acid testing.
  - Excessive screening may lead to unnecessary medical interventions, since most infections resolve on their own or have an extremely slow clinical course.
- Pharmacological treatment
  - Podophyllotoxin
  - Trichloroacetic acid
  - Imiquimod
  - Interferon
- Surgical intervention
  - Cryotherapy
  - Loop electrosurgical excision procedure, or "cold knife cone," is an outpatient procedure. It uses a thin, electrified wire shaped like a cone that is scooped though the abnormal cervical tissue to cut it away.

More frequent testing or testing with nucleic acid amplification assays alone is too sensitive and is not recommended.

III. EPSTEIN-BARR VIRUS. EBV is a highly prevalent virus and the most common cause of infectious mononucleosis. It causes a lifelong, latent infection that is typically asymptomatic but may result in nasopharyngeal and B-cell cancers.

1. **Background:** EBV is a member of the *Herpesviridae* family and is also known as human herpesvirus 4 (HHV-4). It is a member of the oncogenic *Gammaherpesvirinae* subfamily (*see* Box 4.1, Table 4.1).
   - It is an enveloped dsDNA virus (Fig. 10.3).
   - There are two main strains, EBV1 (more common) and EBV2.
2. **Transmission:** through **saliva.** It is passed through kissing, sharing cups or utensils, and intimate contact (infectious mononucleosis is also known as the "kissing disease").
   - In young adults, infection is associated with college and military dormitories and elite athletes.
   - Shedding occurs periodically in oropharyngeal secretions, especially in immunosuppressed hosts (e.g., HIV-infected individuals). Shedding is not symptomatic.
   - EBV infects oral epithelial cells and undergoes the lytic cycle. It can also infect B lymphocytes but causes a latent infection. It does not integrate into the genome but exists episomally.
   - It is highly prevalent (>65% of the population).
3. **Clinical presentation:** EBV can cause several distinct syndromes.
   - **Asymptomatic infection:** more likely when primary exposure occurs in childhood

EBV remains latent for life, like the other herpesviruses. Reactivations are usually mild or asymptomatic, but they may be severe in immunocompromised people.

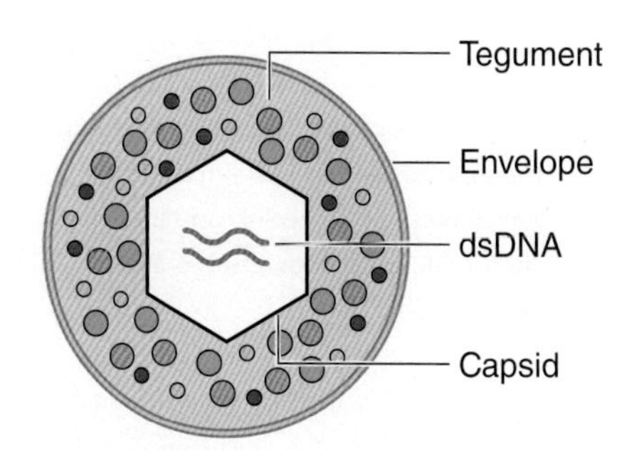

Figure 10.3. EBV.

- **Infectious mononucleosis:** acute, primary infection with EBV that is most common in healthy, young adults (15- to 24-year-olds)
  - Symptoms last for 2 to 3 weeks.
  - Also known as "mono" or "glandular fever"
  - Fever, pharyngitis, and lymphadenopathy
  - Fatigue, splenomegaly, retro-orbital pain, and headaches
  - Administration of ampicillin produces a widespread pruritic, maculopapular rash.
  - Lymphocytosis and atypical mononuclear cells (lymphocytes and monocytes) may be present in the peripheral blood. These cells are not infected but are reactive cytotoxic CD8 cells.
  - Thrombocytopenia
  - Less common: jaundice and rash
  - Rare: autoimmune hemolytic anemia, splenic rupture, and neurologic symptoms (Guillain-Barré syndrome, encephalitis, and others)
  - Fulminant disease: Fulminant hepatitis and hemophagocytic syndrome occur in boys with X-linked lymphoproliferative disease. Infection can be fatal.

EBV causes most cases of mononucleosis. Other pathogens in the differential diagnosis are CMV, HIV, and *Toxoplasma gondii*.

- **Posttransplant lymphoproliferative disease** (PTLD) occurs in transplant patients after immunosuppression.
  - Most common in the first year of transplantation
  - May occur after both primary and reactivated EBV infections
  - Flu-like symptoms
  - PTLD in the CNS: encephalitis, meningitis, and other neurologic abnormalities
- **Nasopharyngeal carcinoma:** endemic in China and Alaskan Inuits
- **Oral hairy leukoplakia**
  - Reactivation of EBV in epithelial cells of the tongue
  - White, "hairy" lesion that does not scrape off (unlike thrush)
  - Mostly associated with HIV infection
- **Burkitt's lymphoma**
  - Endemic in Africa, especially in children. In this geographic location it may be associated with *Plasmodium falciparum* infection.
  - HIV^+ individuals are also at risk.
- **Hodgkin's lymphoma**
- Unlike HSV, VZV, and CMV, EBV does not cause adverse effects to the infant during pregnancy.

4. Diagnosis

- Serology: the primary method of diagnosis for infectious mononucleosis (acute disease) because antibodies are diagnostic and detectable at the time symptoms appear. Serology can be used to detect heterophile antibodies or various anti-EBV antibodies.
  - **Heterophile antibodies:** antibodies that are broadly reactive to animal red blood cells. EBV stimulates production of these antibodies during acute infection.

Serology should be used to diagnose EBV mononucleosis (not PCR or culture). Antibodies appear rapidly and are detectable when the person is acutely symptomatic.

Table 10.2. Interpretation of EBV serologic results

STAGE OF INFECTION	HETEROPHILE ANTIBODIES	VCA IgM	ANTI-EA	VCA IgG	ANTI-EBNA
None (naive)	–	–	–	–	–
Primary, early acute	+	+	–	–	–
Primary, recent	–	+	+	+	–
Past	–	–	–/+	+	+
Reactivation	–/+	–/+	–/+	+	+

- Advantages
  - Highly specific (~100%)
  - Is a rapid agglutination test from serum
- Disadvantages
  - Lower sensitivity (~85 to 90%). Approximately 10% of people with EBV infectious mononucleosis do not produce these antibodies, so testing can be falsely negative.
  - Children <4 years old may not produce these antibodies.
  - Heterophile antibodies may persist for several months, even after resolution of acute infection, so they may not signify current disease.

- EBV-specific antibodies can be used in heterophile antibody-negative cases or to identify the stage of EBV infection (Table 10.2, Fig. 10.4).
  - **Anti-VCA IgM** appears rapidly during early disease and persists for 1 to 2 months. It can persist or recur if the virus reactivates.

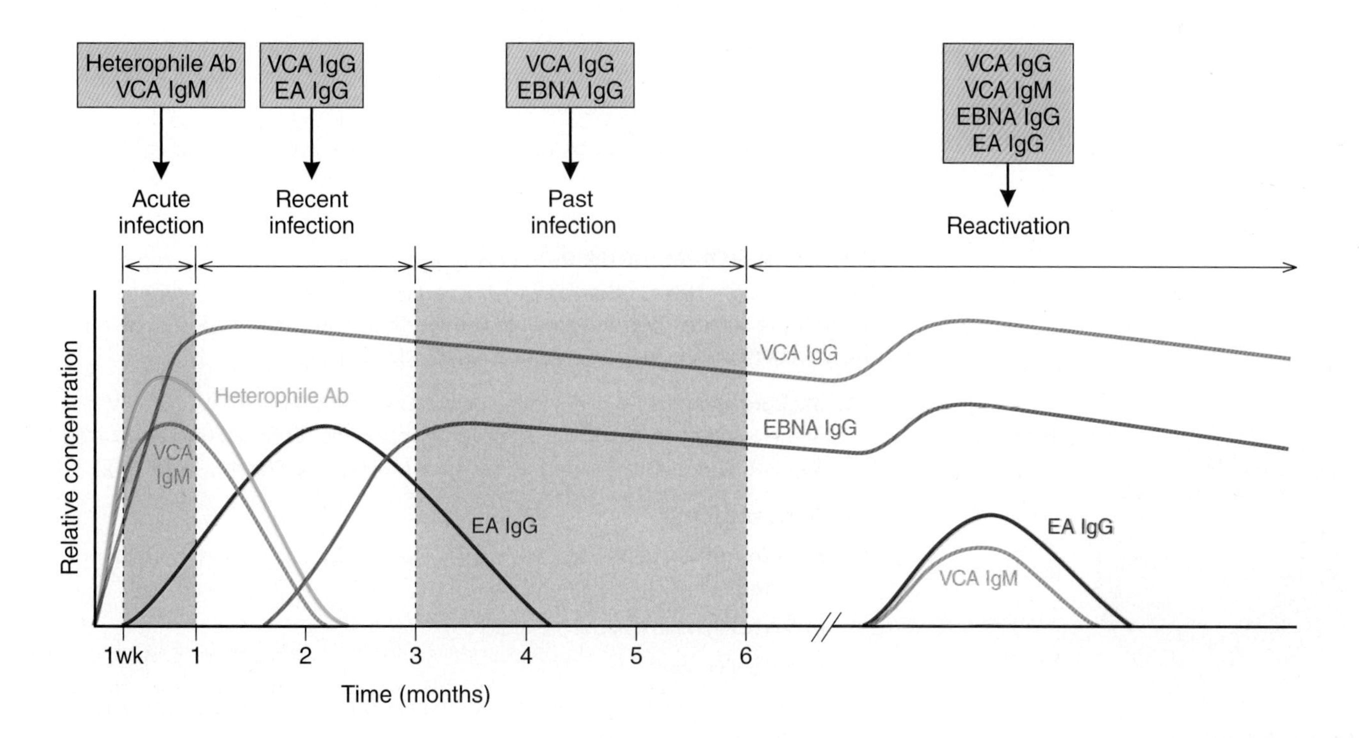

Figure 10.4. Kinetics of EBV serology.

- **Anti-VCA IgG** appears rapidly during early disease and remains lifelong.
- **Anti-EA IgG** appears during early disease and persists for 3 to 6 months. It can appear again if the virus reactivates.
- **Anti-EBNA IgG** appears after clinical disease and can be used to identify past exposure. However, 5 to 10% of patients do not produce these antibodies.

- NAAT: not used for self-limited infections or uncomplicated cases of mononucleosis. PCR assays are used to diagnose and monitor transplant patients for development of PTLD, and can also be used for other severe EBV syndromes too.
  - Specimens: whole blood, respiratory samples, bone marrow, tissue, ocular or joint fluid, and CSF (for CNS presentation)
  - Note that normal virus shedding is occasionally detected in asymptomatic and symptomatic people, so diagnosis should be made based on clinical presentation and a quantitative PCR trend.
- Clinical diagnosis: Oral hairy leukoplakia, for example, has a classic appearance. It cannot be scraped off the way thrush can.
- *In situ* hybridization: can be done on tissue and is used to identify neoplasms or infections in the tissue (like nasopharyngeal carcinoma). **EB**V small **e**ncoding **R**NAs (**EBER**) 1 and 2 are two small, noncoding ("junk") pieces of viral RNA that are produced in latently infected cells.
- Culture: not used. EBV does not grow in commonly used cell lines.

5. **Prevention and treatment**
   - There is currently no vaccine against EBV.
   - Primary infections, or infections in immunocompetent individuals, do not generally require treatment.
   - Antiherpesviral treatments (e.g., acyclovir) have been used but clinical benefit has not been adequately demonstrated. Also these antivirals do not affect latent EBV.
   - Corticosteroids may be used in severe or chronic cases to manage symptoms.

IV. HUMAN HERPESVIRUS 8. HHV-8 typically causes a mild or asymptomatic lifelong infection. Unlike other herpesviruses, HHV-8 is rare in North America but has high prevalence in other parts of the world. It can cause Kaposi's sarcoma, which occurs in high-risk populations such as AIDS patients.

1. **Background:** HHV-8 is an enveloped dsDNA virus (Fig. 10.5). Like EBV, it is in the *Herpesviridae* family, *Gammaherpesvirinae* subfamily. It is also called Kaposi's sarcoma-associated herpesvirus (KSHV) (*see* Box 4.1, Table 4.1).

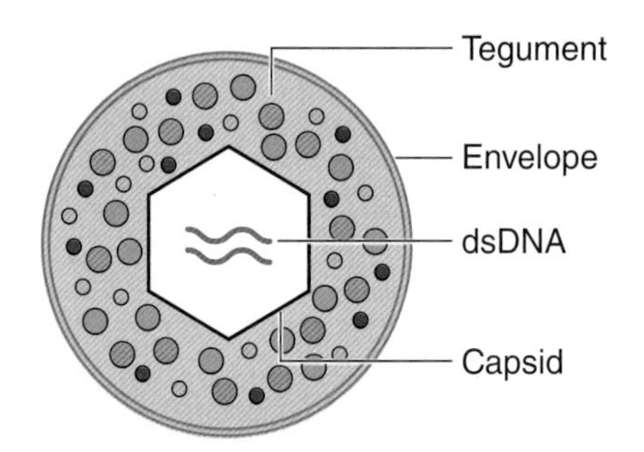

Figure 10.5. HHV-8.

2. **Transmission**
   - Through oral secretions, like saliva. This usually occurs during childhood in regions of high endemicity.
   - Vertically, from mother to child
   - Sexual contact
   - Transmission through blood is inefficient but can occur (e.g., intravenous drug use and transfusions).

Table 10.3. The four types of Kaposi's sarcoma (22)

TYPE	PERSONS AFFECTED	GEOGRAPHIC LOCATION	CLINICAL PRESENTATION
Classic	Older men	**Mediterranean**, East/Central European	Progresses **slowly**, with a relatively low mortality rate.
Endemic (or African)	Males, children	Africa	Can resemble **slowly** progressing classic Kaposi's sarcoma or can cause **rapidly** spreading, visceral, aggressive tumors with lymphadenopathy
Epidemic (or AIDS related)	AIDS patients	Worldwide	Is an **AIDS-defining illness**. It progresses **rapidly**.
Iatrogenic	Transplant or immunosuppressed patients	Worldwide	Aggressive and progresses **rapidly**. It can involve mucosa and visceral organs.

- Risk groups
  - Higher prevalence in the Mediterranean region, Middle East, Africa, Amazonian region, some regions of China
  - AIDS patients (*see* Table 9.1)
  - Men who have sex with men
  - People of Ashkenazi Jewish descent

3. **Clinical presentation:** Primary infection is asymptomatic or mild. Like other herpesviruses, HHV-8 causes a lifelong episomal infection and can reactivate upon immunosuppression. HHV-8 infects endothelial cells, B cells, and spindle cells. It can cause angioproliferation and several types of neoplasms.
   - Nonspecific primary infection: asymptomatic, or symptoms occur mainly in children.
     - Fever and upper respiratory tract symptoms
     - Maculopapular rash
   - **Kaposi's sarcoma**
     - Presents as darkly discolored, purplish or reddish bruise-like lesions on the skin, usually on the legs and feet. These may be flat, raised, or nodular and may ulcerate and bleed.
     - Lesions are usually NOT painful or itchy.
     - Lesions can also occur on mucosal membranes and visceral organs such as the gastrointestinal tract.
     - There are four 4 main forms, which are described in Table 10.3. Importantly, rapid, aggressive Kaposi's sarcoma can have a high fatality rate.
   - **Primary effusion lymphoma:** a form of non-Hodgkin's B-cell lymphoma. It presents as effusions within body cavities, like the pleural, peritoneal, and pericardial spaces. It occurs primarily in AIDS patients.
   - **Multicentric Castleman's disease:** a lymphoproliferative disorder. It occurs primarily in AIDS patients.

4. Diagnosis

- Immunohistochemistry: staining of LANA-1 from a biopsy of a Kaposi's sarcoma lesion
- NAAT: performed on a lesion biopsy or a blood sample, although this test is not frequently available
- Cell culture: not used
- Serology: available at reference laboratories but has limited utility

5. Prevention and Treatment

- There is no vaccine against HHV-8.
- Tumors can be treated with radiotherapy, chemotherapy, or immunotherapy
- Underlying HIV infection can be treated with HAART
- Removal of immunosuppression
- Drugs used to treat other *Herpesviridae*, such as ganciclovir, cidofovir, and foscarnet, may have activity, but efficacy is not known.

Multiple-Choice Questions

1. Which of the following subfamilies of *Herpesviridae* is oncogenic?

a. *Alphaherpesvirinae*
b. *Betaherpesvirinae*
c. *Gammaherpesvirinae*
d. All of the above

2. Pap smear screening of HPV screening should be performed

a. On females ≥15 years old
b. Annually, on sexually active females
c. Every 3 years, on females ≥21 years old
d. Only on symptomatic females

3. HHV-8 is different from other herpesviruses due to which of the following?

a. It does not cause a lifelong infection.
b. It is not widely prevalent in the United States.
c. It causes skin lesions.
d. None of the above

4. Which of the following HPV genotypes is the most likely to cause cervical cancer?

a. HPV2
b. HPV6
c. HPV16
d. HPV68

5. Which of the following groups is NOT typically associated with EBV disease?

a. Neonates
b. HIV-infected individuals
c. Transplant patients
d. Adolescents

Match the following. Use each answer only once.

6. What is the preferred diagnostic test method for each of the following EBV-related conditions?

Mononucleosis	A. Clinical presentation
Oral hairy leukoplakia	B. Serology
Nasopharyngeal carcinoma	C. PCR
PTLD	D. Histology with *in situ* hybridization

True or False

7.	Culture is not used as a diagnostic test for EBV.	**T F**
8.	Culture is not used as a diagnostic test for HPV.	**T F**
9.	HPV is easily transmitted by saliva.	**T F**
10.	Recurrent respiratory papillomatosis is most often acquired through oral sex.	**T F**

11

CHAPTER 11

ZOONOTIC VIRUSES

I. OVERVIEW. Zoonotic viruses are transmitted from animals to humans without an intermediate vector. These viruses are high-risk pathogens because they can cause very severe symptoms, like hemorrhagic fever and CNS infection, and because there are low levels of preexisting immunity to them in humans. See Table 11.1 for a comparison of zoonotic viruses. Viruses that are transmitted from animals to humans by an intermediate arthropod vector (arboviruses) are covered in chapter 12.

1. **Background:** Almost all viruses covered in this chapter are enveloped, (–) ssRNA (Baltimore class V) viruses. There are only two that are dsDNA viruses (class I).
2. **Transmission:** Contact with animals or animal specimens (e.g., pet owners, lab workers, livestock workers) is the most common route of exposure.
   - Animal bites or scratches
   - Ingestion of contaminated meat
   - Inhalation of aerosolized blood (e.g., abattoir workers) and body fluids (e.g., campers exposed to animal urine or feces)
   - Some animals can harbor the virus without getting disease (**reservoirs**), while others are infected and are symptomatic.
3. **Clinical presentation:** Zoonotic viruses can cause several severe symptoms that can result in death.
   - **Central nervous system (CNS) infection**
   - **Severe pulmonary infection**
   - **Viral hemorrhagic fever:** a life-threatening syndrome with hemorrhage and multiorgan involvement. It can be caused by several arboviruses (*see* chapter 12) and zoonotic viruses in the following groups.
     - Bunyavirales (e.g., Rift Valley fever virus, Crimean-Congo hemorrhagic fever virus, and hantaviruses)
     - Paramyxoviruses (e.g., Nipah and Hendra viruses)
     - Arenaviruses (e.g., Lassa virus)
     - Filoviruses (e.g., Ebola and Marburg viruses)
     - Flaviviruses (e.g., dengue, yellow fever, Omsk hemorrhagic fever, and Kyasanur Forest disease viruses)

Rodents are a common reservoir. Rodent-borne viruses (e.g., Lassa virus and Hantaviruses) are sometimes called "roboviruses."

Most zoonotic viruses covered here are highly **neurotropic**.

Differential diagnoses: malaria, typhoid fever, and meningococcemia

For common hemorrhagic viruses remember **"DEATHLY"**: **D**engue virus, **E**bola/Marburg viruses, **A**frican Rift Valley fever virus, **T**ick-borne CCHFV, **H**anta-virus, **L**assa virus, **Y**ellow fever virus

Table 11.1. Comparison of zoonotic viruses

VIRUS(ES)	NUCLEIC ACID	MOST COMMON ANIMAL ASSOCIATION(S)	PERSON-TO-PERSON TRANSMISSION	GEOGRAPHIC LOCATION	CLASSIC PRESENTATION	MORTALITY RATE
Rabies virus	(–) ssRNA	Bats, dogs	No	Worldwide	CNS	**Very high**
Ebola and Marburg viruses	(–) ssRNA	Bats, monkeys	**Yes**	Africa	Abdominal, hemorrhagic fever	**High**
Lassa virus	(–) ssRNA[a]	Rodents	**Yes**	Africa	Hemorrhagic fever	Low, except in neonates
Crimean-Congo hemorrhagic fever virus	(–) ssRNA[a]	Farm animals, ostriches, ticks (*see also* Table 12.1)	**Yes**	Africa, Middle East, Balkans and surrounding areas	Arboviral fever, CNS, hemorrhage	**High**
Lymphocytic choriomeningitis virus	(–) ssRNA[a]	Rodents	No, except vertically	Worldwide	CNS	Low
Hantaviruses	(–) ssRNA[a]	Rodents	No	HPS, USA; HFRS, Korea, China, Western Europe	Pulmonary, renal, hemorrhagic fever	Low to high
Monkeypox virus	dsDNA	Monkeys, prairie dogs, rodents	**Yes**	Africa, USA	Dermatologic	Low
Herpes B virus	dsDNA	Monkeys	No	Worldwide	CNS	**High**
Hendra virus	(–) ssRNA	Horses	No	Australia	Pulmonary, CNS	**High**
Nipah virus	(–) ssRNA	Bats, pigs	**Yes**	South and Southeast Asia	Pulmonary, CNS	**High**
Rift Valley fever virus	(–) ssRNA[a]	Livestock, mosquitoes (*see* chapter 12)	No	Africa	Ocular, CNS, hemorrhagic fever	Low for ocular and CNS disease. **High** for hemorrhagic fever.

[a]Classified as negative sense RNA (Baltimore class V), but the genome can actually be read in both directions (ambisense).

4. Diagnosis

- Serology is used to detect exposure most often because these viruses are relatively rare.
- Public health labs may offer confirmatory molecular testing.
- Culture should not be used in diagnostic labs, in order to reduce the risk of spread posed by live virus propagation.
- Patients and patient specimens can pose a high risk to health care personnel, even during routine care and testing. This is because of factors like spills/breakage of patient specimens, management of waste, improperly followed decontamination procedures, the potential for creating aerosols, instrument contamination, and inappropriate use of personal protective equipment (PPE).
  - Confining suspected specimens to different instruments or doing point-of-care testing can minimize risk.
  - When clinical infection with a high-risk pathogen is suspected, hospital and lab workers should be notified because of occupational risk.

5. Prevention and treatment

- Wash site of inoculation with soap and water thoroughly for 15 minutes. All viruses in this chapter are enveloped and will degrade upon exposure to detergents.
- Supportive care

Washing bites or scratches with soap will reduce the risk of infection because these viruses are enveloped and will degrade upon exposure to detergents.

II. RABIES VIRUS. Rabies virus is the most lethal virus known to infect humans. It is typically transmitted by a bite from a dog or other small animal. The virus takes several months to travel into the CNS, and then it causes irritability, anxiety, abnormal behavior, and other mental status changes. Once symptoms occur, the disease is rapid and almost always fatal (Table 11.1).

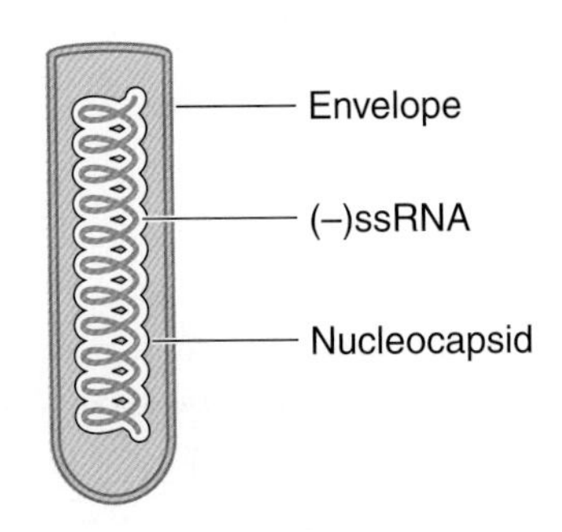

Figure 11.1. Rabies virus.

1. Background

- Family *Rhabdoviridae*. Rabies virus is a bullet-shaped, enveloped, (–) ssRNA virus (Baltimore class V) (Fig. 11.1).
- It is unstable and does not persist in the environment and is very susceptible to drying.

2. Transmission: Rabies virus is present in brain and nervous system tissue and is shed in body fluids like saliva and vaginal and seminal fluid.

- Rabies is mostly associated with **dogs,** bats, and raccoons, but it is also associated with other small mammals (e.g., skunks and foxes).
- Bites are the most common mode of transmission, and the most dangerous.
- Nonbite exposure can occur when infectious material or large amounts of aerosolized virus come in contact with open wounds or mucous membranes (e.g., corneal, nose, or mouth).
    - Scratches from claws contaminated by saliva.
    - Surgical transplant patients receiving tissue from infected patients
    - Lab exposure to infected material
- Rabies virus is not transmitted by feces, urine, or blood.
- Person-to-person transmission is theoretically possible, but so far has occurred only through organ transplantation.
- Geographic location: worldwide

Rabies is acquired through animal **bites and scratches**. It is not transmitted person to person.

3. Clinical presentation

- Rabies virus has a long incubation period, usually between 30 and 90 days (range is ~1 week to more than a year).
- Upon exposure, the virus replicates in muscle cells at the entry site. It ascends to the CNS by retrograde transport through the axons. Because of this, bites further away from the CNS take longer to manifest symptoms (e.g., symptoms occur rapidly when bites occur on the face but manifest slowly when bites occur on the ankles).
- Early symptoms are nonspecific, with malaise, fever, and discomfort at the wound site. This lasts for ~4 days.
- Symptoms advance to acute progressive encephalomyelitis, at which point the patient becomes infectious. There are two forms of the disease.
    - **Encephalitic** or "**furious**" rabies: occurs in ~80% of cases. Characterized by hyperactivity, hallucinations, anxiety and irritability, hydrophobia, muscle spasms, seizures, and paralysis.

Rabies virus typically has a long incubation and symptoms often appear several months after exposure.

Almost 100% of cases are fatal without postexposure prophylaxis, treatment, or vaccination.

- **Paralytic** or "**dumb**" rabies: occurs in ~20% of cases. Characterized by hypersalivation, headaches, and ascending flaccid paralysis.

- Acute disease lasts less than 10 days. The final stages are coma and/or death.

4. Diagnosis

- DFA: staining of biopsied tissue is the gold standard. Sources are brain tissue, skin, and hair follicles from the nape of the neck.
- NAAT: PCR can be used on tissue as well as CSF and saliva
- Histology: nerve cells show characteristic cytoplasmic inclusions called **Negri bodies** (*see* Fig. 13.6). These are found in Purkinje cells and the hippocampus.

The rabies vaccine should be used both preexposure (prophylactic) and postexposure.

5. Prevention and treatment

- Prevention
  - A vaccine is available and highly effective. It should be given to people at high risk of exposure (for example, veterinarians, rabies laboratory workers, travelers to areas where the virus is highly endemic, and some military personnel).
  - Animals are vaccinated as a public health measure.
    - In the United States, vaccination is required for dogs, which reduces exposure to rabies from pets.
    - Animal vaccine for rabies can be given orally. It is inoculated onto bait and disseminated to immunize a large population of wild animals.
- Treatment upon exposure
  - The virus is fragile in the environment and is readily susceptible to detergents, UV light, and drying. The following steps are taken to reduce the risk of infection.
    - Wash site with soap and water for 15 minutes. Wound cleaning reduces infection rates by 90%.
    - In areas of low endemicity, monitor the animal that caused the bite for 10 days.
    - Unvaccinated individuals: Start postexposure vaccination. This includes both the rabies vaccine and passive rabies immunization (human rabies immunoglobulin, which is a concentration of IgG from hyperimmunized donors).
    - Vaccinated individuals: two vaccine booster shots
  - **Milwaukee protocol:** a method of rabies management in which the patient is placed in a coma in order to give the body time to produce antibodies to the virus before it damages the brain.

III. EBOLA AND MARBURG VIRUSES. Ebola and Marburg viruses are found in Africa and can cause deadly hemorrhagic fevers. They spread rapidly through contact with body fluids and have very high morbidity and mortality rates (Table 11.1).

1. Background: Family *Filoviridae*. These viruses are enveloped and filamentous (filo = filament) and contain (–) ssRNA (Fig. 11.2).

- Ebola viruses
  - Four human species: Bundibugyo, Sudan, Zaire, and Tai Forest (used to be Ivory Coast)

Zaire ebolavirus is the most pathogenic strain.

- ▫ One monkey species: Reston (identified in the United States and the Philippines). It does not cause disease in humans.
- Marburg viruses: 2 species

2. Transmission

- Animal to human: contact with blood, feces, and fluids from **primates** and **bats** (fruit bats are thought to be the reservoir). The virus can also be transmitted by ingestion of wild animal meat (**bushmeat**).
- Human to human: direct contact of mucous membranes or broken skin with **human secretions** (blood, saliva, urine, sweat, tears, semen, vomit, and breast milk)
  - ▫ Fomites (e.g., linens) contaminated with these fluids
  - ▫ Ebola virus can remain positive in semen for many months and can be sexually transmitted.
- Geographic location: Africa
- People at high risk
  - ▫ Family members of infected individuals
  - ▫ Health care workers and laboratory workers
  - ▫ Contact with dead persons during burial ceremonies
  - ▫ Contact with infected animals or animal reservoirs

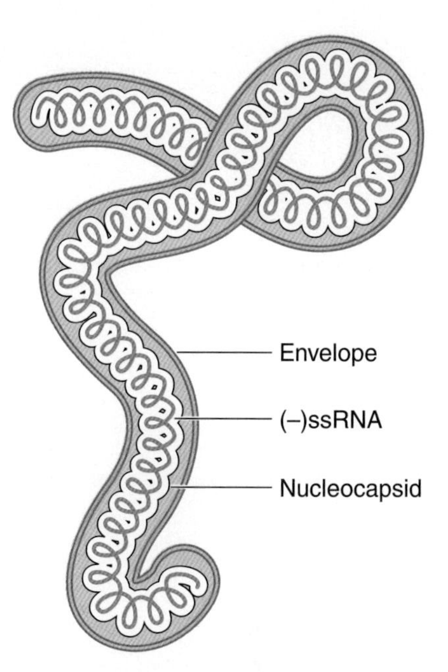

Figure 11.2. Ebola virus.

These viruses are highly transmissible by contact with human body fluids. They are not airborne. Droplet transmission is possible but has not been documented for humans (23, 24).

3. Clinical presentation

- Incubation
  - ▫ Ebola virus: 2 to 21 days
  - ▫ Marburg virus: 5 to 10 days
- Abrupt fever, weakness, fatigue, headache, and myalgia
- Diarrhea (may be bloody), vomiting, and abdominal pain
- Rash and red eyes
- **Internal and external hemorrhage** (e.g., blood from eyes, ears, nose, and rectum)
- Symptoms can progress to organ failure, jaundice, weight loss, shock, seizures, and death.
- For survivors, recovery can take months, with long-term sequelae (joint and vision problems).
- Both viruses have high fatality rates (30 to 90%).

4. Diagnosis

- Serology: ELISA is used to identify IgM and IgG. Serology does not cross-react between Marburg and Ebola viruses.
- NAAT: can be done at some public health labs on tissue, blood, urine, semen, and other body fluid samples. Some multiplex assays capture several hemorrhagic or high-risk pathogens simultaneously.
- Rapid antigen test: provides results within 15 minutes and can be performed in lower-resource settings. However, it is less sensitive (92%) and specific (85%) than PCR. It is approved by the WHO only in emergency contexts (25).

5. Prevention and treatment

- Prevention
  - ▫ No vaccines have been approved by the FDA yet, but several are under investigation.

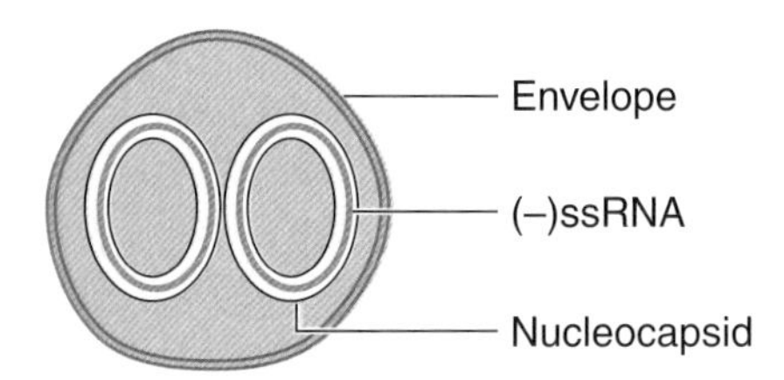

Figure 11.3. Lassa virus.

- An experimental vaccine called **rVSV-ZEBOV** has shown excellent efficacy, and has been distributed during recent outbreaks. It consists of a vesicular stomatitis virus expressing Zaire ebolavirus antigens.
- Treatment
  - Supportive care (intravenous fluid, blood transfusion, etc.). The fatality rate is lower in places with reliable health care facilities.
  - Treatment with serum from convalescent patients has been used.
  - Other therapeutics are being investigated, including new antivirals and antibodies.

IV. LASSA VIRUS. Like Ebola virus, Lassa virus is found in Africa, is highly transmissible through body fluids, and can cause hemorrhagic fever. However, it has significantly lower morbidity and mortality (Table 11.1).

1. **Background:** is an enveloped virus with a segmented, ambisense (–) ssRNA genome (Fig. 11.3). It is a member of the *Arenaviridae* family.
2. **Transmission:** through exposure (ingestion, direct contact of open wounds, or inhalation) to rodents or rodent urine and droppings
   - Reservoir: **multimammate rat**, which sheds virus in the urine and feces. The rat is found in savannahs, forests, and homes.
   - Geographic location: endemic in West Africa
   - It can also be passed from **person to person** through contact with body fluids (e.g., blood, urine, and feces) and tissue.
   - It is not airborne.

Lassa virus has a much lower fatality rate than Ebola virus.

3. **Clinical presentation:** usually produces a mild, asymptomatic, or nonspecific febrile illness, but severe disease occurs in ~20% of cases
   - Incubation: 1 to 3 weeks
   - Symptoms of severe disease include hemorrhaging (eyes, nose, and mouth), vomiting, deafness, respiratory distress, encephalitis, and multiorgan failure.
   - The fatality rate is low, especially compared to those of Ebola and Marburg viruses (1% of all cases).
   - Much higher fatality rate among pregnant women and gestating fetuses
4. **Diagnosis:** usually done by serology
   - Serology: ELISA
   - Immunohistochemistry: can be used for postmortem cases
   - NAAT: in early stage of disease
5. **Prevention and treatment**
   - Reduce transmission by minimizing exposure to rodents.
   - Use PPE in hospitals.
   - Treatment is with supportive care and ribavirin.

V. CRIMEAN-CONGO HEMORRHAGIC FEVER VIRUS. Like Ebola virus, CCHFV can cause hemorrhagic fever with high morbidity and mortality. It has a wide geographic distribution and is typically transmitted through exposure to livestock. However, it is also considered an arbovirus because it is tick borne (Table 11.1; *see also* Table 12.1).

1. **Background:** enveloped, with a tripartite (i.e., segmented into three pieces), ambisense (–) ssRNA genome (Fig. 11.4).

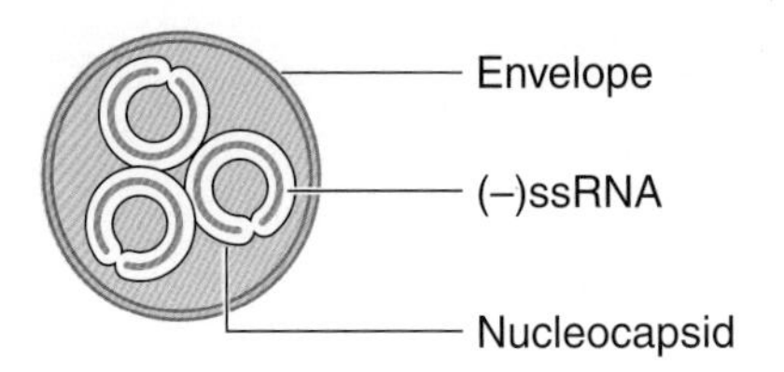

Figure 11.4. CCHFV.

2. **Transmission:** Exposure to animals, infected humans, or ticks can cause disease.
   - Contact with animals, blood, and tissue. The virus can infect a wide range of animals (**farm animals**, livestock, hares, ostriches, and birds).
   - Directly from person to person through contact with blood and tissue
   - **Ticks:** *Hyalomma* sp.
   - Geographic location: Africa, the Middle East, the Balkans, and Asia (*see* Fig. 12.1)
   - Risk factors
     - **Occupational exposure** to animals (abattoir workers, farmers, and veterinarians)
     - Health care workers
3. **Clinical presentation**
   - The incubation period is short, ~1 to 3 days.
   - Nonspecific febrile illness: sudden high fever, myalgia, arthralgia, headache, backache, nausea, and vomiting
   - **Hemorrhagic fever:** petechiae in the mouth, red eyes, blood in stool and sputum, bruising, and bleeding from orifices or injection sites
   - **CNS symptoms:** agitation, mood swings, and confusion
   - Hepatitis, multiorgan failure
   - Recovery is slow.
   - High mortality rate (~30%)

CCHFV is both arboviral (tick borne) and zoonotic (exposure to farm animals)

4. **Diagnosis:** usually done by serology
   - Serology: antibody as well as antigen ELISAs
   - Immunohistochemistry: can be used for postmortem cases
   - NAAT: in early stage of disease
5. **Prevention and treatment**
   - No vaccine is available for humans.
   - Supportive care
   - Ribavirin

VI. HANTAVIRUSES. Hantaviruses are a rare group of rodent-borne viruses that only sporadically cause disease in humans. Some species cause a fatal pulmonary infection, while others cause hemorrhagic fever with renal failure (Table 11.1).

1. **Background:** enveloped viruses with an ambisense, (–) ssRNA, tripartite genome (like CCHFV) (Fig. 11.5)

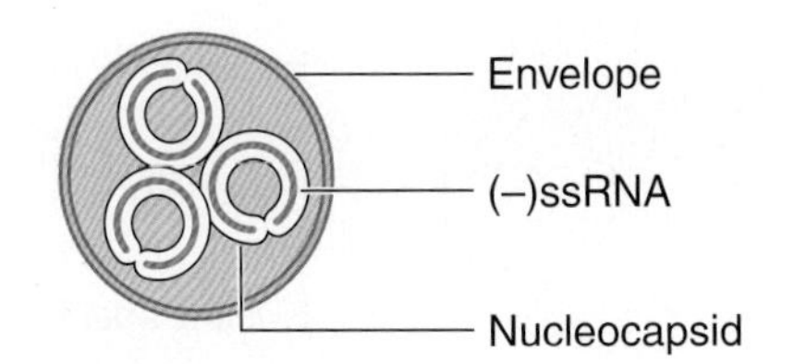

Figure 11.5. Hantavirus.

Box 11.1. Recent reclassification of bunyaviruses

Sometimes hantaviruses and CCHFV are called bunyaviruses because both used to be classified as genera under the family *Bunyaviridae*. They have been recently reclassified as separate families (*Hantaviridae* and *Nairoviridae*, respectively) but they are both still under the order *Bunyavirales*.

Table 11.2. **Four common hantavirus species**

HANTAVIRUS	GEOGRAPHIC REGION	ASSOCIATED SYNDROME	MORTALITY RATE
Sin Nombre virus	United States (especially Colorado, Utah, Arizona, and New Mexico, so it is sometimes called the "**four corners**" virus)	**HPS**	**50%**
Hantaan virus	Korea, China, Russia	HFRS	~10%
Seoul virus	Worldwide, China	HFRS	1%
Puumala virus	Western Europe, Russia	HFRS	1%

2. Transmission

- **Rodents** are the reservoir. Human infection is associated with rodent contact or exposure to rodent urine, saliva, and feces.
- Hantaviruses are not transmitted from person to person

3. Clinical presentation

- Incubation period: ~2 weeks
- Can affect young, healthy people
- There are two syndromes (Table 11.2).
    - **Hantavirus pulmonary syndrome (HPS):** nonspecific initial symptoms of fever, fatigue, and muscle aches. There may be additional abdominal symptoms (pain, diarrhea, and vomiting). There is progression to shortness of breath and fluid-filled lungs. HPS is very rare and is caused by Sin Nombre virus.
    - **Hemorrhagic fever with renal syndrome (HFRS):** a debilitating illness with fever, hemorrhage, renal failure, thrombocytopenia, and hypotension. There is a high incidence in China.

4. Diagnosis: usually by serology. Immunohistochemistry and PCR can be used, but are not commonly available.

5. Prevention and treatment

- Several inactivated vaccines are available, but not in the United States.
- Treatment is with supportive care.
- Ribavirin can be used.

VII. LYMPHOCYTIC CHORIOMENINGITIS VIRUS. Lymphocytic choriomeningitis virus (LCMV) is a common virus among mice but can be transmitted to humans. It can cause CNS infections that are generally mild or asymptomatic but can be severe especially in gestating fetuses (Table 11.1).

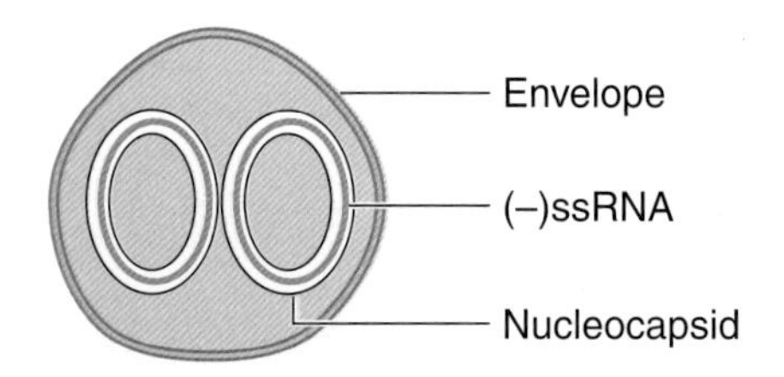

Figure 11.6. **LCMV.**

1. **Background:** enveloped, with an ambisense segmented (–) ssRNA genome (Fig. 11.6). It is a member of the *Arenaviridae* family (like Lassa virus).
2. **Transmission:** rodent-borne and acquired through exposure to rodent urine, feces, and saliva via mucous membranes (nose, eyes, and mouth) or from a rodent bite
    - From **rodents**. People with house mice and owners of pet hamsters and mice are at risk.

- Vertically. The mother gets a primary infection during pregnancy and transmits it to the fetus.
- Not transmitted from person to person
- Geographic location: worldwide, but higher incidence in areas with poor hygiene (i.e., areas with mice)

3. **Clinical presentation:** LCMV causes flu-like symptoms, sometimes with meningitis.
   - First phase: fever, generalized aches and pains, headache, and vomiting. This lasts about 1 week.
   - Second phase: Only occurs in some people. CNS infection occurs, with symptoms of meningitis and/or encephalitis, acute hydrocephalus, spinal inflammation, paralysis, and sensory and motor issues. Symptoms usually resolve without long-term damage.
   - **Fetus/neonate:** fetal death, birth defects, and mental retardation
   - The mortality rate is very low.
4. **Diagnosis:** Testing is not commonly available or performed.
   - Serology for IgM and IgG
   - NAATs like PCR for CSF
5. **Prevention and treatment:** supportive care

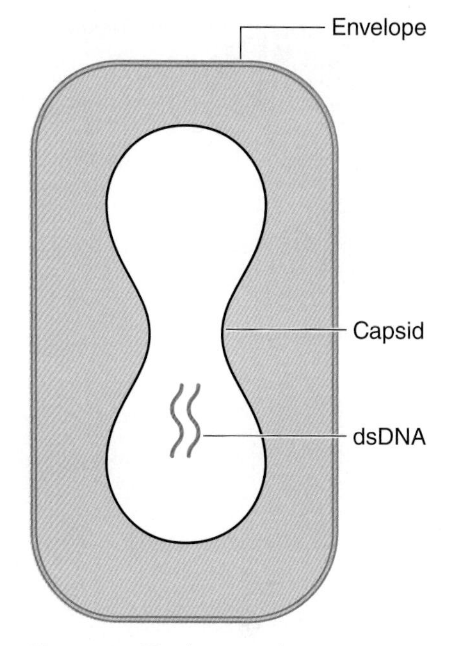

Figure 11.7. Monkeypox virus.

VIII. MONKEYPOX VIRUS. Monkeypox virus causes a smallpox-like disease in monkeys. It can also cause a milder, smallpox-like disease in humans that needs to be differentiated from disease caused by variola virus (Table 11.1).

1. **Background:** enveloped, dsDNA virus in the *Poxviridae* family (Fig. 11.7)
2. **Transmission:** exposure to infected animals like **monkeys** and prairie dogs. The reservoir is probably rodents.
   - Contact with body fluids (bite, scratch, or lesion fluid) of animals
   - Inhalation of aerosolized animal fluids (e.g., abattoirs) or respiratory droplets
   - Ingestion of contaminated meat, possibly
   - Human-to-human transmission can occur.
   - Found in West and Central Africa and the United States
3. **Clinical presentation:** Monkeypox virus causes smallpox-like disease, but it is usually milder or asymptomatic.
   - Fever, chills, headache, and myalgia
   - Causes lymphadenopathy, unlike smallpox virus
   - Smallpox-like rash (umbilicated macules, vesicles, papules, and nodules). Like smallpox, all the lesions are at the same stage.
   - Recovery occurs within 1 month.
   - Mortality rate: low to moderate (10% in Africa and 0% in the United States).
   - Differential diagnoses: smallpox, chickenpox, and herpes B viruses
4. **Diagnosis**
   - Serology for IgM and IgG is the primary method.
   - NAATs like PCR can be done on lesion fluid, blood, tissue, or respiratory swabs.

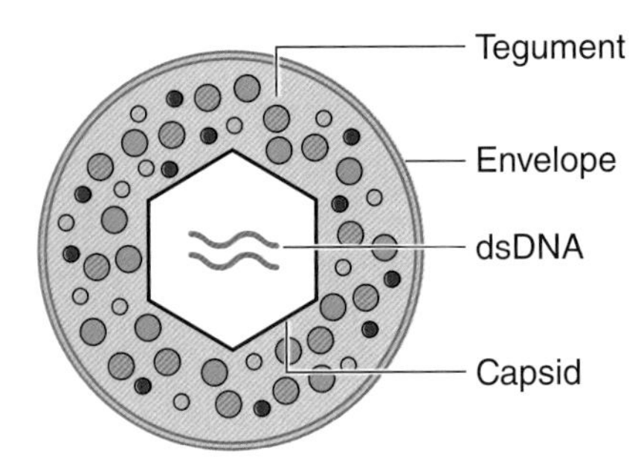

Figure 11.8. **Herpes B virus.**

5. **Prevention and treatment**
   - A live-virus vaccine containing a similar pox virus (vaccinia virus) is available. It is typically used for smallpox and can also be used to protect against monkeypox virus. It can be used both preexposure and postexposure.
   - Supportive care

IX. HERPES B VIRUS. Herpes B virus (also known as **B virus**, monkey B virus, herpesvirus simiae) causes a common, mild, and latent infection in monkeys, like HSV in humans. The rate of transmission to humans is very low despite the number of interactions between monkeys and humans. However, if infection does occur, the disease is severe and often lethal (Table 11.1).

1. **Background:** Herpes B virus is an enveloped, dsDNA virus in the *Alphaherpesvirinae* subfamily (like HSV) that infects monkeys (Fig. 11.8).
2. **Transmission**
   - Traumatic inoculation from infected **monkeys** (bite or scratch).
   - Contact of mucosal surfaces (e.g., eye) or breaks in the skin with monkey body fluids/tissue
   - People at high risk: veterinarians, lab workers, animal handlers
3. **Clinical presentation**
   - Incubation: ~1 month after exposure
   - Vesicular lesions at site of exposure
   - Headaches, fever, and fatigue
   - Nervous system damage: pain and numbness at the site, ascending paralysis, diplopia, brain and motor damage
   - High fatality rate (~40%) a few weeks after symptoms begin
4. **Diagnosis:** Testing for herpes B virus is not commonly available.
   - Serology: highly cross-reactive with common human herpes viruses
   - Culture: gold standard, but must occur in BSL4 area
5. **Prevention and treatment**
   - Prophylactic antivirals: acyclovir and valacyclovir
   - Antivirals after symptoms: acyclovir and ganciclovir

X. HENDRA AND NIPAH VIRUSES. Hendra and Nipah viruses are rare viruses that belong to the genus *Henipavirus*. Nipah virus is found in South and Southeast Asia. It is associated with pigs or with ingestion of date palm sap that has been contaminated by bats. Hendra virus is associated with exposure to horses in Australia. Both Nipah and Hendra viruses can cause encephalitis and have a high fatality rate (Table 11.1).

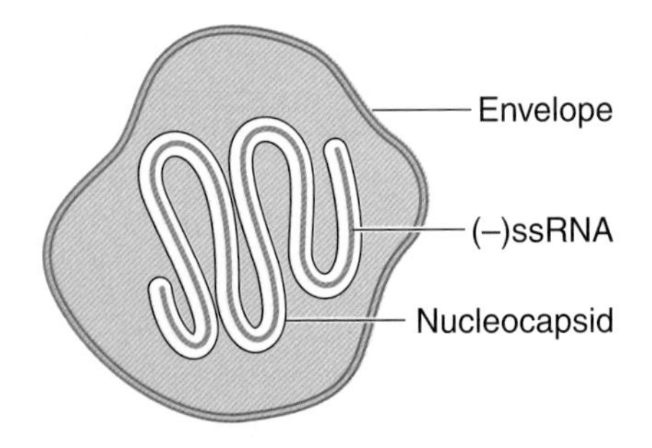

Figure 11.9. **Hendra and Nipah viruses.**

1. **Background:** are enveloped, (–) ssRNA viruses that are part of the *Paramyxoviridae* family, like parainfluenza virus (Fig. 11.9)
2. **Transmission**
   - Nipah virus
     - Exposure to fruit **bats**, infected **pigs**, and **date palm sap** contaminated by fruit bats
     - Can be transmitted from person to person (e.g., caregivers)
     - Geographic location: South and Southeast Asia (Malaysia and Bangladesh)

- Hendra virus
  - **Horses** can be infected from exposure to fruit bats (i.e., **flying foxes**). Humans can be infected through exposure to horses.
  - Recreational and occupational exposure to horses increases risk.
  - Geographic location: Hendra, Australia

3. **Clinical presentation**
   - Incubation: ~1 to 2 weeks
   - Initial symptoms: nonspecific febrile illness, with acute respiratory and pulmonary signs that last 3 to 14 days
   - Can progress to encephalitis and coma
   - High mortality rate (>70%)
   - Long-term sequelae: personality changes, convulsions, reactivation of latent virus resulting in death
4. **Diagnosis**
   - NAATs from respiratory tract (throat and nasal swab) blood, and CSF
   - Immunohistochemistry on tissue can be used in fatal cases.
   - Serology can also be used to identify exposure.
5. **Prevention and treatment:** ribavirin, intravenous immunoglobulin (potentially)

Multiple-Choice Questions

1. **Which of the following viruses is/are transmitted from dogs to humans?**
   a. Canine parvovirus
   b. Rabies virus
   c. Monkeypox virus
   d. All of the above
2. **Which of the following is a severe clinical manifestation associated with hantaviruses?**
   a. Hepatic fever
   b. Encephalitis
   c. Hemorrhagic fever with renal syndrome
   d. Hemorrhagic diarrhea
3. **Bats are a reservoir for the following viruses except**
   a. Ebola virus
   b. Rabies virus
   c. Nipah virus
   d. CCHFV
4. **Which of the following viruses can be acquired through ingestion of bushmeat?**
   a. Ebola virus
   b. HIV
   c. Nipah virus
   d. LCMV

5. Which of the following groups is at risk for herpes B virus infection?

a. Children

b. Elderly

c. Veterinarians

d. Ostrich handlers

6. Which of the following viruses is generally not lethal?

a. LCMV

b. Herpes B virus

c. Hendra virus

d. Marburg virus

7. Which of the following zoonotic viruses is not transmitted directly from person to person?

a. Lassa virus

b. Ebola virus

c. Hantaan virus

d. Nipah virus

Match the following. Use each answer only once.

8. Animals or other risk factors associated with exposure *(May require reading outside of this chapter.)*

Prairie dogs	A. Nipah virus
Deer mouse	B. Monkeypox virus
Bats	C. Ebola virus
Date palm sap	D. MERS virus
Camels	E. Sin Nombre virus

9. Important or predominant geographic locations of zoonotic viruses

Australia	A. CCHFV
Malaysia	B. Ebola virus
Guinea	C. Hendra virus
Utah	D. Nipah virus
Iran	E. Sin Nombre virus

True or False

10. A person infected with rabies virus in the hospital only needs to be placed under standard isolation precautions. Contact, droplet, or respiratory precautions for isolation are not necessary. *(May require reading outside of this chapter.)* **T F**

11. Monkeypox and chickenpox viruses are both poxviruses. *(May require reading outside of this chapter.)* **T F**

12. Ebola virus can be transmitted by the contact, droplet, and airborne routes. **T F**

13. LCMV is transmitted vertically. **T F**

12

CHAPTER 12

ARBOVIRUSES

I. OVERVIEW. Arthropod-borne viruses are also known as arboviruses. These viruses are transmitted by arthropod vectors (e.g., mosquitos and ticks) and may cause disease in humans. Arboviruses are commonly maintained in animal reservoirs. See Table 12.1 for a comparison of important arboviruses and their transmission, distribution, and animal hosts.

1. **Background:** most arboviruses are enveloped and contain ssRNA genomes. Some exceptions exist, including Colorado tick virus, which is a naked dsRNA virus.

2. **Transmission**
   - Most arboviruses do not replicate to high enough titers in humans to be transmitted to a new arthropod vector, so humans are considered **dead-end hosts**.
   - For some arboviruses (e.g., dengue virus), humans can act as a **reservoir**, and the virus achieves a high enough titer in a human host to be taken up by a vector and transmitted to another person.
   - Very rare: transmission via transplant or transfusion
   - Arthropod vectors are segmented and have an exoskeleton, bilateral symmetry, and jointed limbs. The most common arthropod vectors of viruses are mosquitos and ticks. It may be important to differentiate between species of mosquitoes and ticks because they have different geographic distributions and transmit different diseases (Fig. 12.1).
     - **Mosquitoes:** Female, but not male, mosquitoes take a blood meal from the host. Ingested virions replicate in the gastrointestinal tract of the mosquito for ~7 to 14 days (extrinsic incubation period). The mosquito is not infectious during this stage. The virus disseminates through the mosquito and becomes infectious when it reaches the salivary glands. There are several mosquito genera that are particularly important in disease transmission to humans, such as *Aedes*, *Anopheles*, and *Culex* spp. (Table 12.2). Different species transmit different viral pathogens.
     - **Ticks:** Ticks can take blood meals and transmit disease at larval, nymphal, and adult stages (Fig. 12.2). Soft ticks have a leathery body and are more likely to feed on just one host species. Medically relevant

If humans are a reservoir, then infected individuals have the potential for spreading diseases if a vector is present. This is important because it leads to sustained transmission and, potentially, epidemics.

Table 12.1. Transmission and distribution of important arboviruses

IMPORTANT GROUPING	VIRUS	VIRUS ABBREVIATION	VECTOR	ANIMAL RESERVOIR	LIKELIHOOD OF HAVING SYMPTOMATIC INFECTION	DIRECT HUMAN-TO-HUMAN TRANSMISSION (WITHOUT AN ARTHROPOD INTERMEDIATE)	GEOGRAPHIC REGION
Genus: *Flavivirus*	Dengue virus	DENV	Mosquitoes (*Aedes*)	Humans[a] (strains in monkeys are not yet known to infect humans)	**Moderate**	No	Tropics
	Zika virus	ZIKV	Mosquitoes (*Aedes*)	Unknown	Low	**Yes**	Tropics
	Yellow fever virus	YFV	Mosquitoes (*Aedes*)	Monkeys, humans[a]	**Moderate**	No	Africa, South America
	Japanese encephalitis virus	JEV	Mosquitoes (*Culex*)	Birds, pigs	Low	No, except rarely during pregnancy	Far East
	West Nile virus	WNV	Mosquitoes (*Culex*)	Birds	Low	No	Mostly USA, but is found worldwide
	St. Louis encephalitis virus	SLEV	Mosquitoes (*Culex*)	Birds	Low	No	USA
	Powassan virus		Ticks (*Ixodes*)	Rodents	Low	No	Northern regions (Russia, Canada, upper Midwest of USA)
	Tick-borne encephalitis virus	TBEV	Ticks (*Ixodes*)	Rodents. Some trans-mission through ingestion of unpasteurized meat and dairy.	Low	No	Northern regions (Russia, Europe, Asia)
Genus: *Alphavirus*	Chikungunya virus	CHIKV	Mosquitoes (*Aedes*)	Monkeys, rodents, humans[a]	**High**	No	Tropics
	Eastern equine encephalitis virus	EEEV	Mosquitoes (many types)	Birds	Low	No	Eastern half of USA
	Western equine encephalitis virus	WEEV	Mosquitoes (many types)	Birds	Low	No	Western half of USA, South America
	Venezuelan equine encephalitis virus	VEEV	Mosquitoes (many types)	Rodents	Low	No	South and Central America
Order: *Bunyavirales*	Crimean-Congo hemorrhagic fever virus (*see* chapter 11)	CCHFV	Ticks (*Hyalomma*) and animals	Farm animals (cows, sheep, goats), wild animals (ostriches, hares)	Low	**Yes**	Africa, Eastern Europe, China, Middle East, Asia
	Oropouche virus	OROV	Midges				South America
	La Crosse encephalitis virus		Mosquitoes (*Aedes*)				North America
	California encephalitis virus		Mosquitoes (*Aedes*)				North America
	Rift Valley fever virus (*see also* Table 11.1)	RVFV	Mosquitoes (*Aedes, Culex, Anopheles*) and animals	Rodents, livestock	Low	No	Africa
Genus: ***Coltivirus***	**Co**lorado **ti**ck fever virus	CTFV	Ticks (*Dermacentor*)	Rodents		No	Western USA and Canada

[a]Humans can act as a reservoir instead of a dead-end host (i.e., viral titers get high enough in the infected person that an arthropod vector can transmit it from one person to another).

soft ticks: *Ornithodoros*. Hard ticks have a tough shell, called a scutum, and can feed on multiple host species. Medically relevant hard ticks: *Amblyomma*, *Hyalomma*, *Ixodes*, *Dermacentor*, and *Haemphasalis*.

Ticks can take blood meals at all stages of their development. This is important because early stages are tiny and are less likely to be seen by the host.

3. **Clinical presentation:** Most arboviruses cause mild or asymptomatic infection in otherwise healthy adults. If they do produce symptoms, most cause a self-limited, **nonspecific febrile illness with a rash** that can progress to arthritis, meningitis, or encephalitis. Symptoms and severe disease are most common in pediatric and elderly patients.

4. **Diagnostic testing**
   - Arboviruses are often tested together in a panel if they are transmitted by the same vector, if they cause overlapping symptoms, or when they occur in the same geographic location.
     - Example 1: Powassan virus, West Nile virus, eastern equine encephalitis virus (EEEV), western equine encephalitis virus (WEEV), St. Louis encephalitis virus (SLEV)
     - Example 2: dengue virus, chikungunya virus (CHIKV), and Zika virus
   - Serology: the most common method of diagnosis
     - Primary screen
       - IgM: appears early, **~5 days** after infection. It may remain positive for 3–12 months, so the presence of IgM may not always indicate current infection.
       - IgG: appears ~1 to 2 weeks after infection
       - Antibody-based serology can miss early infection (<1 week).

There is significant serologic cross-reaction between arboviruses in the same families (e.g., dengue virus and Zika virus).

Table 12.2. Comparison of common mosquito vectors

TYPE OF MOSQUITO	MOST ACTIVE FEEDING TIME	COMMON FEEDING LOCATION	ANGLE OF ABDOMEN
Aedes sp.	Day (especially morning and evening)	Outdoor	Down, horizontal
Anopheles sp.	Night (dusk to dawn)	Indoor	Points up in the air
Culex sp.	Night	Indoor and outdoor	Horizontal

Different control measures are implemented depending on where and when a vector takes a blood meal. For example, fogging is used for outdoor mosquitoes and bed nets are used for night feeding, indoor mosquitoes.

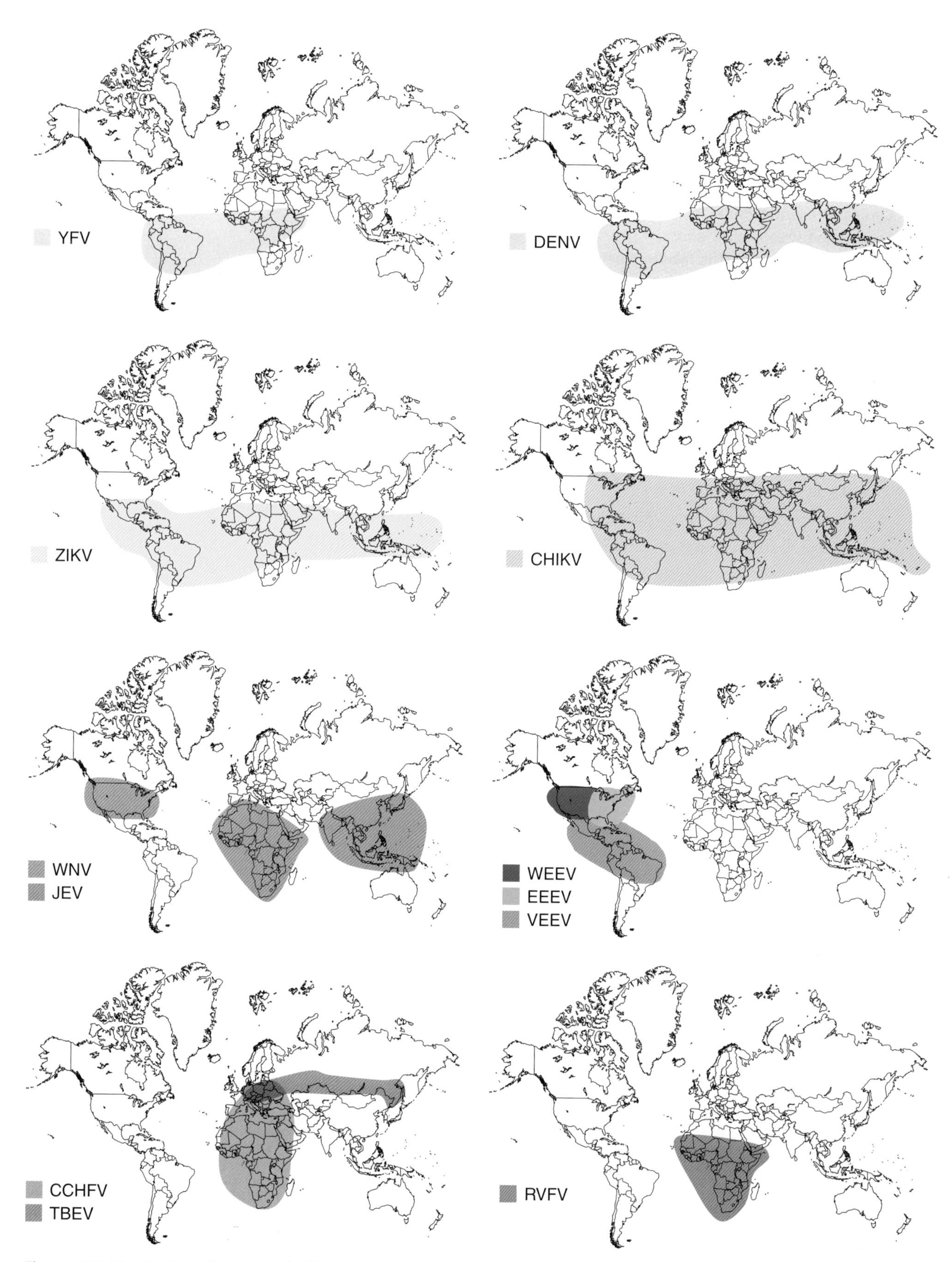

Figure 12.1. Distribution of geographically located arboviruses. YFV, yellow fever virus; DENV, dengue virus; ZIKV, Zika virus; CHIKV, Chikungunya virus; JEV, Japanese encephalitis virus; WNV, West Nile virus, EEEV, Eastern equine encephalitis virus; WEEV, Western equine encephalitis virus; VEEV, Venezuelan equine encephalitis virus; CCHFV, Crimean-Congo hemorrhagic fever virus; TBEV, tick-borne encephalitis virus; RVFV, Rift Valley fever virus.

- Use paired (i.e., acute- and convalescent-phase) sera to demonstrate seroconversion or a 4-fold rise in IgG titer.
    - Confirmatory testing: There is significant cross-reactivity between arboviral antibodies, so confirmatory testing is often needed to increase specificity. Confirmation is typically performed by plaque reduction neutralization testing (PRNT).
- NAATs: typically performed on blood or serum
    - Timing of testing is crucial because the period of viremia is very brief (<5–7 days), so the virus may not be present when specimens are submitted for testing. Therefore, NAATs can be used to assist in the diagnosis of early, active infection but may be negative in later stages.
    - Can also be performed on CSF if central nervous system (CNS) symptoms are present, or on other body fluids and tissue. However, this testing may be insensitive.
- Culture: not performed. Arboviruses do not grow in routinely used cell lines.

Figure 12.2. Actual size of a tick at different stages of its lifecycle.

Arboviruses are detectable from the blood for a very short period. So PCR on blood should generally be done on samples collected <5 days from symptom onset; otherwise, the test may be negative.

5. Prevention

- Vaccination: vaccines do not exist for most arboviruses, but do exist for some (*see* Table 18.1).
- Vector control: Controlling the population of arboviral vectors and avoiding bites are the primary ways of preventing transmission of arboviruses. Several personal and public health strategies can be used for **vector control** (Table 12.3).

II. DENGUE VIRUS. Dengue virus may cause a severe febrile illness with rash and debilitating joint and muscle pain. It can progress to hemorrhagic disease, shock, and death. It is transmitted by *Aedes* mosquitoes and is endemic in tropical countries (Table 12.1).

1. **Background:** enveloped, (+) ssRNA in the genus *Flavivirus* (Fig. 12.3). There are four serotypes, DENV 1, 2, 3, and 4.

2. Transmission

- Mosquitoes: *Aedes aegypti* and *Aedes albopictus*
- Monkeys are the primary reservoir, but a mosquito can also transmit the virus from other infected humans. The virus is not transmitted directly (i.e., without a mosquito vector) between humans.
- Geographic location: tropics and subtropics (Fig. 12.1)

3. **Clinical presentation:** ~50% of infections are symptomatic, and can present from mild undifferentiated fever to severe hemorrhagic disease and shock.

- Incubation: <1 week
- **Undifferentiated fever:** mild, nonspecific febrile illness. This usually occurs in young children or those with primary infection.
- **Dengue fever (DF)**
    - 2 to 7 days of sudden, high fever, headache, and retro-orbital pain
    - Intense joint, bone, and muscle pain (dengue is also known as "bone-crusher" or **"breakbone" disease**)
    - Rash over skin, palms, and soles
    - Increased vascular permeability may cause mild hemorrhage from nose or gums and easy bruising.
    - Nausea and vomiting

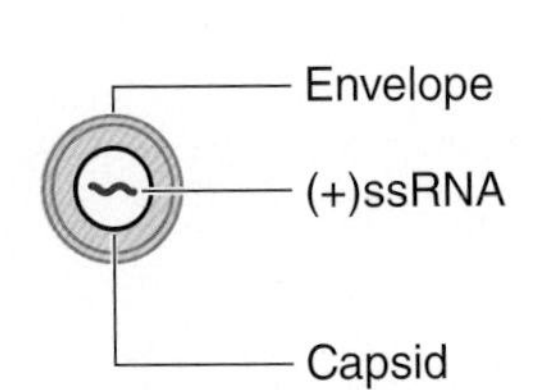

Figure 12.3. Dengue virus.

Table 12.3. **Examples of vector control strategies against mosquitoes and ticks**

VECTOR CONTROL STRATEGY	AGAINST MOSQUITOES	AGAINST TICKS
Chemical repellents applied directly to skin	**DEET**, picaridin, oil of lemon eucalyptus, and IR3535[a]	**DEET**, picaridin, oil of lemon eucalyptus, and IR3535[a]
Chemical repellents applied to clothing	**Permethrin**	**Permethrin**
Disseminated chemical repellents	Aerosol of pyrethrins and pyrethroids ("fogging")	Spray of acaricides
Impermeable barriers	Long-sleeved clothing, window screens, and bed nets	Tape pant legs to prevent ticks from crawling under clothing.
Habitat modification	Reduce breeding areas, improve drainage, and eliminate areas of stagnant water, like improperly stored tires.	Remove leaf litter, install mulch barriers, and trim tall grass. Annual controlled burning may be useful in some areas.
Physical inspection and removal		Perform a full-body inspection. Use forceps to pull ticks out so that the mouthparts are removed cleanly.
Dissemination via animal vectors		Animals are baited with food and get brushed with insecticides that kill ticks. This is used mostly for mice (tick tubes or bait boxes) and deer (4-poster bait stations).

[a]DEET, diethyltoluamide; IR3535, ethyl 3-acetylbutylaminopropanoate

- **Severe dengue:** dengue fever with severe symptoms such as bleeding, plasma leakage and organ involvement.
  - Hepatomegaly and vomiting
  - **Dengue hemorrhagic fever.** Symptoms include bleeding from mucosal sites, petechiae, and thrombocytopenia. Risk of dengue hemorrhagic fever increases upon secondary infection with different dengue serotypes (antibody-dependent enhancement).
  - **Dengue shock syndrome.** Significant plasma leakage leads to fluid accumulation, hypotension, and shock.
  - Neurologic disease (meningitis and encephalitis)
  - With supportive therapy, this stage can progress to reabsorption of excess fluids.

The risk of hemorrhage and severe disease increases when a previously exposed person is infected with a different serotype of dengue virus (antibody-dependent enhancement).

- **Antibody-dependent enhancement:** exposure to one dengue virus serotype does not provide cross-protection against the other serotypes. Instead, a secondary infection with a different serotype may increase the risk of dengue hemorrhagic fever (*see* chapter 18).

4. Diagnosis

- Serology: the preferred method of diagnosis.
  - Unlike other arboviral serologies, dengue virus antigen detection can be used together with antibody testing. **NS1** is a dengue virus protein that is present in the first 7 days after exposure, so it can be used as a marker of early infection (Fig. 12.4).
  - ELISAs are common.
  - Lateral-flow assays are available for rapid testing (~15 minutes).
- NAAT: used on blood and serum. The period of viremia is short, so NAATs should only be used in the first week following infection in order to detect early disease (Fig. 12.4).

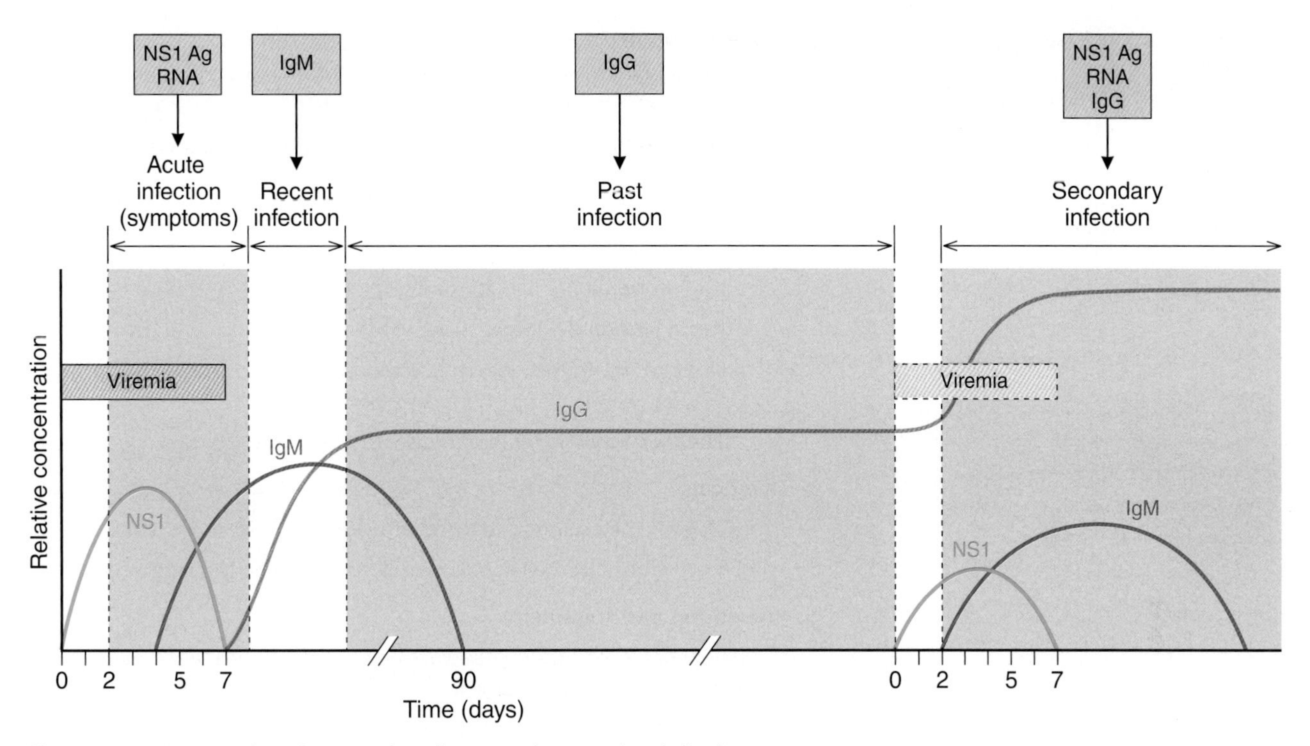

Figure 12.4. Kinetics of markers used to diagnose dengue virus infection.

5. **Prevention and treatment**
   - There is one vaccine available that contains all 4 serotypes, but it has low efficacy and is approved only in a few countries where there is a high burden of disease, due to the risk of antibody-dependent enhancement.
   - Supportive care, no specific treatment

III. YELLOW FEVER VIRUS: Yellow fever virus is transmitted by *Aedes* mosquitoes and is endemic to Africa and South America. Unlike other arboviruses, it causes jaundice and gastrointestinal symptoms. Disease can be very severe, but at-risk individuals can be protected with a vaccine (Table 12.1).

1. **Background:** enveloped, (+) ssRNA *Flavivirus* (like dengue virus) (Fig. 12.5)
2. **Transmission**
   - Mosquitoes (*Aedes* and *Haemagogus* spp.) can transmit yellow fever virus from either humans or monkey reservoirs.
     - Sylvatic cycle: transmitted between monkeys in the jungle
     - Savannah cycle: transmission along jungle borders from monkey or human reservoirs
     - Urban cycle: transmission between human reservoirs
   - The virus is not transmitted directly from person-to-person
   - Geographic location: South America and Africa (*see* Fig. 12.1)
   - Risk factors for infection
     - Travel to areas of endemicity
     - Forest workers

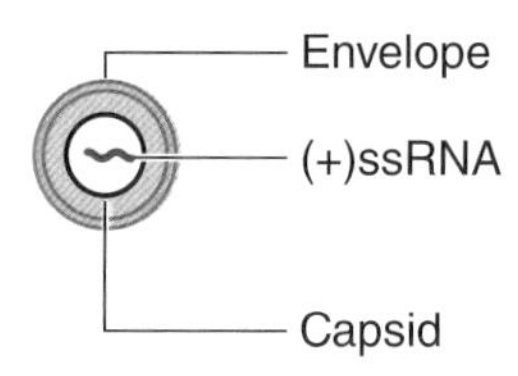

Figure 12.5. Yellow fever virus.

3. **Clinical presentation:** About 45% of infections are symptomatic, and they are typically self-limited and resolve within 3 to 4 days. Infection can occur in two phases.
   - **Acute infection:** fevers, chills, headache, muscle pain, and severe lower back pain
   - **Severe disease:** In about 15% of patients, the infection appears to subside but then progresses within 24 hours.
     - Nausea, vomiting, and abdominal pain
     - Rapid **jaundice** ("yellow" fever virus)
     - Hemorrhage (bleeding from the mouth, nose, eyes, or stomach). Blood may be seen in vomit and feces ("black vomit").
     - The mortality rate for severe disease may approach 50%.

Yellow fever virus presents with jaundice and gastrointestinal symptoms. This is different from nonspecific fever and rash exhibited by other arboviruses.

4. **Diagnosis**
   - Serology is the preferred method of diagnosis.
   - PCR on serum may be used to detect viremia in early infection.
5. **Prevention and treatment**
   - Prevention: A live, attenuated vaccine is available and recommended for travelers to areas of endemicity.
   - Treatment: supportive care, no specific treatment

Chikungunya virus and dengue virus are transmitted by the same vector, occur in the same geographic locations, and have similar clinical presentations.

Dengue = "breakbone fever"

Chikungunya = "to become contorted"

IV. CHIKUNGUNYA VIRUS. Like dengue virus, Chikungunya virus causes debilitating joint pain that can last for months. It is also transmitted by *Aedes* mosquitoes and endemic to the tropics. However, it is not in the same family, it does not cause hemorrhagic disease, and mortality rates are low (Table 12.1).

1. **Background:** enveloped, (+) ssRNA virus (Fig. 12.6). Unlike dengue virus, it is in the *Togaviridae* family and *Alphavirus* genus.
2. **Transmission:** overlaps significantly with that of dengue virus
   - Mosquitoes: *A. aegypti* and *A. albopictus*
   - Monkeys and rodents are the reservoir, but a mosquito can also transmit the virus from human reservoirs.
   - Geographic location: tropics and subtropics (Fig. 12.1)
   - Is not transmitted directly between humans without a mosquito vector
3. **Clinical presentation:** Most infections are symptomatic (>70%).
   - Fever, myalgia, headache, and rash for ~1 week
   - Debilitating **polyarthralgia** (chikungunya means "to become contorted"). This can persist for months to years.
   - Other severe complications (including neurologic involvement) may occur.
   - Morbidity is high, but the mortality rate is low.
4. **Diagnosis**
   - Serology is the preferred diagnostic method.
   - NAAT on serum may be used to detect viremia in early infection.
5. **Prevention and treatment**
   - No vaccine available
   - Supportive care, no specific treatment

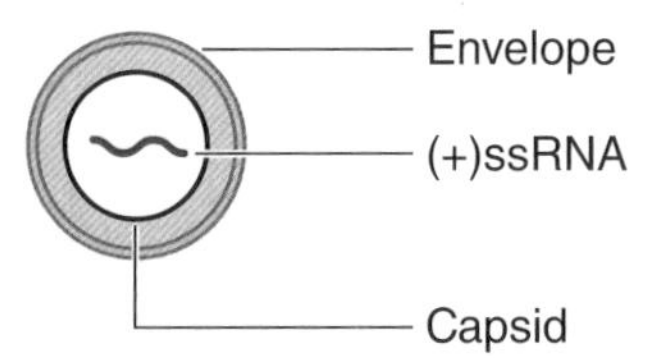

Figure 12.6. Chikungunya virus.

V. WEST NILE VIRUS. West Nile virus is transmitted by *Culex* mosquitoes and is endemic in North America. It is generally asymptomatic, but it can cause a self-limited, nonspecific febrile illness. In rare cases, it can cause infection of the CNS, which can result in long-term sequelae and death (Table 12.1).

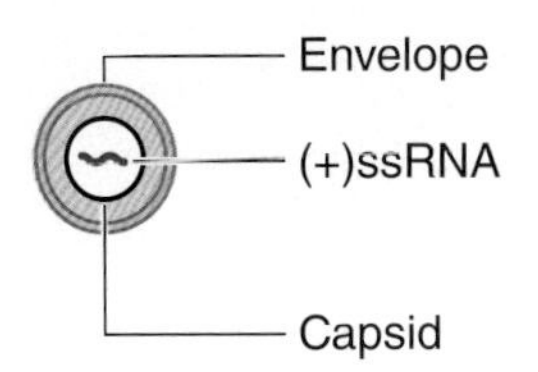

Figure 12.7. West Nile virus.

1. **Background:** enveloped, with a (+) ssRNA genome and in the genus *Flavivirus* (like dengue virus) (Fig. 12.7)
2. **Transmission**
   - Mosquitoes: *Culex* spp.
   - The West Nile virus life cycle is maintained between mosquitoes and birds, but sometimes humans are infected as dead-end hosts.
   - Dead birds can act like **sentinel animals** and can be a sign of circulating West Nile virus.
   - Is is not transmitted directly between humans.
   - Geographic location: worldwide, but now endemic in North America (Fig. 12.1)

Sentinel animals: when infection in animals signals that infection may also be occurring in humans (e.g., canary in a coal mine).

3. **Clinical presentation**
   - Only ~20% of infections are symptomatic. Symptoms are most common in pediatric and elderly patients.
   - Nonspecific febrile illness lasting ~3 to 6 days
   - High fever, headache, rash, backache, nausea, vomiting, lethargy, myalgia, and retro-orbital pain
   - **Neuroinvasive disease:** occurs in only <1% of infections
     - Mild fever, headache, stiff neck, disorientation (i.e., meningitis and encephalitis), convulsions, and conjunctivitis
     - Paralysis may occur.
     - Mortality rate: 10%
4. **Diagnosis**
   - Serology is the preferred method. It can be done on both serum and CSF.
   - NAAT on serum during the first week following infection
5. **Prevention and treatment**
   - No vaccine available
   - Supportive care, no specific treatment

VI. ZIKA VIRUS. In most cases, Zika virus causes an asymptomatic or mild febrile illness. It is found in the tropics and South America because it is usually transmitted by *Aedes* mosquitoes. However, infections during pregnancy can be transmitted congenitally and may cause significant birth defects, like microcephaly (Table 12.1).

1. **Background:** enveloped, (+) ssRNA *Flavivirus* (like dengue virus and chikungunya virus) (Fig. 12.8)
2. **Transmission**
   - Mosquitoes: *A. aegypti* and *A. albopictus*
   - Direct human-to-human transmission can occur via the following routes.
     - Sexual intercourse
     - Transplacentally

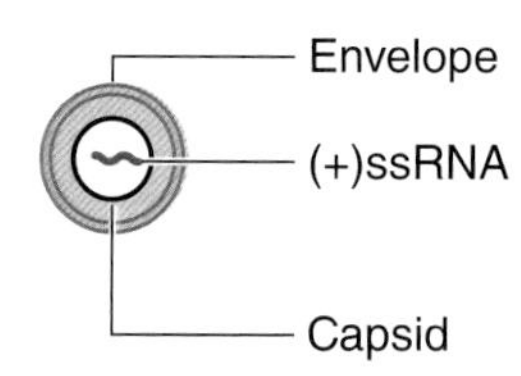

Figure 12.8. Zika virus.

- Geographic location: tropics and subtropics, especially South America (Fig. 12.1)
- Risk factors
  - Pregnancy, for congenital infection
  - Travel to areas of endemicity or outbreak

3. **Clinical presentation**
   - Only ~20% of Zika virus infections are symptomatic, and it causes almost no fatalities.
   - Asymptomatic or mild, nonspecific viral syndrome in adults and children
   - Can cause fever, diffuse rash, joint pain, myalgia, and eye pain (similar to mild dengue). It is also associated with **Guillain-Barré syndrome**.
   - **Zika congenital syndrome**: Transmission to the fetus during pregnancy can result in microcephaly, intracranial calcifications and other effects on the brain, redundant scalp skin, and deformed joints.

Zika virus infection during pregnancy can cause birth defects.

4. **Diagnosis:** Testing is recommended for pregnant women, their sexual partners, and neonates to identify potential congenital infection.
   - Serology from serum is preferred.
   - NAAT on serum may be used to detect viremia in early infection. Virus may be present for longer (several months) in urine, semen, and placental tissue.
5. **Prevention and treatment**
   - No vaccine is available.
   - Specific treatment is not available or needed for adults.
   - Supportive care is provided for affected babies.

VII. EASTERN AND WESTERN ENCEPHALITIS VIRUSES. EEEV and WEEV are rare viruses that are transmitted by mosquitoes and are endemic in the eastern and western regions of the United States. They generally do not cause disease but may cause nonspecific febrile symptoms. In some cases, they can cause neurologic infection with long-term sequelae or even death (Table 12.1).

1. **Background:** enveloped, (+) ssRNA viruses in the *Togaviridae* family and *Alphavirus* genus (like chikungunya virus) (Fig. 12.9)
2. **Transmission**
   - Mosquitoes: several types (*Culiseta*, *Aedes*, *Culex*, and others)
   - The life cycle is between mosquitoes and birds, but sometimes humans and horses are infected as dead-end hosts.
   - Sentinel animals: **horses**
   - Geographic location (Fig. 12.1)
     - EEEV: eastern half of the United States
     - WEEV: western half of North America, South America
   - Risk factors
     - Outdoor activities
     - Travel/residence in rural areas of endemicity, especially swampy areas

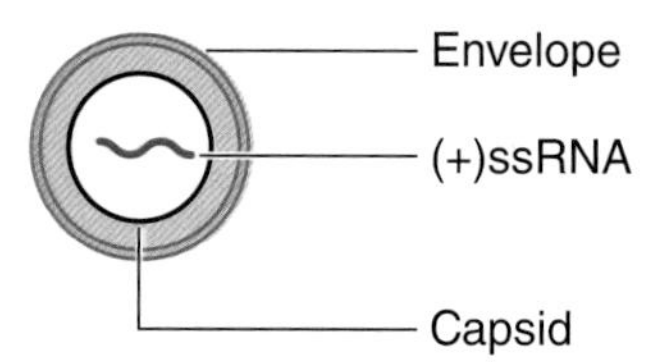

Figure 12.9. EEEV and WEEV.

3. Clinical presentation

- Less than 5% of infections are symptomatic.
- Nonspecific febrile illness: myalgia, fever, and arthralgia
- **Encephalitis:** headache, fever, encephalitis, confusion, behavioral changes, seizures, coma, and permanent neurologic damage
- **Severe, long-term sequelae** in survivors is possible and may include:
  - Seizures, paralysis, cognitive impairment
  - Death after a few years
- Mortality rate
  - EEEV: ~33%
  - WEEV: ~3%

4. Diagnosis

- Serology is the preferred method. Can be done on both serum and CSF.
- NAAT on serum may be used to detect viremia in early infection.

EEEV and WEEV cause rare and **sporadic** cases, while VEEV can cause large **outbreaks**.

5. Prevention and treatment

- No vaccine available for humans
- Supportive care; no specific treatment

VIII. OTHER IMPORTANT ARBOVIRUSES. (*see also* Table 12.1)

1. **Japanese encephalitis virus:** Like West Nile virus, JEV is a flavivirus and causes similar symptoms, but it occurs in the Far East and can be prevented with a vaccine (Fig. 12.1).

2. **Powassan virus:** Like West Nile virus, Powassan virus is also a flavivirus and most infections are asymptomatic. However, it can cause more severe encephalitis with permanent neurologic sequelae. Also, it is transmitted by ticks instead of mosquitos, is very rare, and occurs only in northern latitudes.

3. **Tick-borne encephalitis virus:** Like Powassan virus, tick-borne encephalitis virus is a flavivirus that is transmitted by ticks, frequently causes CNS involvement like encephalitis, and is found in northern regions of the world (especially Russia) (Fig. 12.1). However, it can also be acquired through ingestion of unpasteurized meats and dairy and can be prevented by a vaccine.

4. **Venezuelan equine encephalitis virus:** VEEV is an alphavirus that infects horses, like EEEV or WEEV. However, it causes outbreaks rather than sporadic cases, is endemic in South and Central America instead of the United States, and has a reservoir in rodents rather than birds (Fig. 12.1).

5. **Rift Valley fever virus:** Like CCHFV, RVFV is a bunyavirus that is primarily transmitted to humans via contact with animals but can also be transmitted by arthropods (mosquitoes). It is found in Africa and usually causes asymptomatic to mild disease but can cause three severe types of infections (Fig. 12.1, Table 12.1; *see also* Table 11.1):
   - Hemorrhagic fever. This is rare but has a high mortality rate.
   - Ocular disease, which can result in permanent blindness.
   - Encephalitis, which can result in persistent neurologic deficits.

Multiple-Choice Questions

1. **The partner of a 3-months-pregnant female returned to the United States after a 6-week business trip to Brazil. He recalled having a mild fever and conjunctivitis for approximately 3 days during the first week of his stay. Which of the following is true regarding Zika virus infection?**
   a. The partner's semen may still be infectious.
   b. The partner's saliva may still be infectious.
   c. There is high risk of maternal death if infection is transmitted to the mother.
   d. There is no risk to the fetus after this length of time.

2. **Which of the following viruses is the least likely to cross-react with dengue virus in a serologic assay?**
   a. Chikungunya virus
   b. Zika virus
   c. West Nile virus
   d. Japanese encephalitis virus

3. **Which of the following viruses can be transmitted directly from person to person?**
   a. West Nile virus
   b. Yellow fever virus
   c. Zika virus
   d. Dengue virus

4. **Which of the following tick-borne viruses is a rare cause of encephalitis in northern Wisconsin?**
   a. CCHFV
   b. Powassan virus
   c. West Nile virus
   d. La Crosse encephalitis virus

5. **Which of the following vectors does not transmit viruses?** *(May require reading outside of this chapter.)*
   a. *Culex* sp.
   b. Midges
   c. *Cimex lectularius*
   d. *Ixodes scapularis*

6. **A vaccine exists for which of the following viruses?**
   a. Colorado tick fever virus
   b. West Nile virus
   c. Chikungunya virus
   d. Japanese encephalitis virus

7. **Dengue virus, chikungunya virus, and Zika virus are transmitted by which of the following vectors?**
   a. A day biting mosquito
   b. A night biting mosquito
   c. The larval stage of a tick
   d. The adult stage of a tick

8. Which of the following arboviruses causes vomiting and jaundice?

a. Venezuelan equine encephalitis virus

b. Dengue virus

c. Chikungunya virus

d. Yellow fever virus

9. Which of the following tests is most likely to demonstrate a negative result 10 days after onset of encephalitis caused by West Nile virus?

a. PCR from blood

b. Serology for IgM in blood

c. Serology for IgG in blood

d. Serology of IgM in CSF

10. Which of the following viruses can achieve high enough levels in humans that a vector can further transmit disease (i.e., cause infections where humans are NOT dead-end hosts)?

a. West Nile virus

b. Japanese encephalitis virus

c. Powassan virus

d. Dengue virus

True or False

11. Almost all arboviruses have an RNA-based genome. **T F**

12. Birds are a reservoir for several arboviruses. **T F**

13. Secondary infection with a different serotype increases the risk of dengue hemorrhagic fever. **T F**

14. Tissue histology is the gold standard for diagnosis of arboviral infection. **T F**

15. Ticks cannot transmit disease during the nymphal stage. **T F**

SECTION III

DIAGNOSTIC ASSAYS AND TECHNIQUES

13

CHAPTER 13

CULTURE AND TISSUE-BASED DIAGNOSTIC TECHNIQUES

I. OVERVIEW. Viral culture can be used to identify the presence of live viruses. It can be used for a wide range of specimen types, from tissue to body fluids, and can identify many types of viruses from a single specimen. On the other hand, it requires significant technical expertise, may have low sensitivity, is highly labor-intensive, and may take a long time because viruses must replicate to detectable quantities.

Viral culture identifies **live** (actively replicating) viruses.

II. CONVENTIONAL VIRAL CULTURE. Conventional viral culture is used to grow and amplify viruses in the laboratory. This procedure involves inoculating the specimen onto a single layer of eukaryotic cells grown *in vitro*, also known as a **cellular monolayer** (Fig. 13.1).

1. **Cell culture medium:** The cell monolayer is bathed in nutrients as well as substances that inhibit bacterial growth. It usually consists of the following components.
   - Buffered medium: a liquid medium typically containing salts, amino acids, glucose, and vitamins. A phenol red indicator is often added which produces an obvious color change when the pH of the medium (usually between 7 and 8) is out of range. Bright pink = too basic; yellow = too acidic.
   - Fetal bovine serum: a highly enriched supplement that is added to the medium at 0 to 10% of the total medium volume to enhance survival and growth of the cell monolayer.
   - Antibiotics/antimycotics: Bacteria present in specimens can grow in the culture medium and destroy the cell monolayer. A solution of antibiotics and antimycotics is typically added at 1% of the total volume to inhibit bacterial and fungal contamination. To reduce bacterial contamination and overgrowth, certain specimens (like respiratory or fecal specimens) are also passed through filters with a 0.45-µm pore size.

Mycoplasma species are 0.2 to 0.3 µm in diameter and can pass through bacterial filters. They are a known cause of contamination in viral cultures.

2. **Solid phase:** Cell culture can be performed on different kinds of surfaces.
   - **Tubes:** Cell monolayers are grown on the side of plastic or glass tubes. Each tube is capped individually to minimize the risk of contamination from spills, splashes, or stray drops. Tube cultures are bathed in medium

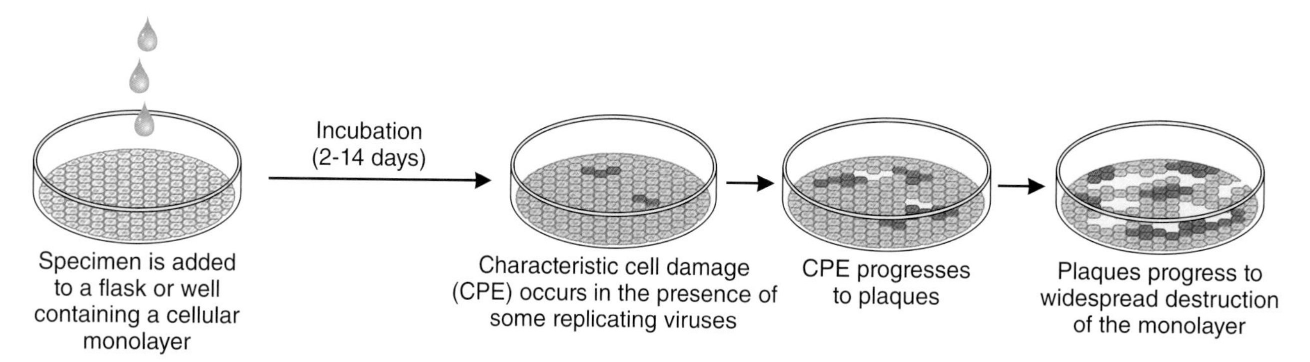

Figure 13.1. Process of viral culture.

and incubated. They are often agitated continuously to promote gas exchange and maintain a balanced pH.

- **Plastic plates:** contain multiple wells and are convenient for plating multiple specimens, saving space, and rapid screening of viral growth. On the other hand, there is a greater risk of contamination since each well is not covered individually.

Different cell types are capable of growing different viruses.

Rule of thumb: DNA viruses grow better in MRC-5 cells. RNA viruses grow better in RMK cells.

3. **Cells:** There are several different cell types that can be used to form the cell monolayer. Cells that can be **passaged** (perpetuated in culture) many times are easy to use in a laboratory because they will stay viable and healthy over time. Different cell types will grow different viruses, so the specimen is generally inoculated onto multiple cell types depending on the sample type and the potential infecting viruses. Table 13.1 lists the viruses that grow in some of the most commonly used cell lines.
   - **Primary cells** are harvested directly from animal or human tissue. These are the most like normal cells, but they can be passaged only one or two times. Like normal tissues, they may even have cellular polarity (different appearance and/or function of the apical and basal sides of the cell). Example: rhesus monkey kidney cells (RMK)
   - **Immortalized cell lines** behave a lot like cancer cells. They can divide indefinitely without losing viability and are easy to use. On the other hand, they are the most unlike normal cells. Examples: lung carcinoma (A549), cervical carcinoma (HeLa), and laryngeal carcinoma (HEp-2) cells
   - **Human diploid cells** are somewhere in between primary and immortalized cells. They still have the same number of chromosomes as normal human cells, but they can be passaged many times. They lose viability after 10 to 50 passages. Example: human embryonic lung (MRC-5) cells
   - **Reporter cells** are immortalized cells that contain an integrated reporter gene. If the cells are infected with a certain type of virus, gene expression is turned on, which reports the presence of the virus quickly and easily. Example: ELVIS cells. These cells contain the *lacZ* gene from *Escherichia coli*, which is not expressed under normal conditions. When the cells are infected with either HSV-1 or -2, *lacZ* expression is activated and β-galactosidase is produced. A substrate called X-Gal (5-bromo-4-chloro-3-indolyl-β-D-galactopyranoside) is added, which is cleaved by β-galactosidase and changes the cells from colorless to blue.

Table 13.1. Cells used for growing viruses (26)[a]

VIRUS	PRIMARY CELLS (RMK)	HUMAN DIPLOID CELLS (MRC-5)	IMMORTALIZED CELLS (A549)	CLASSIC CPE
RNA viruses				
Influenza	+	0	0	Hemadsorption positive (CPE is not unique)
Parainfluenza	+	0	0	
Mumps	+	0	+	
RSV	+	~	0	Syncytia
Rhinovirus	~	+	0	Small, round, refractile cells
Enterovirus	+	+	0	Tear-shaped, small, round, refractile cells
DNA viruses				
Adenovirus	~	+	+	Cluster of grapes
CMV	0	+	0	Foci of rounded cells
HSV	~	+	+	Large, globby cells
VZV	0	+	+	Small, round cells

[a]+, grows well; 0, grows poorly or not at all; ~, moderate or variable growth.

- **Cocultures** are when several cell types are grown together in a single monolayer so that multiple types of viruses can be grown and identified simultaneously.

4. **Incubation:** Viruses have different rates of growth (Table 13.2). Some grow within 1 to 2 days, while others take several weeks. A specimen cannot be reported as negative until the full incubation period has elapsed. This is important because results may not be available within a clinically actionable time.

Viral cultures are incubated for 2–3 weeks.

5. **Identifying viruses:** Virus particles are grown and amplified in culture, but they cannot be seen directly using light microscopy. Instead, they are identified by characteristic effects they have on cellular morphology (CPE). Then a more specific test is used to confirm the identity of the viral pathogen.
   - **Cytopathic effect (CPE):** The morphologic changes and cell death produced in virally infected cells are called CPE. Some viruses produce very characteristic types of CPE as shown in Fig. 13.2, but it takes substantial technical expertise to identify them.

Table 13.2. Rate of growth of viruses

PACE	RATE OF GROWTH	VIRUS(ES)
Fast	1–3 days	**HSV**
Moderate	2–10 days	Influenza virus, enterovirus (including coxsackievirus, rhinovirus, and echovirus), adenovirus, mumps virus, measles virus, parainfluenza virus, RSV
Slow	7–14 days	**VZV, CMV**

- CPE is not very specific, since multiple viruses can cause similar cellular changes. Causes of false-positive CPE include the following.
  - Cell lines derived from hosts infected with endogenous viruses or other pathogens can show false-positive CPE. For example, simian viruses can cause false-positive CPE (Fig. 13.2). This virus does not infect humans but can infect the monkeys from which primary cell lines are generated. This should be suspected when uninoculated negative-control cells also show CPE.
  - Molecules in the specimens may be toxic to the cells and cause cellular changes similar to CPE. Toxicity is often associated with fecal, urine, bile, and blood specimens. Dilution of the original specimen can reduce the amount of toxic substances in the sample, although it will also reduce viral load and decrease sensitivity.
- Because of its low specificity, CPE must be confirmed by additional methods. These can include PCR, rapid tests to detect viral antigens, or antigen staining with fluorescently tagged antibodies.

6. **Occupational hazards** (*see* chapter 17)
   - Amplification of live, infectious virus can increase the risk of transmission to laboratory personnel. Viral cultures are performed in a biosafety cabinet with appropriate PPE and containment (see chapter 17). If there is a risk that particularly pathogenic viruses (like SARS or Ebola virus) are present in a specimen, routine culture (even for other viruses) should be minimized.
   - The eukaryotic cells used for viral culture are produced from living animals and may contain unexpected, pathogenic contaminants. For instance, cell lines derived from a monkey that was infected with *Coccidioides immitis* were used by many laboratories and caused two laboratory exposures.

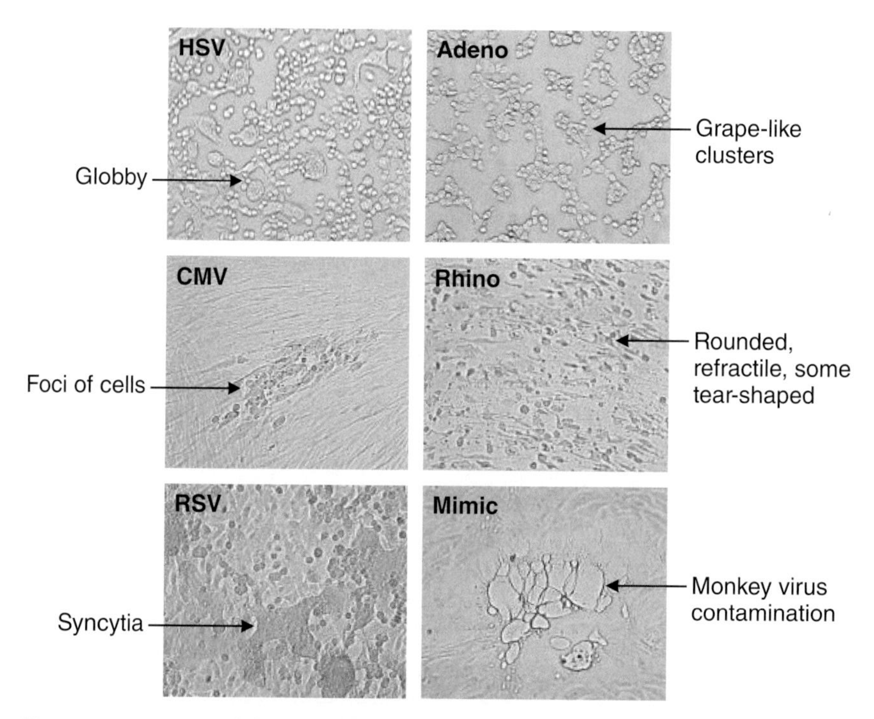

Figure 13.2. **Images of characteristic CPE.** Cell types (with infecting viruses): A549 (HSV and adenovirus), MRC-5 (CMV and rhinovirus), Hep-2 (RSV), and RMK (monkey virus contaminant). Adapted with permission from reference 58.

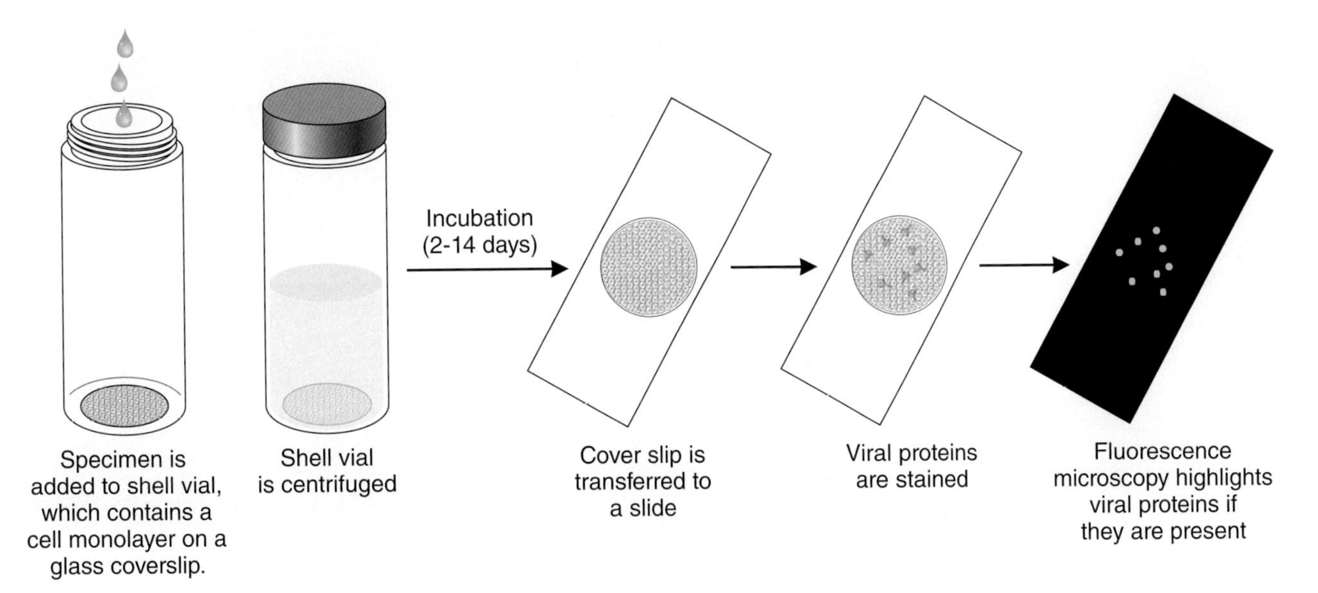

Figure 13.3. Process of shell vial culture.

III. SHELL VIAL ASSAY. Shell vial assays are more sensitive and rapid than conventional culture because they detect the production of viral proteins, which happens as soon as viral replication begins. On the other hand, viral culture relies on the appearance of CPE, which occurs only after a sufficient quantity of virus has been produced to cause visible changes in the cell monolayer. Shell vials are performed as follows (Fig. 13.3).

Shell vials can be read in 2–3 days.

1. Shell vials are individual vials that have a glass coverslip on the bottom. The top side of the coverslip is coated with a cell monolayer.
2. The sample is added to the monolayer. The shell vial is then centrifuged, which stresses the cells and increases infectivity.
3. Viruses infect cells and begin replication within 24 to 72 hours. Viral proteins are produced, even though complete viral particles may not yet be fully assembled.
4. The coverslip is removed from its shell and transferred to a glass slide, where acetone is added to inactivate the viruses and fix the cells.
5. Fluorescently tagged antibodies are added that bind to specific viral proteins. A fluorescence microscope is used to read the slides for any staining that is characteristic of the virus.

IV. HEMADSORPTION. Hemadsorption is a test used to confirm the growth of viruses in cell culture that produce hemagglutinin, such as influenza virus, parainfluenza virus, measles virus, mumps virus, and rubella virus. To perform this test, guinea pig red blood cells are added to the infected cell monolayer. If hemagglutinin is present, red blood cells will **adsorb** (or stick) to the cells in clumps (Fig. 13.4).

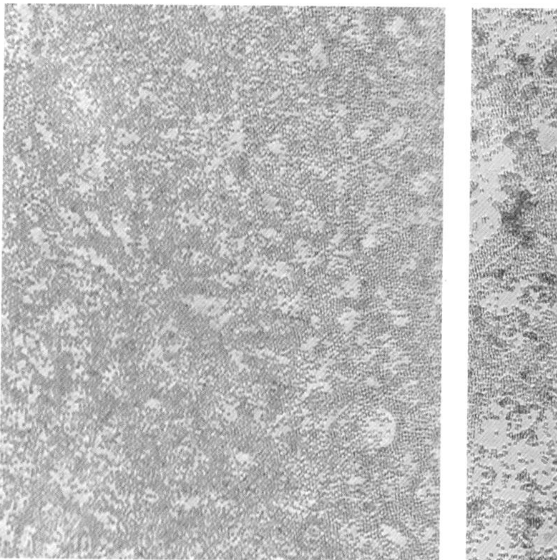
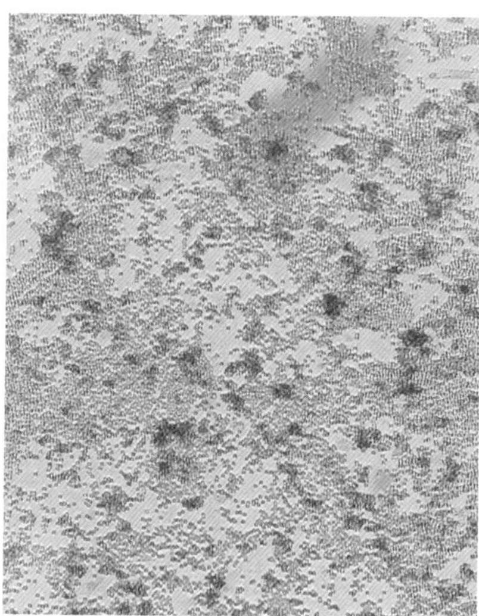

Figure 13.4. **Hemadsorption of red blood cells on a cell monolayer.** (Left) No virus present. (Right) Influenza virus present. Guinea pig red blood cells are adsorbed in clumps to the monolayer. Photos courtesy of Dr. Ute Werringloer, Northwell Health.

V. VIRAL QUANTIFICATION.

1. **Plaque-forming units (PFU):** As viruses replicate, new virions spread out radially to infect neighboring cells. A cluster of necrotic cells in a monolayer is called a **plaque**. Like a bacterial colony, a plaque is assumed to have originated from a single virion that has multiplied. PFU can be counted to quantify the amount of virus that is present in a certain volume of specimen (i.e., PFU per milliliter).

Viral plaque-forming units (PFU) are similar to bacterial colony-forming units (CFU).

2. **50% tissue culture infectious dose ($TCID_{50}$):** The $TCID_{50}$ assay is used to estimate the amount of infectious virus (**infectious titer**) present in a suspension.
   - A cell monolayer is grown on each well of a plate.
   - A suspension of virus is serially diluted (usually 10-fold dilutions) and then used to infect the cell monolayers. This serial dilution is done in many replicates.
   - The dilution at which there is no virus (usually detected by absence of CPE) in 50% of the replicates is considered the $TCID_{50}$.

VI. HISTOPATHOLOGY AND CYTOPATHOLOGY. These are techniques used to detect viral infection by directly examining stained cells (cytology) or sections of tissue (histology) (Table 13.3; *see also* Fig. 13.6 and 13.7)

1. **Changes to cellular morphology:** Some viruses produce characteristic morphologic changes to cells. Some changes are highly specific for viruses, while others may overlap with other diseases or infections. In this case, the etiology should be confirmed with a more specific test. For example, **Tzanck smears** are rapid tests that can be performed on lesion scrapings or fluid. The sample is fixed, stained, and visualized for multinucleated giant cells, which may indicate the presence of HSV or VZV. However, this test is no longer recommended because of its relatively poor sensitivity and specificity.

Table 13.3. **Characteristic histologic or cytologic features of virally infected cells (Fig. 13.6 and 13.7)**

VIRUS	COMMONLY USED STAIN	HISTOLOGIC/CYTOLOGIC FEATURES	LOCATION OF INCLUSIONS
CMV	H&E	Enlarged cell and nucleus ("**owl's eye**")	Cytoplasmic and nuclear
HSV	H&E, Tzanck	3 M's of cell morphology: **margination** (precipitation of chromatin around the nuclear margin), **molding** (cells start molding around or hugging neighboring cells), and **multinucleation** (cells merge together). This may be seen as multinucleated giant cells.	Nuclear
VZV	H&E, Tzanck	Multinucleated giant cells	Nuclear
Adenovirus	H&E	**Smudge cells** (infected cells have large, round nuclei that look smudgy)	
Rabies virus	Sellar	**Negri bodies**	Cytoplasmic
Molluscum contagiosum	Giemsa	**Molluscum bodies**	Cytoplasmic
HTLV-1	Wright	**Flower nuclei** in T cells	
RSV	H&E	Syncytia	
HPV	Pap, Giemsa	**Koilocytic atypia** on cytology (squamous epithelial cells that have dark, enlarged nuclei with a perinuclear halo and an irregular nuclear membrane); cervical intraepithelial neoplasia (**CIN**) in histology appears as **disordered**, **undifferentiated**, and **dividing** cells	
Parvovirus B19	H&E	Enlarged, ground-glass nuclei with chromatin precipitation in erythroid precursors	
BK virus	Pap	**Decoy cells** on urine cytology	Nuclear
JC virus	H&E	Bizarre (i.e., irregular) astrocytes present; oligodendrocytes have enlarged nuclei	Nuclear
	Luxol fast blue	Foci of **demyelination** in white matter of brain	

2. **Viral inclusions:** Depending on the type of virus, newly assembled virus particles can accumulate in the cytoplasm or the nucleus. Aggregates of viruses and viral proteins can sometimes be seen as viral inclusion bodies.

3. **Immunohistochemistry (IHC):** Tagged antibodies specific to certain viral proteins are added to a section of tissue. If these antibodies bind, they show the presence of specific viruses. This can be done for many viruses, including herpesviruses as well as nonculturable viruses like JC and BK viruses (covered in chapter 14).

4. ***In situ* hybridization (ISH):** similar to immunohistochemistry but detects nucleic acids instead of proteins. So instead of using a tagged antibody, the specimen is probed with a tagged nucleic acid complementary to the target viral genome. When it binds, it shows the presence of viral genome. EBV-encoded small RNA (**EBER**) is a commonly used *in situ* hybridization target (Fig. 13.5).

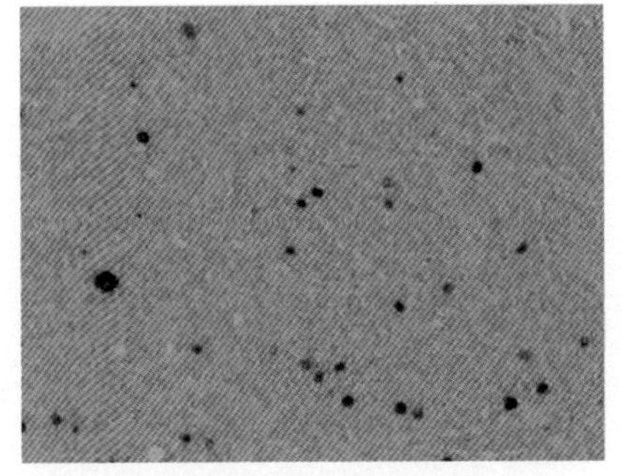

Figure 13.5. **EBER-positive lymphoid cells in a patient with T-cell lymphoma.** Reprinted from reference 59, with permission.

VII. ELECTRON MICROSCOPY. Individual viruses can be seen using electron microscopes but not using optical techniques, like light microscopy. Electron microscopy can be used directly on specimens, and until molecular methods became more prevalent, this was the only technique available to identify viruses that do not grow well *in vitro* (e.g., norovirus). However, today electron microscopy is generally

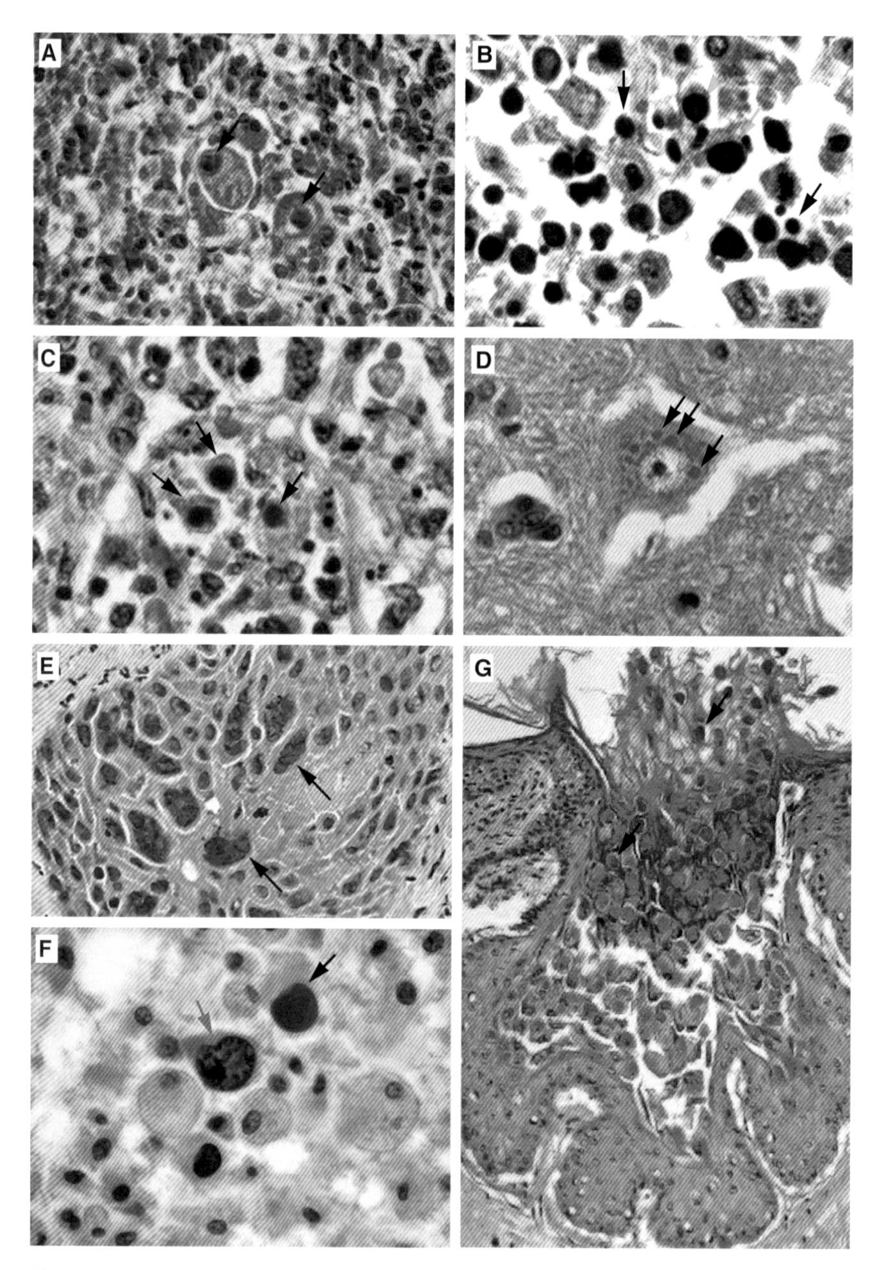

Figure 13.6. **Histologic identification of viruses by characteristic inclusions or cell morphology. (A) CMV:** Shown are owl's eye cells with nuclear (black arrows) and cytoplasmic (green arrow) inclusions. Photo courtesy of CDC-PHIL (ID#22200). **(B) Parvovirus B19:** Shown are enlarged ground glass nuclei (green arrows) versus normal erythroid precursor cell nuclei (black arrows) in fetal liver. **(C) Adenovirus:** Shown are smudge cells in lung tissue. Photos courtesy of Dr. Morris Edelman, Northwell Health. **(D) Rabies virus:** Shown are Negri bodies in the cytoplasm of a nerve cell. Photo courtesy of CDC-PHIL (ID#3376). **(E) HSV:** Shown are multinucleation, margination, and molding. Photo courtesy of Dr. Bobbi Pritt, Mayo Clinic. **(F) JC virus:** Shown are an oligodendrocyte with a ground-glass intranuclear appearance (black arrow) and a bizarre astrocyte (blue arrow). Photo courtesy of Jian Yi Li, Northwell Health. **(G) Molluscum contagiosum virus:** Shown is a skin lesion with molluscum bodies. Photo courtesy of CDC-PHIL (ID#860).

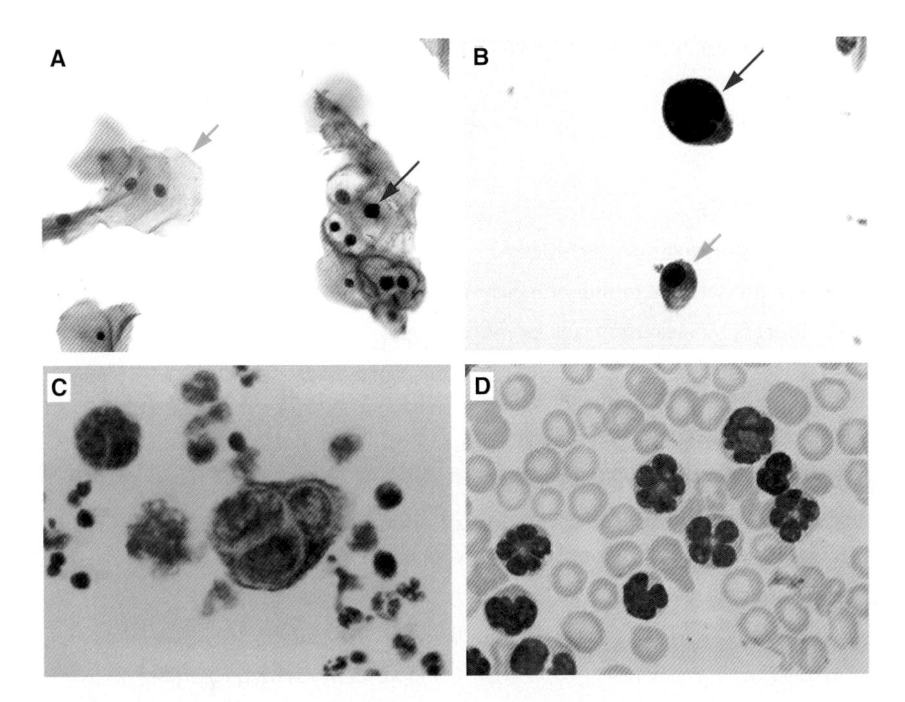

Figure 13.7. **Cytologic identification of viruses. (A) HPV:** Shown is a Pap smear with normal squamous epithelial cells (green arrow) versus koilocytic atypia (red arrow). **(B) Parvovirus B19:** Shown are enlarged (red arrow) versus normal (green arrow) nuclei of erythroid precursors in a peripheral blood smear. Photos courtesy of Dr. Cecilia Gimenez, Northwell Health. **(C) HSV:** Shown is a Tzanck smear of a lesion scraping with a multinucleated giant cell. Photo courtesy of CDC-PHIL (ID#14428). **(D) HTLV:** Shown are flower CD4 T cells in a peripheral blood smear. Photo reprinted from reference 60, with permission.

not used in diagnostic laboratories because it is highly manual, expensive, labor-intensive, and slow. On the other hand, it is useful for identifying new viruses (e.g., electron microscopy was used to identify SARS-CoV).

Multiple-Choice Questions

1. **What is a significant benefit of viral culture compared to PCR or serologic methods?**
   a. It is rapid.
   b. It is highly specific.
   c. It identifies the presence of "live" (actively replicating) virus.
   d. It requires minimal technical expertise.
2. **What is the most important difference between shell vials and conventional cultures?**
   a. Shell vials identify the presence of viral genomes, while culture identifies the presence of viral proteins.
   b. Specimen and cells are centrifuged in shell vials but not in conventional culture.
   c. Shell vial is used only for fast-growing viruses.
   d. Shell vial is used only for slow-growing viruses.

3. Which of the following viruses can be identified by ELVIS cells?

a. HSV-1/2

b. VZV

c. CMV

d. Rhinovirus

4. *In situ* hybridization techniques

a. Are the same as immunohistochemistry

b. Identify viruses using cellular morphology

c. Detect proteins

d. Detect nucleic acids

Match the following. Use each answer only once.

5. Characteristic identifying features of adenovirus by different methods.
(May require reading outside of this chapter.)

Conventional viral culture	A. Crystalline lattice
Shell vials	B. Grape-like clusters
Histology	C. Fluorescent staining of proteins
Electron microscopy	D. Smudge cells

6. Types of CPE

Syncytia	A. CMV
Positive hemadsorption	B. HMPV
Foci of cells	C. Mumps virus
None	D. RSV

7. Characteristic histologic/cytologic features

Koilocytic atypia	A. HPV
Multinucleated giant cells	B. JC virus
Presence of bizarre astrocytes	C. Rabies virus
Negri bodies	D. VZV

True or False

8. CPE is highly specific for individual viruses and should be used independently for identification. **T F**

9. RMK cells grow primarily RNA viruses. **T F**

10. HSV can grow in traditional cell culture within 1 to 2 days. **T F**

14

CHAPTER 14

DIAGNOSTIC TECHNIQUES BASED ON IMMUNOLOGICAL INTERACTIONS

I. OVERVIEW. Immunologic assays are tests that make use of antigen-antibody interactions. **Antigens** are foreign substances (usually proteins) that trigger the immune system. They have antigenic sites called **epitopes**. Antibodies are Y shaped proteins produced by modified B cells called plasma cells. Antibodies are produced in response to antigens and can bind to them highly specifically. Differences in types of antibodies, labels, and solid phases allow immunologic assays to have different functions, turnaround time, advantages, and limitations (*see* Table 14.3 at the end of the chapter).

Antibodies are also called immunoglobulins (Ig).

1. **Antibodies used for binding antigens:** The **Fab** segments of antibodies are responsible for binding to epitopes.
   - **Polyclonal** antibodies: highly similar antibodies produced by different plasma cells. They target the same antigen but recognize slightly different epitopes.
   - **Monoclonal** antibodies: a purified group of antibodies that will recognize only a single type of epitope. This increases the specificity for a target (Fig. 14.1).
2. **Labels:** The **Fc** segment of an antibody can be bound to a label or tag. This acts as a flag when an antigen is detected by the antibody (Fig. 14.2).
   - **Enzyme label:** an enzyme will cleave a substrate to produce a detectable product, like light or a colored compound.
   - **Fluorescent label:** a fluorophore will absorb light at a specific wavelength (i.e., color) and then emit light at a different wavelength. Special equipment needs to be used for this technique. For example, fluorescent microscopes have lasers to activate the fluorophore and different colored filters to see the emitted light. Fluorescein isothiocyanate (FITC) is a commonly used fluorophore that absorbs blue light and emits green light.
   - **Colored beads:** These are synthetic particles that are inexpensive and large. Their color can be read visually when enough of these conjugated antibodies aggregate together.

Enzyme labels = EIAs
Fluorescent labels = Immunofluorescence assays
Colored beads = Lateral flow assays

3. **Solid phase:** in many diagnostic assays, antibodies or antigens are bound to a solid surface so that multiple reactions can take place without the critical components being washed away. Solid phases include microtiter wells (small wells), beads, or membrane strips.

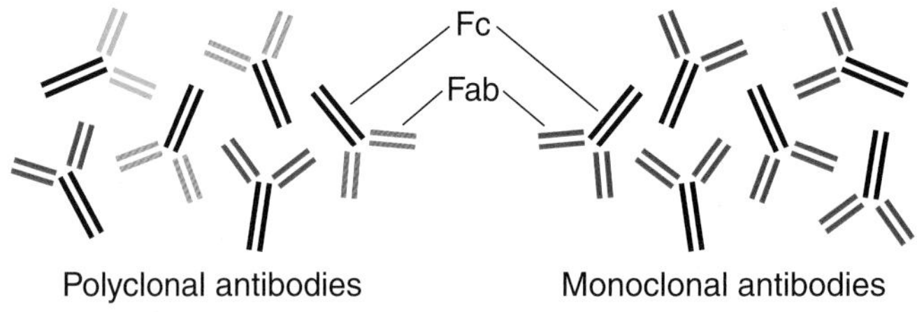

Fc
Fab
Polyclonal antibodies
Monoclonal antibodies

Figure 14.1. Polyclonal and monoclonal antibodies. Polyclonal antibodies detect similar epitopes. Monoclonal antibodies detect the same epitope.

II. ENZYME IMMUNOASSAYS. EIAs are techniques that detect a protein of interest using antibodies that are conjugated to an enzyme label.

1. **Enzyme-linked immunosorbent assay (ELISA):** is a frequently used method that is usually performed in 96-well plates so that many specimens can be tested at the same time. The most commonly used enzyme label in ELISAs is **horseradish peroxidase (HRP)**, which can cleave a colorless substrate into a colored compound. The presence of a color change in a microtiter well is usually read by measuring a decrease in absorbance (i.e., a color change means that less light can pass through the well). The ELISA procedure has many steps and can be labor-intensive if done manually, but some laboratories have instruments that automate addition of the reagents. There are many types of ELISA formats, but three of the most common are described here. (Fig. 14.3 and Table 14.1).

ELISAs are typically performed in a **microtiter plate**. The output is typically a **color change** (but it can be other things, like luminescence).

- **Direct ELISA:** The target of interest is detected with a single antibody binding step because the primary antibodies used for detecting a protein of interest are tagged. This method is usually used to directly detect an antigen that is present in a patient specimen.
    - The patient specimen is fixed to a solid phase, like a slide.
    - An antigen-specific antibody that is directly conjugated to an enzyme label is added. After incubation, the unbound antibodies are washed off.
    - The substrate is added and is cleaved by the enzyme label. The amount of color produced correlates with the amount of antibody bound to the target protein.
- **Indirect ELISA:** The target of interest is detected after multiple antibody binding steps (i.e., the secondary antibodies are tagged instead of the primary antibodies).

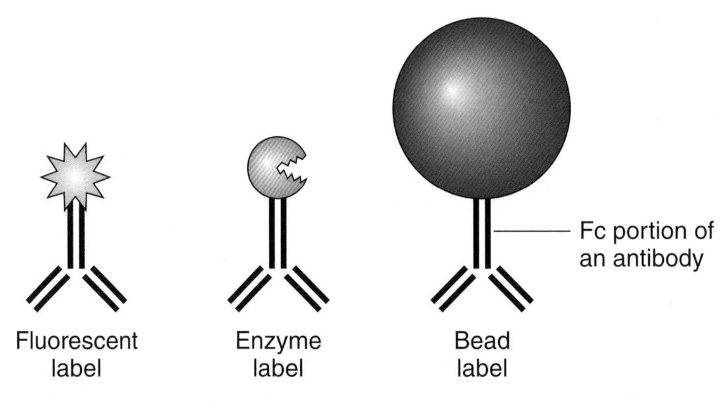

Figure 14.2. Different kinds of antibody labels used in immunologic assays.

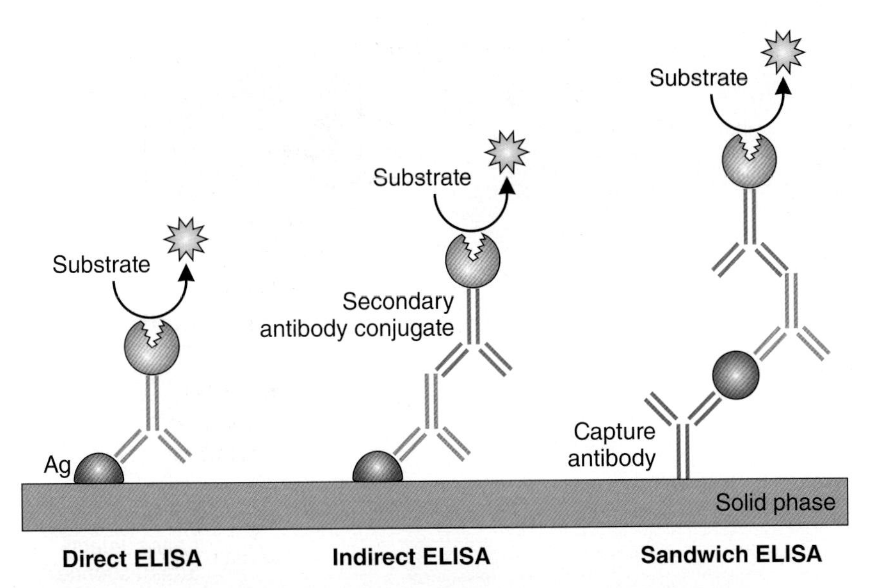

Figure 14.3. Comparison of enzyme-linked immunosorbent assays (ELISA) techniques. Direct ELISA: a primary, labeled antibody binds to the antigen. Indirect ELISA: a primary antibody binds to the antigen. A secondary, labeled antibody binds to the primary antibody. Sandwich (capture) ELISA: a capture antibody is bound to a solid surface and captures the antigen of interest before detection with primary and secondary antibodies.

- A purified antigen is attached to a solid phase, like a slide or microtiter well.
- Patient specimen is added. Any antibodies contained within the specimen can bind to the antigen. The unbound antibodies are washed off after incubation.
- An enzyme-labeled secondary antibody that is specific to the primary antibody is added to the specimen. After incubation, the unbound antibodies are washed off.
- The substrate is added. The amount of color produced by the enzyme label correlates with the amount of antigen present in the specimen.

Table 14.1. Comparison of ELISA techniques

TYPE OF ELISA	ADVANTAGES	DISADVANTAGES
Direct	Relatively rapid (only one antibody binding step)	Lower sensitivity because the signal is not amplified. The enzyme-labeled antibody is specific to each assay and is expensive to make.
Indirect	Higher sensitivity (multiple secondary antibodies can bind to the primary antibody so that the final signal is amplified)	Extra step in the procedure Potentially reduced specificity (the secondary antibody may cross-react with other antigens)
Sandwich (or capture)	High specificity even from a complex specimen (because it is directly detected by two reagents: one for capture and another for detection)	Optimization of the antibody pair (capture and detection) may be challenging.

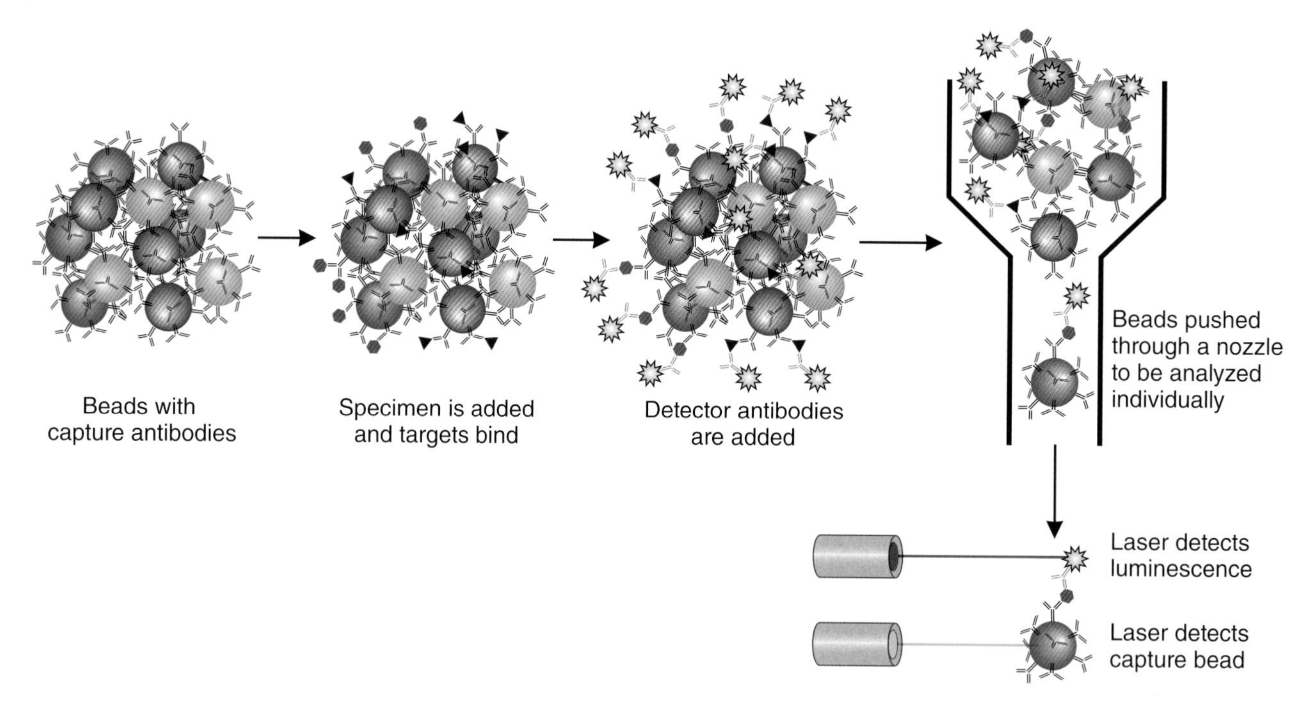

Figure 14.4. Principle of chemiluminescence immunoassays (CMIA, CLIA, CIA).

- **Sandwich (or capture) ELISA:** The target of interest is specifically captured onto a solid phase by an antibody prior to a direct or indirect ELISA. This method is used to increase the specificity of detection for the target of interest.
  - A capture antibody is attached to a solid phase, like a microtiter well.
  - A patient's specimen is added. The protein of interest (either an antigen or an antibody) will be captured onto the plate while the unbound proteins in the sample are washed off. This step increases the specificity of the assay.
  - The target of interest is then either detected in a single step by a directly conjugated antibody or in two steps using a target-specific primary antibody and a labeled secondary antibody.
  - The substrate is added. The amount of color change correlates with the amount of antigen present in the specimen.

2. **Chemiluminescent immunoassays (CIAs, CMIAs or CLIAs)**: are similar to sandwich ELISAs, but they use enzyme-labeled antibodies that cleave substrates to produce light. Also, the solid phase is often magnetic or nonmagnetic microbeads (Fig. 14.4). This test can be high-throughput, rapid, and automated, and as much as ~10,000 times more sensitive than an ELISA.
   - **Capture antibodies** are bound to a solid phase, like a microbead or a 96-well plate.
   - The patient specimen is added so that target antibodies or antigens can bind. Unbound material is washed off. When using magnetic beads, magnets hold the beads in place so that unbound material can be washed off easily.
   - A second antibody specific to the protein of interest and conjugated to an enzyme label is added.

- The substrate is added, which is cleaved to produce light.
- The amount of light produced correlates with the amount of target protein present in the specimen.
- If the assay is designed using microbeads, the beads are passed through a reader in a high-throughput manner. These readers use lasers to detect light from the bead as well as the color of the bead. By attaching different beads to different capture antibodies, bead-based CMIAs can be used to bind and differentiate multiple targets from a single specimen (i.e., multiplexing).

The output of chemiluminescent EIAs is **light**. These assays can be performed on **beads**.

3. **Immunohistochemistry:** a technique used to detect antigens directly from tissue samples while preserving tissue morphology (*see* chapter 13).
   - A slice of patient tissue, fresh or paraffin embedded, is fixed to a slide.
   - Monoclonal antibodies specific to the antigen of interest are added to the tissue.
     - Direct staining: enzyme-labeled detector antibodies are bound directly to the antigen of interest (one antibody binding step).
     - Indirect staining: Unlabeled primary antibodies are used to bind to the antigen of interest. Secondary labeled antibodies are added that are specific to the first antibody at one end and are conjugated to an enzyme label at the other end.
   - The enzyme substrate is added. When the enzyme cleaves the substrate it produces a colored product that stains the cells. The excess substrate is washed off.
   - This technique is useful because pathogens can be seen in context of the tissue. Also the stain is permanent and can be examined under a regular **light microscope**.
4. **Immunoblot assays:** these tests use tagged antibodies to detect viral proteins on a membrane. They are labor-intensive and relatively insensitive, since a lot of protein must be present in order to be detected.
   - **Western blot**
     - Viral proteins are separated by gel electrophoresis and transferred onto a membrane. Note that viral protein preparations are complex, may contain other proteins, and may not separate well. This can reduce the specificity or make the final result difficult to interpret (Fig. 14.5).
     - Patient serum (containing antibodies) is incubated with the membrane. If antibodies are present, they will bind to a specific band of viral proteins. Any unbound material is washed off.
     - An enzyme-labeled antibody is added that cleaves a substrate into a colored product.
   - **Line blot,** line immunoblot, or line immunoassay: very similar to the Western blot procedure, but instead of using a relatively "dirty" protein preparation, selected proteins (either synthetic or recombinant) that are known to be of diagnostic value are applied directly onto a membrane. Patient serum is incubated with the membrane. If antibodies are present, they will bind to specific regions. This minimizes background and nonspecific binding.
   - Typical interpretation
     - Positive: A certain number of antigen bands must be present.

Note: the line blot assay is not the same as a line probe assay (covered in chapter 15). The two assays look similar and work on similar principles, but the former is based on antibodies binding to target proteins, while the latter is based on oligonucleotide probes binding to target nucleic acids.

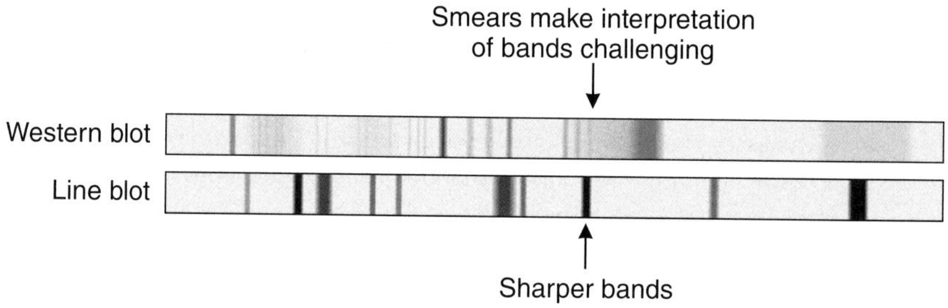

Smears make interpretation of bands challenging

Western blot

Line blot

Sharper bands

Figure 14.5. Immunoblot assays. Western blots contain purified viral preparations which can produce a "dirtier" banding pattern. Line blots contain a select number of purified proteins and produce sharper results.

- Negative: No bands are positive except the control band.
- Indeterminate: Some bands are present, but the result does not fulfill the criteria for a positive result.
- Unreadable: The strip has nonspecific binding that makes it difficult to read. Bands may be very faint, which makes these kinds of assays subjective and difficult to interpret (Fig. 14.5).

Immunoblot EIAs are performed on a **membrane.** Detector antibodies typically produce a **color change.**

III. IMMUNOFLUORESCENCE ASSAYS. Immunofluorescence assays use antibodies that are conjugated to a fluorescent label and must be read with a **fluorescence microscope**.

1. **Direct fluorescence antibody assay (DFA):** detects viral antigens directly from a specimen using one antibody binding step (Fig. 14.6).
   - A patient specimen is fixed directly on a slide.
   - Fluorescently labeled antibodies are added to the slide. This step is often performed in a humidified chamber to prevent drying.
2. **Indirect immunofluorescence assays (IFA):** detect antigens or antibodies that a patient has formed against a virus. It is indirect because there are two antibody binding steps.
   - Detecting antigens
     - A patient specimen is fixed to a slide.

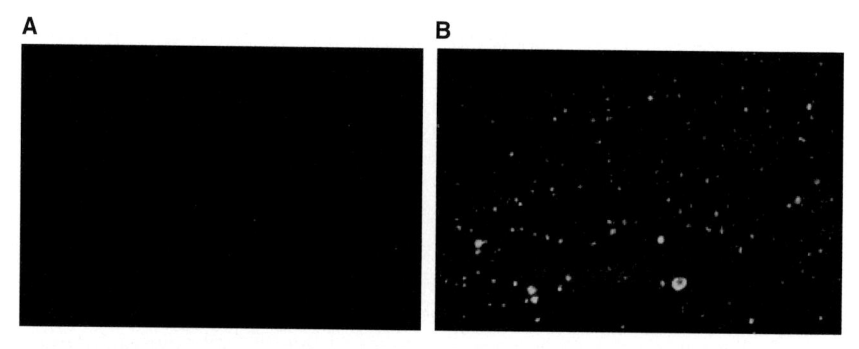

Figure 14.6. Direct fluorescence immunoassay for rabies virus antigens in brain tissue. (Left) Negative. (Right) Positive. Photos courtesy of CDC, NCEZID, DHCPP (61).

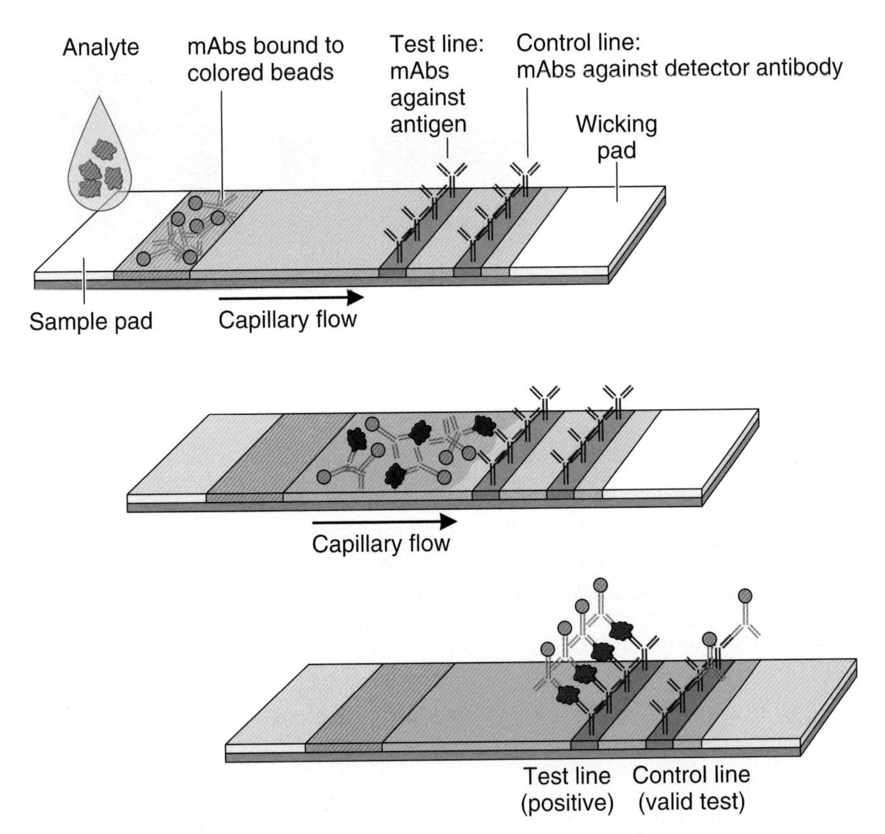

Figure 14.7. Principle of lateral-flow immunoassays. The specimen is drawn laterally through the strip by capillary action. In the first zone, labeled antibodies will bind the antigen of interest. In the second zone, capture antibodies will capture the antibody-antigen complexes. In the third zone, capture antibodies will capture unbound antibodies (control). Binding in the latter two regions can be read visually.

  - An unlabeled primary antibody is added that binds specifically to the target of interest (first antibody binding step).
  - A secondary, fluorescently labeled antibody is added that is specific to the primary antibody (second antibody binding step).
- Detecting antibodies
  - Viral antigens are fixed to a slide, usually as a preparation of virally-infected cells.
  - Patient serum is added. Antibodies present in the serum will bind if they are specific to the antigen (first antibody binding step).
  - Fluorescently labeled antibodies are added that are specific to the antibody of interest (second antibody binding step).

Fluorescent tags will dim over time, so these reactions should be read immediately for maximum sensitivity. On the other hand, enzyme-induced color changes will be stable or increase in intensity over time.

IV. IMMUNOCHROMATOGRAPHIC ASSAYS. In immunochromatographic assays, detector antibodies are conjugated to a colored bead. When enough of these detector antibodies have bound to an antigen and have aggregated, they will produce a color that can be read visually (Fig. 14.7).

1. A small amount of sample is added to one end of a nitrocellulose strip. The sample fluid is drawn laterally through the strip by **capillary action**. Because of this design, these assays are also called **lateral-flow assays**.

Lateral-flow assays are performed on a **strip** and work via capillary action. They are read visibly by looking for a **color change.**

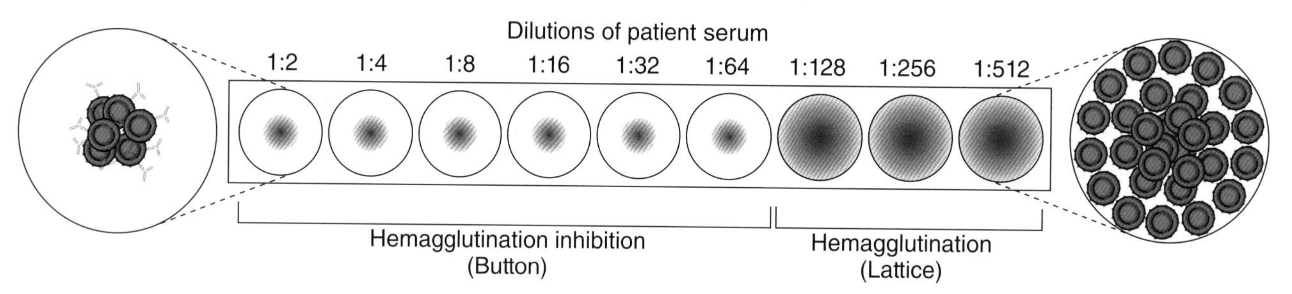

Figure 14.8. Principle of hemagglutination inhibition titers. Wells containing high levels of antiviral antibodies block the virus from agglutinating the RBCs so they fall to the bottom of the well and form a button. In wells with low levels of antiviral antibodies, virus hemagglutinin is exposed and will cause the RBCs to agglutinate into a diffuse lattice.

2. Sample fluid is first drawn through an area containing antibodies that bind specifically to the antigen of interest. These antibodies are labeled, usually with gold nanoparticles or colored latex beads.
3. Sample fluid will then migrate into the next zone, which contains immobilized capture antibodies.
   - In the "test" area, the immobilized antibodies are also specific to the antigen of interest. Antigen-antibody complexes will be captured at this section.
   - In the "control" area, the immobilized antibodies will capture excess bead-labeled antibodies. The control region is placed beyond the test region to ensure that the sample fluid flows through the entire length of the strip.
4. Once immobilized, the aggregate of colored beads will produce a colored band at the control region and the test region (if the antigen of interest is present) that is read visually.
5. These assays are rapid and can produce results within a few minutes.

Low levels of antibodies can reflect mild or remote exposure. High levels can reflect recent or multiple exposures.

V. QUANTIFYING THE AMOUNT OF ANTIBODY. It is sometimes useful to determine whether a patient has high or low levels of antibodies in their serum.

1. **Hemagglutination inhibition assay:** used to detect the amount of antibodies formed against a hemagglutinin-producing virus (e.g., influenza and parainfluenza viruses) (Fig. 14.8).
   - Patient serum is serially diluted in a microtiter plate (e.g., 2-fold dilutions).
   - The same amount of hemagglutinin-producing virus is added to each well.
   - Red blood cells are then added to each well.
   - Wells that have antiviral antibodies will bind to the virions and prevent them from agglutinating (i.e., sticking together) the red blood cells. Nonagglutinated cells will fall to the bottom and form a button.
   - If no antibodies are present, then the hemagglutinin will be exposed and the red blood cells will clump together. Agglutination of red blood cells will form a lattice and, macroscopically, the mixture will look diffuse.
2. **Plaque reduction neutralization test (PRNT):** is used to determine the amount of **neutralizing antibodies** against a virus. This is a very specific

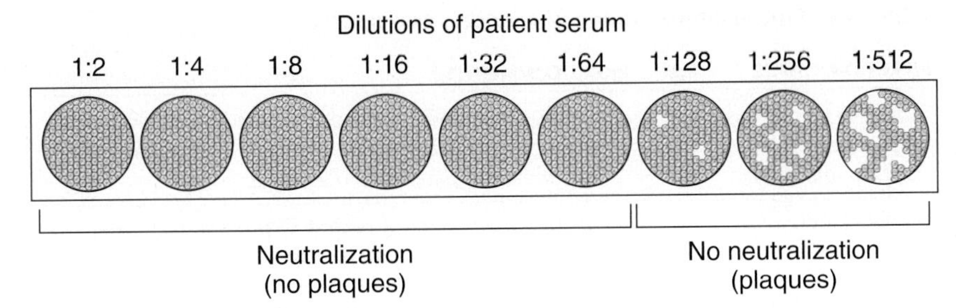

Figure 14.9. Principle of the plaque reduction neutralization test (PRNT). Wells containing low levels of antiviral antibodies will result in nonneutralized virus causing cell death in the cell monolayer (plaques).

test, so it is sometimes used to confirm antibody screens between closely related viruses (e.g., arboviruses) (Fig. 14.9).

- Patient serum is diluted serially in a microtiter plate (e.g., 2-fold dilutions).
- The same amount of virus is added to each well. If antibodies are present in the patient specimen, they will bind the virus.
- The virus antibody suspensions are then added to wells containing cell monolayers. These are overlaid with agar.
- If the well has enough antiviral antibodies, the virus is neutralized, or blocked from killing cells. So no plaques will form.
- If no antibodies are present, then the virus will be able to infect the cells and produce a plaque.

VI. SEROLOGIC ASSAYS. Immunologic assays used on serum specimens can be used to detect circulating viral proteins or subsequent antibody responses to infection. By understanding the kinetics of an infection, serologic assays can help to identify acute, resolved, or ongoing viral infections (Table 14.2).

1. **Initial exposure:** The virus enters host cells, replicates, and produces viral proteins. Some viruses produce proteins that are detectable from serum within days of exposure (e.g., dengue virus and HIV).

2. **Primary antibody response:** The first time a host is exposed to viral antigens it takes several weeks for IgM and IgG antibodies to form against their epitopes.
   - IgM is the first antibody to be produced. It takes ~1 to 2 weeks for it to become detectable and typically lasts **~3 months.** As a result, it is often used as a marker of acute or recent infection.
   - IgG is produced ~1 to 2 weeks after IgM (~2 to 4 weeks after exposure) but persists for life. Since it is produced after IgM, **seroconversion** from IgM to IgG indicates resolving or past infection.

IgM is detectable ~7-14 days after exposure and lasts ~3 months.

IgG is produced ~2-4 weeks after exposure and lasts for life.

3. **Subsequent antibody responses:** IgG persists in circulation for decades and provides long-term immunity against reinfection with pathogens. This **memory response** also provides immediate protection upon secondary exposure because memory B cells are able to rapidly ramp up IgG (Fig. 14.10).

4. **Interpreting serologic results:** IgM and IgG do not always appear and disappear when they are supposed to. For example, a positive IgM result may be a false positive, or it may be lingering from a past, resolved infection.

Table 14.2. **Typical interpretation of serologic test results**

INFECTION STATUS	IgM	IgG	COMMENT(S)[a]
Naive (no exposure or very early in infection)	–	–	This result may also occur if testing was performed too soon after infection. If clinical suspicion is high, retest the patient in 1–2 weeks.
Early infection (~1–2 weeks after exposure)	+	–	IgM is a marker of **recent exposure.** Note that IgM can be highly **cross-reactive** and may result in false positives. Retest patients in 2 weeks to confirm that the result is a true positive and to document seroconversion to IgG.
Recent or past infection	+	+	May indicate recent infection and seroconversion. However, IgM may **persist** for 3–6 months after infection, so it may be present even after resolution of the infection. Clinical correlation is required.
Resolved or past infection	–	+	IgG is a marker of **past exposure** because it takes several weeks to appear but persists for life.
Resolved or past infection with a secondary exposure	–/+	+	A **4-fold increase in IgG** over baseline IgG levels can indicate a secondary exposure.
Vaccinated	–	+	IgM may be positive if vaccination occurred recently.

[a] Antibody generation and persistence may vary depending on the host or on the type of virus.

A negative IgG result may mean that the patient was not previously exposed, the patient is unable to mount an antibody response due to immunosuppression, or the antibody response has not yet been generated due to recent infection. Therefore, a positive or negative result does not necessarily rule an infection in or out. Instead, the trends from multiple serologic test results should be monitored.

A **4-fold change** in antibody titer indicates a true secondary exposure.

- **Paired acute- and convalescent-phase specimens** are the best way to evaluate serologic results. Specimens are taken from the patient at acute and convalescent time points to confirm that an infection occurred and verify that it has resolved. The first specimen is collected as soon as possible following suspected infection. A second specimen is taken after 2 to 3 weeks to detect seroconversion to either IgM and/or IgG. In secondary infections, IgG that is already present against the pathogen is boosted. A 4-fold increase in IgG between the acute and convalescent sera is typically considered suggestive of a true secondary infection.
- **Avidity** is the total strength of an antibody-antigen interaction. Avidity testing is an uncommon secondary test that can sometimes be used to identify whether an IgG response is recent or mature.
  - The overall strength of binding between an IgG antibody and its antigen increases over time. This is because as IgG continues to encounter the agent, the binding sites mature and are optimized for the specific epitopes. As a result, high antibody avidity typically indicates that the patient was infected at least 3 to 4 months prior, while

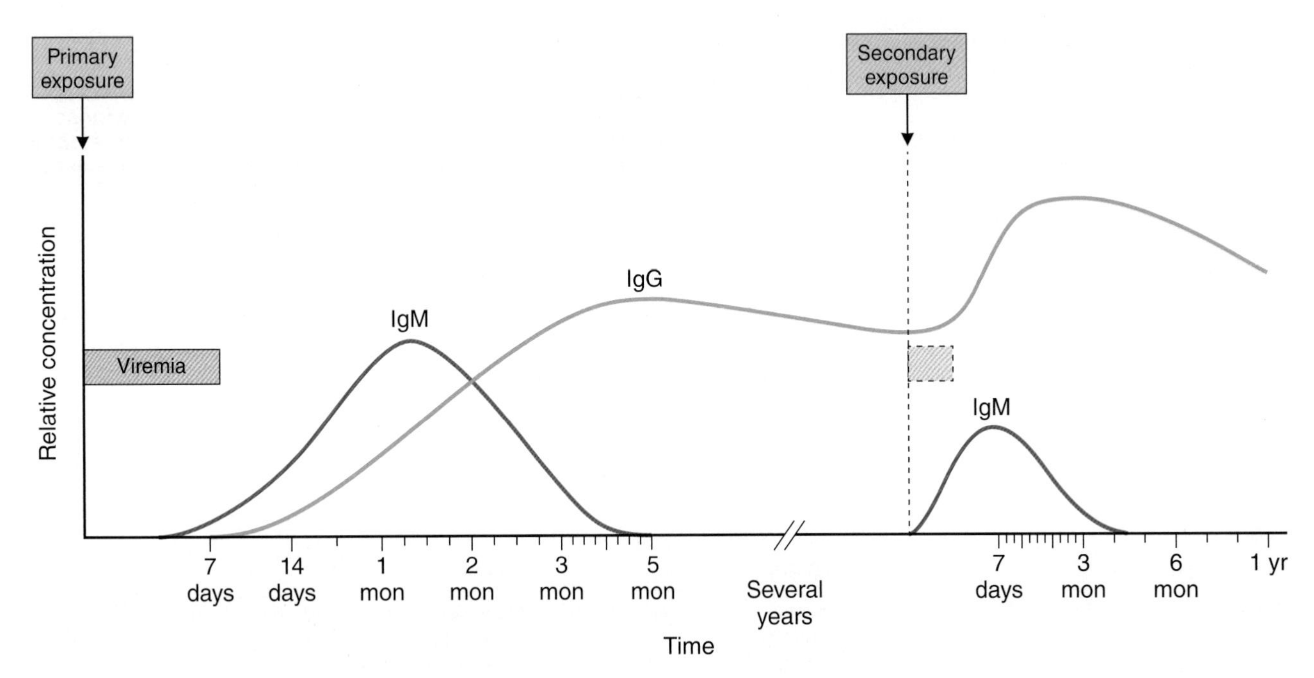

Figure 14.10. Kinetics of antibody responses. Upon initial exposure, IgM antibodies are produced first, followed by IgG. After the secondary exposure, IgG antibodies are rapidly produced in large amounts.

low-avidity antibodies suggest more recent infection (<3 months prior).

- Note that IgM binds antigens weakly compared to IgG but it can have high avidity for a target because it is pentameric (i.e., has 10 binding sites; *see* Figure 18.1).

High avidity IgG likely means past infection, low avidity IgG likely indicates recent infection.

5. Advantages of serologic assays

- Unlike NAATs or cell culture, serologic assays can identify the stage of viral infection, such as acute or resolved infection.
- IgM and IgG can be detected from samples that are relatively easy to acquire (e.g., plasma and serum). Antigens can also be detected in the blood, as well as wherever the virus is present (urine, tissue, CSF, respiratory secretions, stool, etc.).

6. Limitations of serologic assays

- Serology generally cannot be used to diagnose very early primary infections because the first antibodies are produced after lag time of 1 to 2 weeks. Therefore, serology is not the preferred method of diagnosis for acute diseases that require immediate intervention.
- Serology may also produce false-positive and -negative results.
  - False negatives: Patients who have been exposed to a virus may appear to be antibody negative if they are tested too early (prior to antibodies being produced) or if they have weakened immune systems and do not show robust antibody titers.
  - False positives: Reactivation may induce antibody titers. Antibodies produced against one virus can cross-react against other, similar viruses. Cross-reaction can also occur with rheumatoid factor and during pregnancy. These factors reduce the specificity of diagnosis.

Serology is not generally useful in diagnosing very early acute infection because the first antibodies are typically produced after 7 days.

Table 14.3. Comparison of immunologic assays[a]

ASSAY NAME	ABBREVIATION(S)	TYPE OF LABEL	PURPOSE	MOST COMMON FORMAT OR SOLID PHASE	ADVANTAGES	LIMITATIONS	EXAMPLES OF PATHOGENS DETECTED BY THIS METHOD
Immunohistochemistry	IHC	Enzyme	Detects antigens in **tissue sections**	Slide	• Maintains tissue morphology • Stains are permanent and do not fade • Can be performed on formalin-fixed, paraffin-embedded tissue	• Manual • Antibodies not commercially available for all viruses	Herpesviruses, JC virus, BK virus, HPV, adenovirus, RSV, parainfluenza virus, parvovirus
Enzyme-linked immunosorbent assay	ELISA	Enzyme	Detects circulating antibodies or viral antigens in fluid specimens	Microtiter well (usually 96-well plate)	• Used on fluid specimens like blood and CSF • Can be **automated**	• Has many steps • Can be time-consuming and slow	Mumps virus IgG, HIV (3rd-generation assay)
Chemiluminescence assay	CIA, CMIA, CLIA	Enzyme	Detects circulating antibodies or viral antigens in fluid specimens	Microbead	• **More sensitive** than an ELISA • **High throughput** • **Automated**	• More costly than an ELISA	EBV, HIV (4th-generation assay)
Direct/indirect fluorescence assay	DFA/IFA	Fluorescent	Detects viral antigens/antibodies in tissue, body fluids, other secretions	Slide	• Fairly rapid • Direct detection from specimen (no culture or amplification steps needed)	• Manual • Signal fades over time	Many respiratory viruses, herpesviruses, enteroviruses
Immunochromatographic (lateral-flow) assay	LFA	Colored bead	Detects circulating antibodies or viral antigens in fluid specimens	Strip	• **Very simple and rapid** to perform (i.e., point of care) • Inexpensive	• May have lower sensitivity (e.g., rapid influenza tests). On the other hand, rapid HIV LFAs are very sensitive)	Influenza virus, HIV
Western blot	WB	Enzyme	Detects antigens or antibodies	Membrane	• More specific than ELISAs	• Labor-intensive and time-consuming • Highly subjective • Low sensitivity	HIV confirmation (early generation assays)
Line immunoblot		Enzyme	Detects antigens or antibodies	Membrane	• Specific and cleaner than Western blotting	• Labor-intensive • Relatively subjective	HIV

[a]These assays can be used to detect viral antigens produced during an infection or antiviral antibodies produced in response to an infection.

Multiple-Choice Questions

1. **How is immunohistochemistry (IHC) different from *in situ* hybridization (ISH)?** *(May require reading outside of this chapter.)*
   a. IHC is done on tissue, while ISH is done on serum.
   b. IHC is for binding proteins while ISH is for binding nucleic acids.

c. Staining by ISH cannot be read with a light microscope.

d. They are the same; IHC and ISH are interchangeable terms.

2. Which of the following can cause a false-positive serology?

a. Infection with a similar virus

b. Rheumatoid arthritis

c. Reactivation of a latent virus

d. All of the above

3. What makes serology for Zika virus difficult? *(May require reading outside of this chapter.)*

a. Zika virus does not stimulate antibodies.

b. A positive serology often represents cross-reaction with similar viruses.

c. Serology can be done only on respiratory viruses.

d. Zika virus has a short viremic period.

4. Which of the following immunologic assays is performed using a microtiter well as the solid phase?

a. ELISA

b. Immunohistochemistry

c. Chemiluminescent immunoassay

d. Immunoblotting

5. Chemiluminescent immunoassays are useful because they

a. Are highly sensitive

b. Use antibodies to bind to nucleic acids instead of proteins

c. Are typically done at the bedside

d. Are highly manual

6. In which of the following situations is serology more useful than PCR?

a. For diagnosis during acute disease

b. When the suspected viral etiology is common in the population

c. To determine if an asymptomatic individual has been exposed to a virus

d. If available, PCR is always the preferred test.

True or False

7. Polyclonal antibodies are more specific than monoclonal antibodies. **T F**

8. Lateral-flow assays are rapid and are often used as point-of-care tests. **T F**

9. Serology is the primary screening tool for many blood-borne viruses. *(May require reading outside of this chapter.)* **T F**

10. High avidity represents low-specificity antibodies. **T F**

11. Immunofluorescence assays are a type of enzyme immunoassay. **T F**

15

CHAPTER 15

MOLECULAR TECHNIQUES: NUCLEIC ACID AMPLIFICATION

I. OVERVIEW. Nucleic acid amplification tests (NAATs) are used to amplify minute amounts of nucleic acids (as little as one copy) based on a specific gene sequence. Because of this, they are highly sensitive and specific techniques to detect the presence of viruses, especially compared to viral culture, antigen, or serologic methods.

II. STRUCTURE OF NUCLEIC ACIDS. The structure of nucleic acids is the key to understanding how they are amplified. This is because a single strand of RNA or DNA is a template for many more strands.

1. **Nucleotide (NTP) and deoxynucleotide (dNTP) triphosphates** are the building blocks of nucleic acids. They consist of a sugar (with 5 carbon atoms), three phosphate groups, and a base (Fig. 15.1).
   - **Sugar:** The sugar in RNA is **ribose**. The sugar in DNA is **deoxyribose** (which is a ribose that is missing a hydroxyl group).
   - **Base:** There are only four bases used in forming strands of DNA. RNA uses the same four bases except for thymine, which is replaced with uracil.
     - Adenine (A)
     - Thymine (T) in DNA and uracil (U) in RNA
     - Guanine (G)
     - Cytosine (C)
   - **Phosphates:** Nucleotides are joined together, or ligated, by enzymes called **polymerases** to form long chains. As each nucleotide is incorporated, it loses two phosphates (called a pyrophosphate) to become a nucleoside. This leaves a strand of DNA or RNA consisting of a phosphate-ribose backbone with all the bases pointing to one side (Fig. 15.1).

   Nucleotides are sometimes called bases, even though they are not the same thing. This is because the base (e.g., A, C, G, and T) component is the most distinctive portion of a nucleotide.

2. **Nucleic acids can be double stranded.** Bases on one strand can pair up with bases on another strand of nucleotides, like a zipper, using hydrogen bonds. Adenine always pairs with thymine (or uracil, in dsRNA) with two hydrogen bonds. Guanine always pairs with cytosine with three hydrogen bonds (Table 15.1).

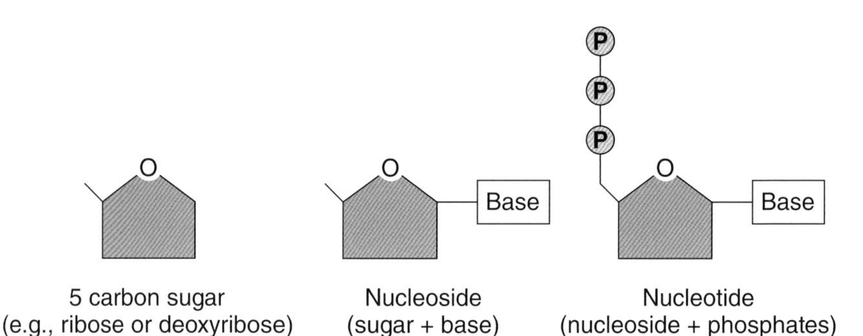

Figure 15.1. Building blocks of nucleic acids.

3. **Directionality:** Enzymes read, copy, and translate nucleic acids in one direction (Fig. 15.2).
   - The beginning of a nucleic acid strand is called the 5′ end. This is because the first nucleotide's ribose has a free 5′ carbon that is not connected to another nucleotide.
   - New nucleotides are added onto the 3′ carbon of the ribose, so nucleic acid strand is said to **grow from the 5′ to 3′ direction.** This directionality is abbreviated as 5′ → 3′.
   - **Sense** (also known as the **template** or **coding**) strand: has the gene in the correct 5′ → 3′ orientation (i.e., the orientation of the strand makes "sense" because it reads in the correct direction for protein synthesis). For example, mRNA codes for proteins so it is always in the sense direction.
   - **Antisense** (also known as **complementary**) strand: has the complementary sequence of bases in the 3′ → 5′ orientation. The antisense strand is the template for mRNA.

   Example

Template DNA (sense):	5′-CCTGGTA-3′
Complement (antisense):	**3′-GGACCAT-5′**
mRNA:	5′-CCUGGUA-3′

The complementary strand is the template/blueprint for making mRNA

DNA and RNA synthesis always occurs in the **5′ → 3′ direction.**

III. SAMPLE PROCESSING PRIOR TO AMPLIFICATION. Nucleic acids are extracted and purified from the specimen so that amplification can occur cleanly. This can be done manually, which is labor-intensive and time-consuming, or using automated instruments, which is more common in clinical laboratories.

1. **Lysis:** Cells infected with viruses are lysed open to expose all the nucleic acids. There are several methods that can be used to lyse cells.
   - Mechanical disruption: Cells are vortexed or beaten with glass beads to cause them to burst.

Table 15.1. Hydrogen bonds formed during base pairing

BASES THAT PAIR	NUMBER OF HYDROGEN BONDS	ABBREVIATION
Guanine with cytosine	**3**	G≡C
DNA: adenine with thymine	**2**	A=T
RNA: adenine with uracil	**2**	A=U

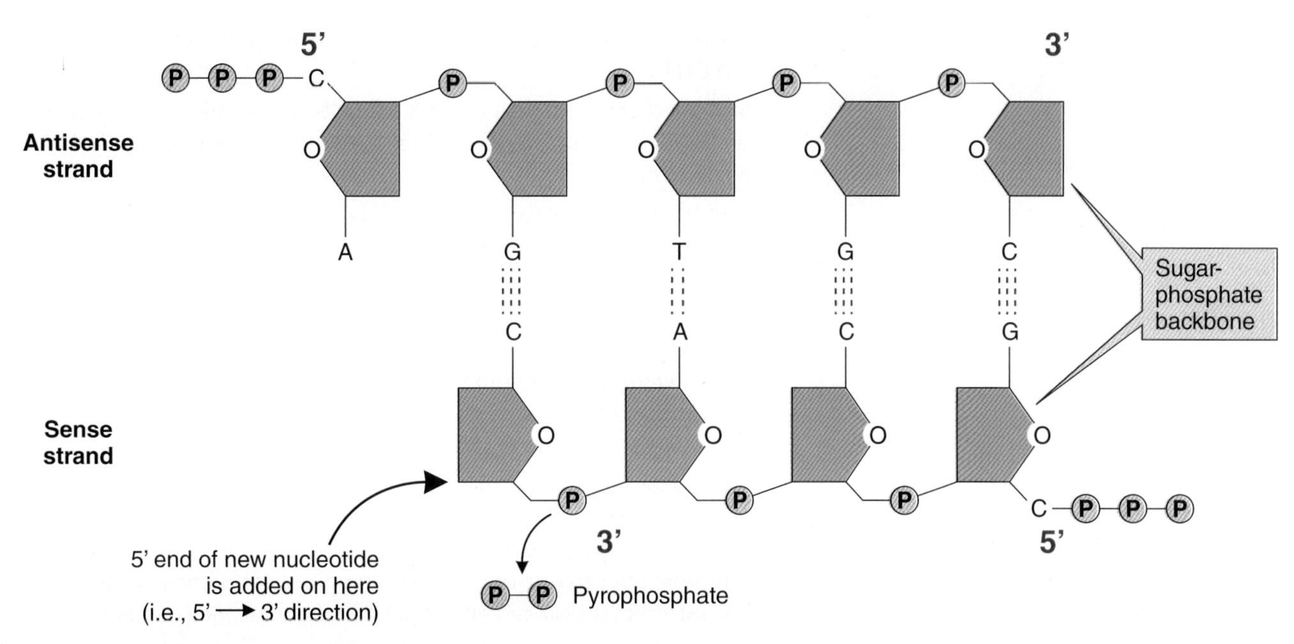

Figure 15.2. Directionality of nucleic acid synthesis. Nucleotides are only added in the 5' → 3' direction. The 5' end of a new nucleotide is added to the 3' end of the growing chain.

- Detergents: added to dissolve membrane lipids
- Proteinase K: added to degrade excess protein

2. **Extraction:** Nucleic acids are purified from the lysis suspension.
   - There are different methods of extraction.
     - Manual methods: Ethanol is added, which causes the DNA to precipitate out of solution. The precipitate is separated out from most of the debris by centrifuging it into a pellet. Phenol and chloroform are added to the pellet. Any remaining debris will dissolve into the phenol, which separates out into a distinct layer of liquid. The water layer on top now contains highly purified DNA and RNA.
     - Automated methods: Nucleic acids are bound onto magnetic beads. The debris is washed off and the nucleic acids are released from the beads by applying heat.
   - Extraction helps remove inhibitors. These are substances that inhibit PCR by interfering with the template, the DNA polymerase, or the magnesium cofactor. Some specimens have higher levels of inhibitors and are therefore more difficult to amplify accurately. Examples of inhibitors found in clinical samples include the following.
     - Stool: bile salts and complex polysaccharides
     - Blood: heme, hemoglobin, and lactoferrin
     - Urine: urea
     - Other: heparin

IV. POLYMERASE CHAIN REACTION. Conventional PCR is a common method of amplifying DNA exponentially.

1. **Reagents:** There are 6 reagents that are necessary for performing PCR.
   - **Template**, or target, DNA: contains the specific section of DNA that needs to be amplified

- **dNTPs:** nucleotides containing deoxyribose which are used for making new DNA
- **Primers:** short sequences of DNA about 20 nucleotides long. They are carefully designed to be complementary to the sequences that flank the region of interest.
  - PCR primers are used in pairs. One primer hybridizes to the DNA immediately upstream of the target gene. This is the forward primer. The reverse primer will hybridize at the end of the gene, but on the complementary strand. Well-designed primers are highly specific to the target region.
  - The **melting temperature (T_m)** of primers is the temperature at which half the primers have melted off and half the primers remain bound to the target DNA. This is often set around ~60 to 64°C.
  - A primer that has too many G and C bases will have more hydrogen bonds and will need much more energy to break—or melt—it off the template DNA. So, primers with a high GC content will have a melting temperature that is too high. The appropriate number of G and C nucleotides is between 40 and 60% of the total primer length.
  - Primers that are too short, or have too many of one base, will have lower specificity.
  - Primers should not have sections that are complementary to each other; otherwise they will hybridize and form **primer dimers** instead of binding to the template.

Taq polymerase comes from *Thermus aquaticus*, a thermophile that can make new DNA under hot conditions. This is important because DNA synthesis in PCR happens at a high temperature.

- **DNA polymerase:** an enzyme that is able to synthesize new strands of DNA by adding nucleotides onto a bound primer. ***Taq* polymerase** is the most commonly used DNA polymerase because it is thermostable. This means that it can be used under high heat conditions needed for denaturing DNA without being denatured itself.
- **Magnesium:** a cofactor for DNA polymerase. Without this cofactor, the polymerase will perform poorly and exponential PCR amplification will not occur.
- **Buffer**: maintains a neutral pH in the reaction in order to prevent degradation of the nucleic acids.

A master mix is a premixed solution of dNTPs, primers, DNA polymerase, and magnesium. The template DNA can be added to aliquots of the master mix.

2. **PCR amplification:** There are 3 main steps in PCR amplification called denaturation, annealing, and elongation. These steps are repeated for multiple cycles in order to amplify a target region of interest. Each step is done at a different temperature, so the instrument used for PCR is called a **thermocycler**. Below is a typical protocol, although temperature and incubation times may vary widely (Fig. 15.3).
   - **Denaturation:** Template DNA is double stranded, so the two strands are denatured (i.e., separated or melted apart) by being heated to ~95°C for about 30 seconds.
   - **Annealing:** The temperature of the reaction mixture is reduced for about 30 seconds in order to let the forward and reverse primers bind (anneal) to their target regions. The forward primer binds upstream of the sequence on the complementary strand, and the reverse primer binds downstream of the sequence on the template strand.
     - The annealing temperature is usually around 56°C; however, it may range from 35 to 65°C, with the specific temperature being optimized for the C and G content and length of the amplicon. It is usually set ~5°C lower than the T_m.

Denaturing is usually at 95°C.

Annealing is usually at ~56°C.

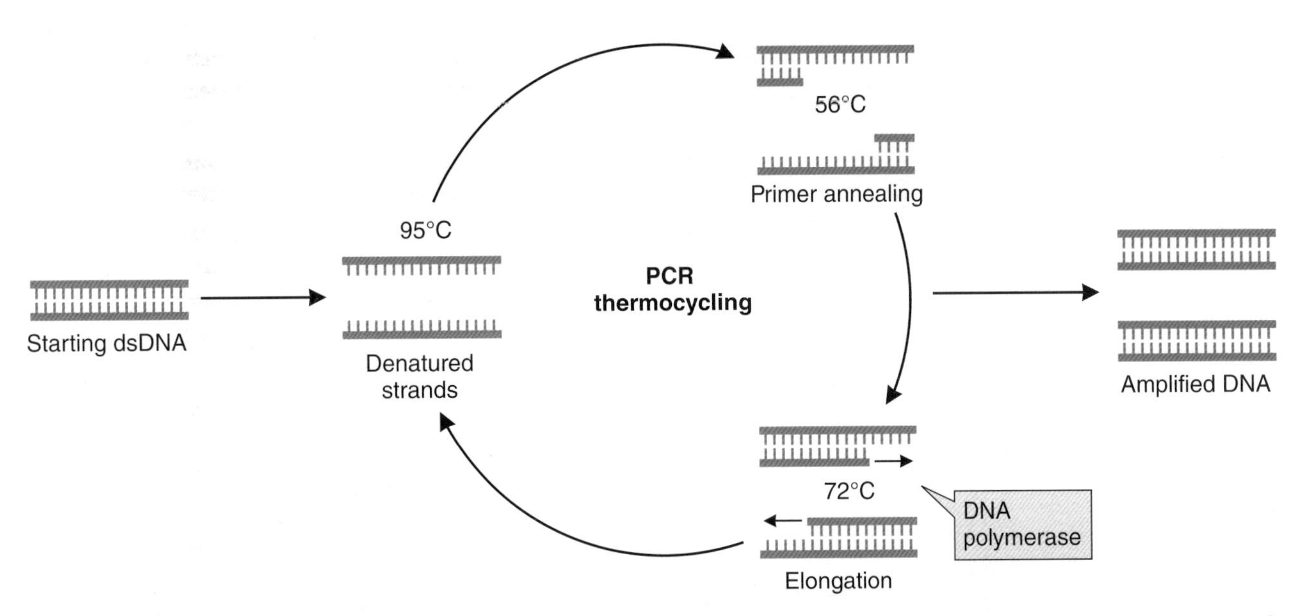

Figure 15.3. Conventional PCR. Denaturation: DNA is separated into single strands under high heat (95°C); Annealing: primers anneal (~56°C); Elongation: DNA polymerase synthesizes new DNA (72°C) by adding nucleotides onto the ends of the primers; Thermocycling: the process is repeated for several cycles.

  - If the annealing temperature is set too low, the primers will bind nonspecifically. If the annealing temperature is set too high, the primers will not be able to bind tightly enough and will fall off during elongation, and amplification will not occur properly.
- **Elongation:** The temperature is increased to ~72°C. This is the optimal temperature for *Taq* polymerase, which will use the annealed primer as a starting point to add on new nucleotides. The size of a target sequence is typically between 100 and 2,000 bases long. Larger sections take longer to generate and are at risk for more replication errors.
- **Thermocycling:** The denaturing-annealing-elongating steps are repeated between 20 and 30 times.
  - During every cycle, the amount of amplified DNA, or **amplicons**, doubles (i.e., exponential increase in amplicon).
  - The final amount of copies will be $2^{\text{number of cycles}}$.

Elongation is usually at 72°C.

Clinical laboratories do not typically use endpoint PCR because detection of amplified products is labor-intensive and time-consuming. Instead, they use automated amplification methods, like real-time PCR.

3. **Amplicon detection:** In classical PCR, the amount of DNA that is produced is measured only at the end of the reaction. So, it is sometimes called **end-point PCR** to differentiate it from real-time PCR.
   - The copied DNA is detected by agarose **gel electrophoresis** (Fig. 15.4).
   - This method is manual, time-consuming, and insensitive.

V. REAL-TIME PCR. Real-time PCR is a modification of PCR in which the number of amplicons being produced can be measured in real time, instead of just at the end. To do this, dyes or fluorescent signals are used to label DNA, which is detected by a laser during each cycle (Fig. 15.5). There are two significant advantages to this. First, this process is automated. Second, the quantity of DNA present in the original sample can be estimated based on how early a fluorescent signal is detected. This specific type of real-time PCR is also known as **quantitative PCR (qPCR)**.

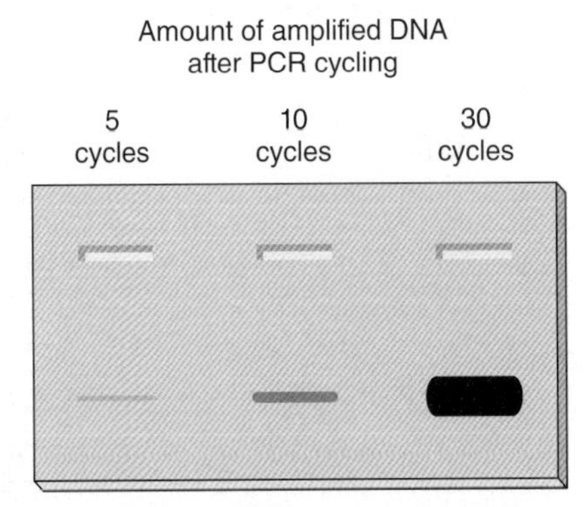

Figure 15.4. Detection of DNA on a gel. New DNA doubles with every PCR cycle. In classical PCR, also called end-point PCR, DNA is run on a gel after all the cycles are completed.

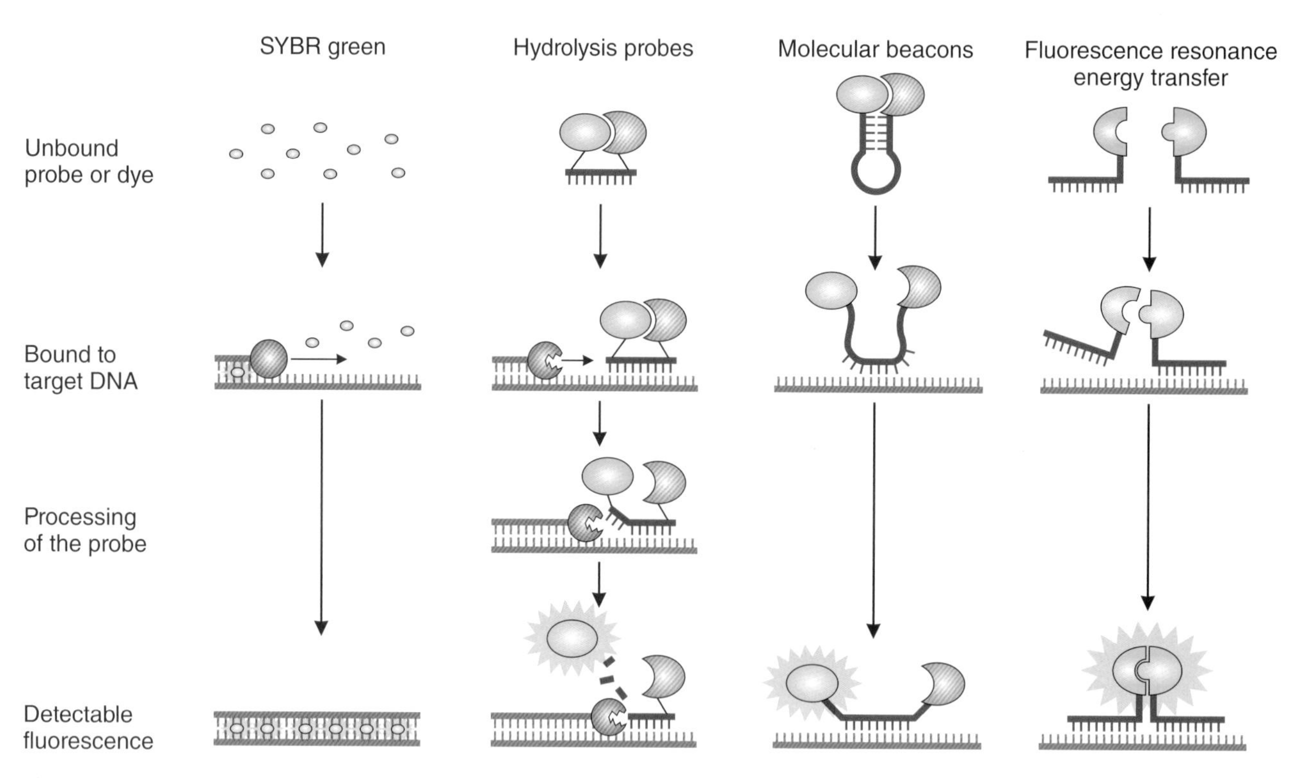

Figure 15.5. Principles of fluorescently tagged probes and dyes.

Probe-based PCR is more specific than intercalating dyes because both primers and probes have to bind specifically to a target sequence.

SYBR green dye fluoresces when intercalated into dsDNA

1. **Detection by fluorescence:** Dyes or fluorescently labeled probes are added to the reaction mixture. **Dyes** bind to DNA nonspecifically. **Probes** are short sequences of DNA (~20 bp) that are complementary to sections of the amplified region. Probes bind DNA at very specific locations so they improve the specificity of amplicon detection.
   - **SYBR green:** a commonly used dye that **intercalates** (inserts) within the base pairs of double-stranded DNA, such as new amplicons. SYBR green fluoresces only when it is intercalated into dsDNA, so the amount of fluorescence is the greatest at the end of each elongation step.
   - **Hydrolysis probes:** single probes that are bound to two tags. These probes release fluorescence when they are hydrolyzed and the tags are separated from each other.
     - Hydrolysis (or TaqMan) probes are tagged with a **fluorophore** at the 5′ end and a **quencher** at the 3′ end. Probes are short so that the two tags are in very close proximity to each other. As a result, the quencher extinguishes (quenches) any fluorescence released by the fluorophore.
     - Probes will anneal to the region of interest at a high temperature. They are designed to have a T_m about 10°C higher than the T_m of the primers. As a result, they are more specific.
     - During the elongation step of PCR the temperature is raised to 72°C and *Taq* polymerase adds new nucleotides onto a growing complementary strand. When it bumps into the annealed probe, the exonuclease activity of the polymerase "chews" through the probe in order to continue elongating the complementary strand.

- ▫ The "chewing" dismantles the probe so that the fluorophore is separated from the quencher and it is able to fluoresce.
- ▫ Hydrolysis probes are not reusable; they are consumed in order for fluorescence to occur.

TaqMan works like Pac-Man. The polymerase has to "chew" through the probe in order for fluorescence to be released.

- **Molecular beacons:** Like TaqMan probes, molecular beacons use a single probe with a fluorophore and a quencher at each end. Also, fluorescence is released when the fluorophore and quencher are separated. However, unlike TaqMan probes, the tags are kept in close proximity because molecular beacons are long enough to fold over. The tags are separated naturally when the probe binds to its target of interest, rather than by "chewing" up the probe.
  - ▫ Molecular beacons have sequences at the beginning and the end of the probe that are complementary to each other. Because of these, beacon probes fold over and hybridize to themselves in a hairpin loop. This brings the fluorophore and quencher in close proximity to each other and quenches any fluorescence when the probes are in their inactive state.
  - ▫ In the presence of target DNA the probe unfolds so that the interior section of the probe can anneal to its complementary sequence. This separates the quencher and fluorophore from each other and fluorescence is released.
  - ▫ Unlike hydrolysis probes, molecular beacon probes are not consumed and can be reused in each cycle.

Molecular beacons are sometimes called scorpion probes because they unfurl like a scorpion's tail.

- **Fluorescence resonance energy transfer (FRET):** These are probes that are used in pairs and release fluorescence when they are together.
  - ▫ A pair of probes is designed so that they anneal adjacently to each other within the target region. One probe is tagged with a **donor** fluorophore (3′ end), and the other is tagged with an **acceptor** fluorophore (5′ end).
  - ▫ When the pair of probes binds to the target region of DNA, they bring the fluorophore tags close together. These fluorophores are then activated by light so that an energy transfer from the donor to the acceptor can take place and a fluorescent signal is released.
  - ▫ Like molecular beacons, these probes are not consumed and can be reused.

Unlike hydrolysis and molecular beacon probes, FRET probes generate, instead of quench, fluorescence when the two tags are adjacent to each other.

2. **Analysis of fluorescence output:** real-time measurement of fluorescence is a powerful tool because it can be used to extrapolate the quantity of virus in the original specimen (Fig. 15.6).
   - Since the amount of amplicon doubles after each PCR cycle, the amount of fluorescence generated from the amplicons through fluorescently labeled dyes or probes also doubles. This is detected as an exponential increase in fluorescent output.
   - The **cycle threshold (C_T)** value is the cycle number at which the detection of a target crosses the background fluorescence level. Above this point, the sample is considered positive for the target.
   - **Qualitative** assays: the result is reported as either positive or negative based on the presence or absence of an amplified product. These kinds of assays are useful when the presence or absence of a virus alone will affect management of the patient. For example, these assays are useful in acute infections, or in significant specimen types like CSF.

Rule of thumb: a difference of 1 C_T between two curves is equivalent to about a 2-fold difference in copies of DNA. A difference of 3 C_T is about a 10-fold difference in copies.

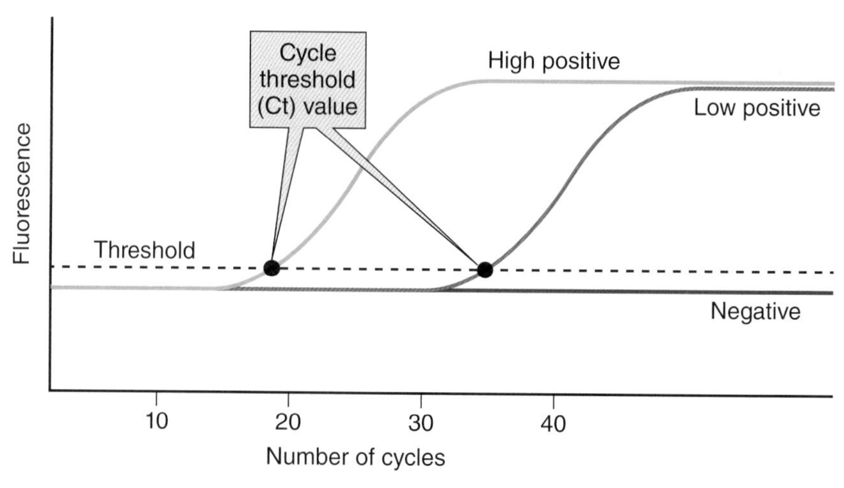

Figure 15.6. Diagram of quantitative PCR (qPCR) output. Fluorescence detected under the threshold (black line) is considered background.

Low C_T value = large amount of virus in the original sample.

High C_T value = small amount of virus in the original sample.

Qualitative assays give a positive or negative result. Quantitative assays give a quantitative viral load.

- **Quantitative** assays: specifically quantify the amount of virus present in the original sample. This can be done by comparing the C_T value in a specimen to a known standard curve, such as described below.
  - A **standard curve** is created with known concentrations of virus. For example, 10^0, 10^1, 10^2, 10^3, 10^4, and 10^5 viral copies/ml. The C_T value from the unknown sample is compared to the standard curve. This results in quantification of virus that was originally present in the unknown sample.
  - Low C_T values = large amount of starting template present in the original sample (i.e., the virus became detectable after just a few rounds of amplification).
  - High C_T values = small amount of starting template in the original samples (i.e., many rounds of amplification were needed for the virus to become detectable).
- Quantitative assays are used to measure the **viral load,** or the burden of virus in a specimen. This is measured by quantitative assays, and can be a useful monitor of viral infection over time. For instance, a low or undetectable viral load may indicate mild or controlled infection, while an increasing or high viral load may indicate progressive or severe infection. Viral load can be expressed in three ways.
  - **Copies/ml:** The number of genome copies detected per milliliter of specimen. Importantly, this unit is not standardized and the same specimen may result in significantly different "copies/ml" when tested in different laboratories and by different assays. Because of this, the same laboratory and same assay should be used to monitor a patient's copies/ml over time.
  - **IU/ml:** The number of international units per milliliter of specimen. This unit is highly standardized. It is based on standard virus preparations that are disseminated by the World Health Organization (WHO). Individual laboratory assays use this preparation to calibrate their assay and standardize their values. Only some viruses have WHO international standards and can be reported in IU/ml. The following viruses have international standards for PCR: HIV, HAV, HBV, HCV, parvovirus B19, HPV16, HPV18, CMV, BK virus, and EBV (27, 28).

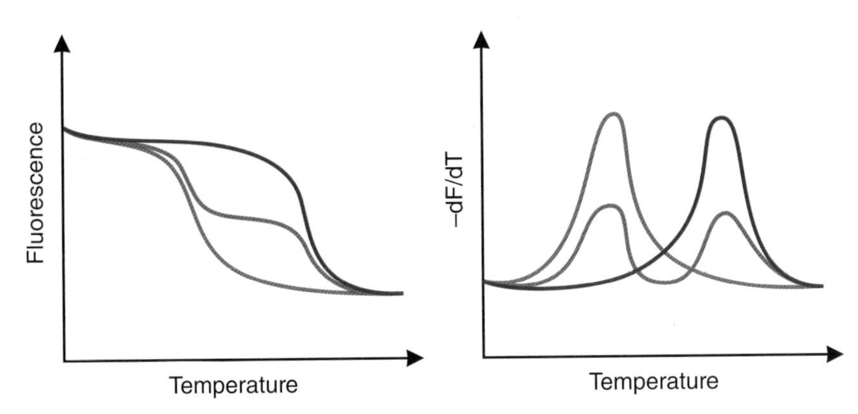

Figure 15.7. Melt curve analysis. Fluorescence output (left) decreases as the temperature increases. The red line indicates a single amplicon with a high T_m, the blue line indicates a single amplicon with a low T_m, and the purple line shows a reaction containing two amplicons, one with a high and one with a low T_m. The first derivative of the melt curves (right panel) is used more frequently because it is easier to interpret.

- **Log values:** This is the 10 "to the power of" number. It is used because it is easy to read when viral load numbers are large. For example: "5 logs" is clearer than 100,000 IU/ml (10 to the power of 5). It also keeps differences in viral load values in perspective. Consider the following example, where an increase in 50,000 IU/ml seems like a lot, but it is actually a difference of only 0.12 log.

  Viral load on day 0: 150,000 IU/ml = $10^{5.18}$ = 5.18 logs.

  Viral load on day 14: 200,000 IU/ml = $10^{5.301}$ = 5.30 logs.

Rule of thumb: an increase or decrease of half a log (0.5) is typically considered significant.

3. **Melt curve analysis:** The **melting temperature (*T_m*)** of an amplified target is the temperature at which 50% of the amplicons have melted (denatured). To determine this value, the amplified DNA is heated up. As the temperature rises, the double-stranded amplicons denature, causing a drop in fluorescence (Fig. 15.7, left).
   - The T_m is the temperature at which there is a 50% drop in fluorescence.
   - The graph for decreasing fluorescence can be hard to read. Usually the first derivative of the curve is shown instead, which converts the 50% fluorescence midpoint into an obvious peak (Fig. 15.7, right).
   - Melt curve analysis can be used to differentiate similar amplicons, such as different species in the same genus. For example, primers are developed that can anneal to all 4 serotypes of dengue virus. If the amplified sections of each serotype have different sequences (and therefore slightly different numbers of A, G, C, and T nucleotides), they will melt at different temperatures and can be differentiated based on their distinct T_ms.
   - Melt curve analysis can be done only if the fluorescent dye or probe has not been consumed.

Melt curve analysis cannot be performed when hydrolysis probes are used (these probes are consumed).

VI. REVERSE TRANSCRIPTION-POLYMERASE CHAIN REACTION. RT-PCR is used to perform PCR on an RNA template instead of a DNA template. To do this, the RNA first needs to be converted into DNA. This is called reverse transcription. Then regular PCR is performed on the DNA copy.

1. **Reagents:** There are 7 main reagents needed for RT-PCR.
   - Template RNA
   - dNTPs: building blocks necessary to make new nucleic acids

Box 15.1. Important PCR terminology

qPCR = real-time PCR ("real-time PCR" is often shortened to "quantitative PCR," or qPCR. Do not get this confused with qualitative PCR).

RT-PCR = reverse transcription PCR (do not get this confused with real-time PCR).

qRT-PCR = quantitative reverse transcription PCR (RT-PCR can also be quantitative).

RT-PCR uses the same reagents as conventional PCR except RNase inhibitors are added for stability and reverse transcriptase is used instead of DNA polymerase.

- Reverse transcriptase: a polymerase that binds to an RNA template and adds dNTPs to create a cDNA strand
- RNase inhibitors: These must be added to the reaction tubes because RNA is very fragile. RNases are ubiquitous on skin, reagents, and supplies, and will degrade the template RNA unless they are inhibited.
- Magnesium: a cofactor for reverse transcriptase
- Primers: need to be specific to viral gene sequences. For amplification of cellular RNA, primers can be strings of poly(T) that hybridize to poly(A) tails on mRNA.
- Buffer: maintains neutral pH

2. **Reverse transcription:** Formation of cDNA from RNA is **isothermal** (it occurs at just one temperature). Protocols may vary, but in a typical procedure the reagents are incubated together for ~30 to 45 minutes at 48°C (Fig. 15.8).
   - The primer will bind to the RNA template.
   - Reverse transciptase will form the complementary DNA strand.
   - RNase H will degrade the original RNA template. RNase activity is intrinsic to some reverse transcriptase enzymes, or else RNase H can be added to the reagents above.
   - Reverse transciptase will synthesize the template strand as DNA.
3. **Method of RT-PCR:** There are two main ways that RT-PCR can be performed.
   - One-step RT-PCR: Reverse transcription and PCR amplification happen in the same tube. In this type of reaction, the same gene-specific primers are used for both reactions.

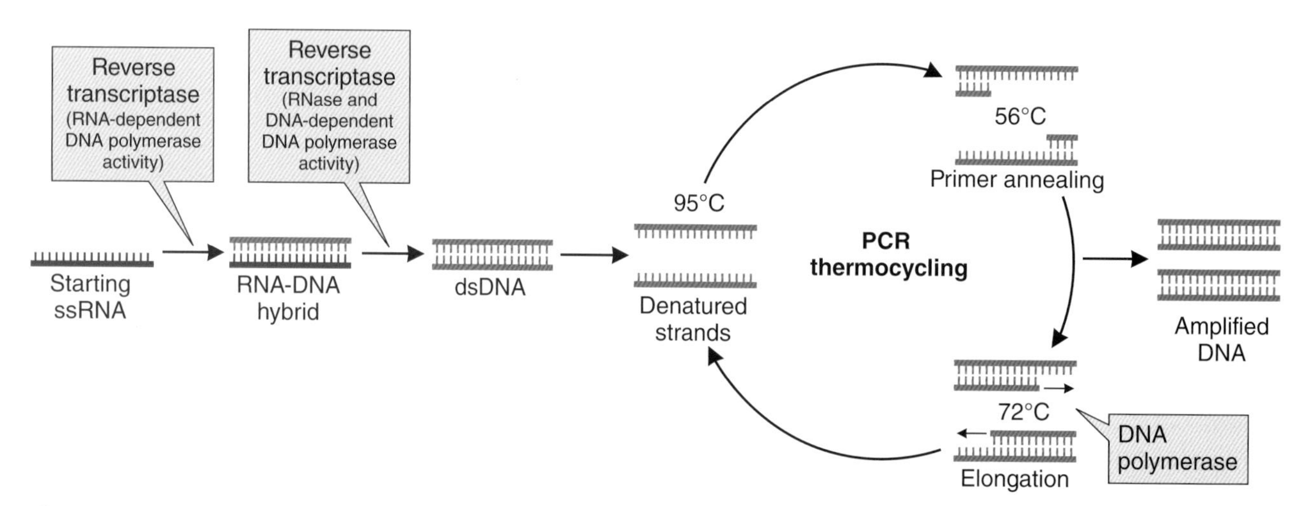

Figure 15.8. **Reverse transcription followed by regular PCR (RT-PCR).** RT-PCR can be done in one or two steps. In both methods ssRNA is converted to an RNA-DNA hybrid and then double-stranded cDNA with reverse transcriptase. The cDNA enters regular PCR for amplification.

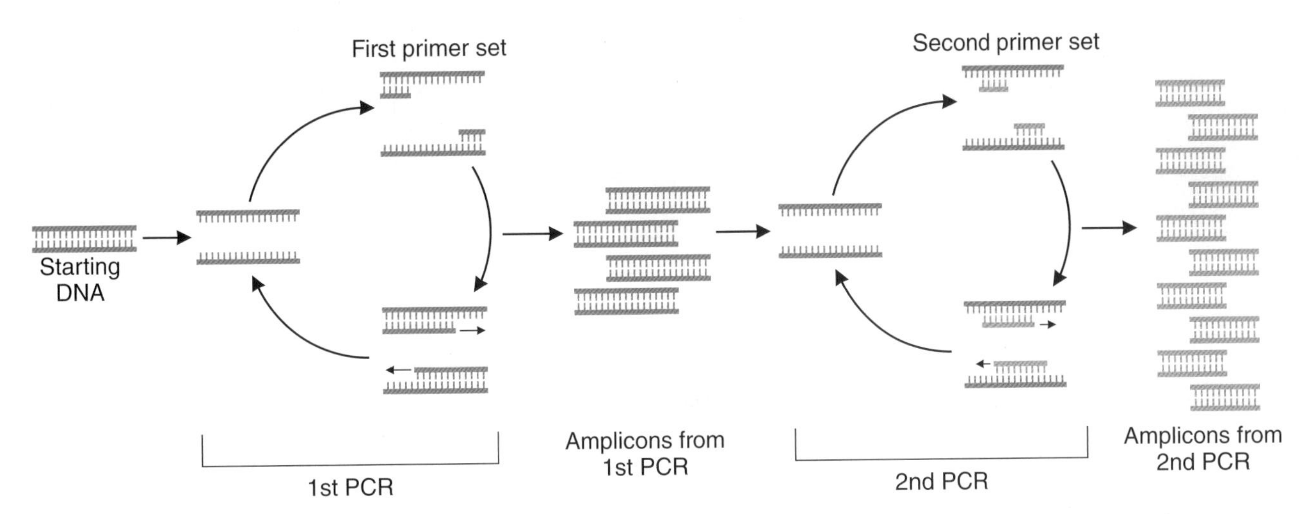

Figure 15.9. Nested PCR. A regular PCR reaction is performed. Amplicons from the first primer set (shown in purple) are then amplified in another PCR reaction using primers nested within the first amplicon.

- Two-step RT-PCR: Reverse transcription is performed. Then cDNA is transferred to another tube, where it is used as the template DNA for PCR amplification with DNA polymerase. Different sets of primers can be used in the first and second reaction tubes.

VII. OTHER FORMS OF NUCLEIC ACID AMPLIFICATION. There are many other methods of nucleic acid amplification, and not all of them are considered PCR. Some of the more common alternative methods are as follows.

1. **Nested PCR:** A PCR reaction is performed with an initial set of PCR primers. A second PCR reaction is done with a second set of PCR primers that fall within the original amplicon sequence. This will produce a final product that is nested within the original target sequence (Fig. 15.9).
   - Advantage: improves specificity because further amplification occurs only on correctly amplified amplicons. Improves sensitivity, due to two rounds of PCR amplification.
   - Disadvantage: high risk of amplicon contamination when it is performed manually. The first PCR amplification tube has to be opened in order to transfer the amplicons into the second PCR tube (i.e., an open system).
2. **Droplet digital PCR (ddPCR):** Another less commonly used method of quantitative PCR. In this assay DNA is quantified in absolute numbers.
   - The nucleic acid within a sample is fragmented. Each fragment is partitioned into a separate water droplet by creating an emulsion in oil.
   - Individual PCR reactions are then performed simultaneously in each droplet. Fluorescently labeled hydrolysis probes are used to detect when target-specific amplification occurs.
   - Each droplet is read by a flow cytometer to determine the presence or absence of fluorescence.
   - The presence of target DNA is determined by counting the number of PCR-positive to PCR-negative droplets. In addition, a **Poisson** mathematical correction is applied in case some droplets contained more than one fragment of starting DNA. The final quantification represents the **absolute** copy number, so it does not have to be calculated against a standard curve.

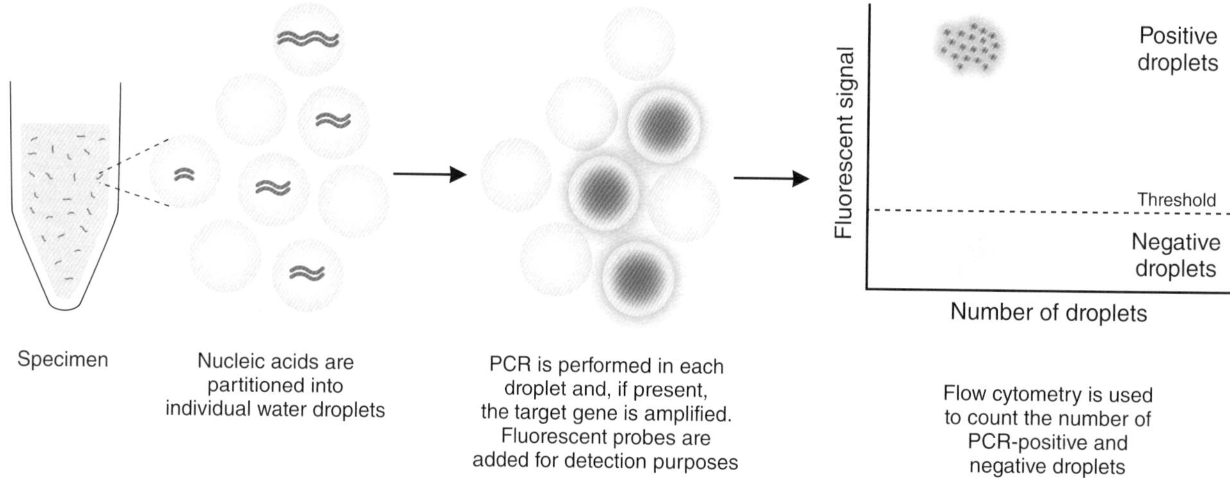

Figure 15.10. Droplet digital PCR (ddPCR). Individual fragments of DNA are partitioned and amplified in separate droplets. Fluorescence in PCR-positive droplets represents the presence of a target gene. Quantification of positive and negative droplets can be used to determine the absolute copy number of the target gene.

Multiplex PCR and melt curve analyses are similar because they can both be used to identify multiple targets. However, multiplex PCR uses multiple primer/probe sets to amplify different sequences. Melt curve analysis occurs after amplification of one or more target sequences and differentiates them based on their variable T_ms.

3. **Multiplex PCR:** Multiple targets are amplified simultaneously within the same reaction tube. To do this, primer sets specific to different targets of interest are combined in the master mix.
   - Advantages: very useful for **syndromic testing**, which is when many pathogens can cause clinically similar syndromes (e.g., respiratory or gastrointestinal symptoms). When the differential is broad, or when specimen is limited (e.g., CSF), it is useful for clinicians to order a single test that will identify multiple pathogens from a single specimen.
   - Disadvantages: Typically, true multiplex assays can amplify only a few targets (~3 or 4) while still maintaining high specificity and sensitivity.
     - Primers must be designed more carefully. They should not form primer dimers with other primers, and they should have approximately the same annealing temperature.
     - Reagents in the master mix are depleted faster because multiple amplifications are occurring simultaneously.
     - The amplified targets need to be differentiated in order to be identified. For example, amplicons should have different sizes (if they will be visualized on a gel) or should have differently colored fluorescent signals. However, these signals can overlap with each other, which reduces specificity.
4. **Transcription-mediated amplification (TMA):** This assay increases amplification by using RNA polymerase instead of DNA polymerase. RNA polymerase is able to make many copies of RNA from a single template of DNA. Reverse transcriptase is added to convert the amplified RNA back to DNA so that the process can repeat many times. (Fig. 15.11)
   - Reverse transcriptase converts starting RNA into DNA.
   - **RNA polymerase** is used to amplify the template nucleic acid. This is because RNA polymerase is designed to transcribe 100 to 1,000 copies of RNA from a DNA template during each cycle (i.e., 100^n to $1{,}000^n$ amplicons produced over n cycles. On the other hand, DNA polymerase is designed to replicate DNA, so it can only double the amount of template during each cycle (i.e., 2^n amplicons produced over n cycles).

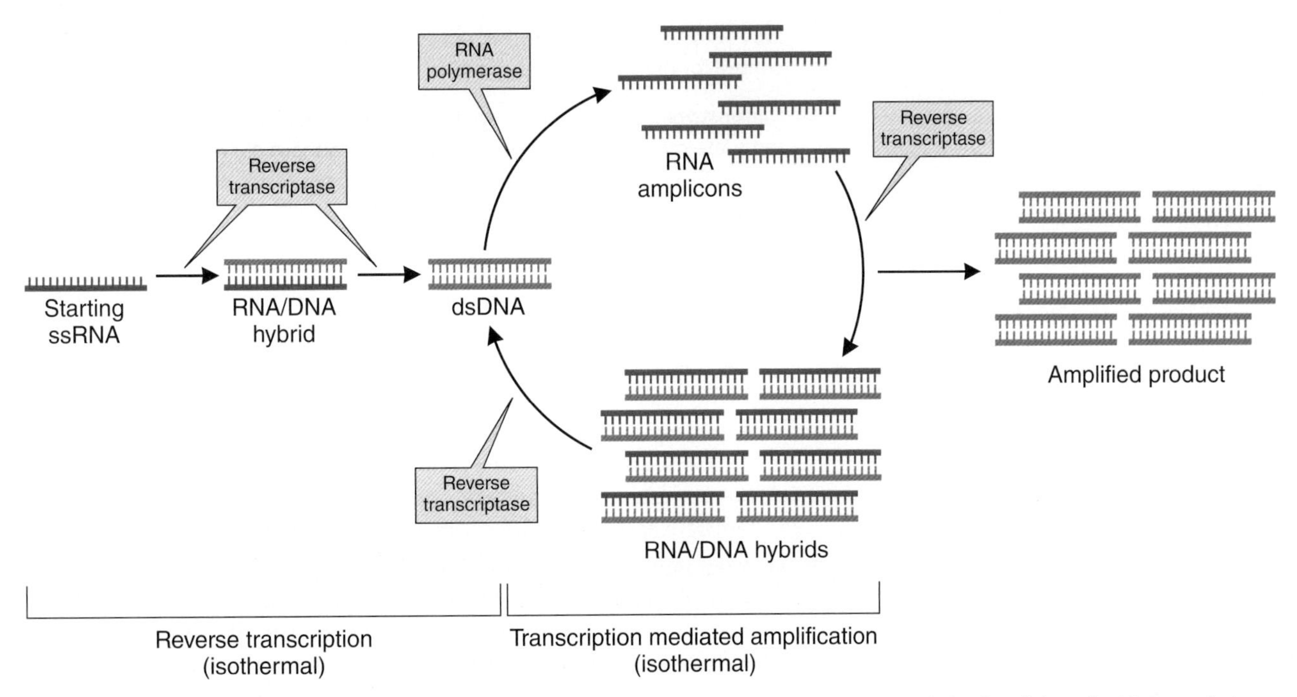

Figure 15.11. Transcription-mediated amplification (TMA). ssRNA is converted to an RNA-DNA hybrid and then double-stranded cDNA with reverse transcriptase. RNA polymerase transcribes many copies of RNA. The RNA is then converted into cDNA so that the cycle can be repeated.

- Reverse transcriptase converts all of the RNA amplicons back into DNA templates so that the process starts again.
- Amplification is **isothermal**. RNA polymerase can produce transcripts without denaturing DNA strands or primer annealing, so no temperature cycling is needed.

PCR doubles the DNA with every cycle. TMA makes hundreds of copies of RNA in every cycle.

5. **Line probe assay:** This technique is useful because it can be used to differentiate between many similar strains of virus or bacterial nucleic acids without sequencing (e.g., HCV genotyping) (Fig. 15.12).
   - PCR is used to amplify target sequences.
   - The amplicons are conjugated to an enzyme.
   - The amplicons are added to a strip with multiple types of oligonucleotide probes immobilized onto it. The conjugated nucleic acid sequence will hybridize to its complementary probe.
   - A substrate is added. The enzyme will cleave the substrate and produce a visible change in color at the location of hybridization.

VIII. ENSURING PCR QUALITY

1. **Internal controls:** A control sequence is added to the same reaction tube as the target gene. Control specific primers are added as well so that both the control and target sequences can be amplified together. This controls for the efficacy of amplification reagents and conditions within each tube.
2. **External controls:** A positive or negative control sequence is added to a separate reaction tube from the target sequence so it can go through all PCR processing steps, including lysis and extraction. During amplification, the control

tube and target tube are thermocycled at the same time. External samples control for preprocessing steps in addition to the amplification reaction.

3. **Calibrators:** A control with known quantity (i.e., copy number) of starting template material is run to ensure that the standard curve is correct and that the NAAT is performing consistently.

4. **Negative controls**
   - A no-template control is the most common negative control. The reaction tube contains all of the essential PCR components EXCEPT the template material. If any amplification occurs, it indicates that contaminating material is present. This contaminating material may cause false-positive results in actual patient tubes.
   - A nontarget control is where the reaction tube contains all of the essential NAAT components and a nontarget sequence. It functions like the no-template control but also controls for errors during addition of the template and any nonspecificity of primer/probe sets.
   - A no-amplification (or no-enzyme) control contains all of the essential NAAT components EXCEPT the polymerase. No amplification can occur in this tube, so if the results show high levels of fluorescence, it could indicate issues with the probe (i.e., probe degradation).

5. **Positive controls:** contain all of the essential NAAT components, including a known template DNA. If the NAAT does not amplify the positive control, then patient results may be falsely negative because thermocycling did not occur properly, inhibitors were present in the sample, or a critical reagent was not added.

Positive controls identify potential false-negative results. Negative controls identify potential false-positive results.

6. **Type of material used as the control**
   - Purified DNA: This can be purified viral DNA or plasmid DNA containing viral genes. These are safe to handle because they are noninfectious. Also, the DNA is relatively pure, so its copies per milliliter can be quantified very precisely. But because it is a lot cleaner than a specimen (i.e., there are no inhibitors, no complicated nucleic acid structure, etc.), it is not always an accurate representation of how well the viral genes in a specimen are being amplified.
   - Virus particles: These mimic a virus-containing sample most closely but are infectious and can be hazardous to lab personnel. Also, virus particles can degrade and are less quantifiable and consistent.
   - Armored RNA: This is a pseudovirion that is made by coating an RNA sequence with bacteriophage proteins. As a result, the RNA is stable and protected from degradation by RNases and the sequences are highly consistent so they can be quantified precisely. Because of this, they can be used as a quantitative standard.

Internal controls can detect issues occurring within each PCR tube (such as inhibitors).

External controls can detect issues occurring with the overall process (such as issues during preprocessing).

IX. CONTAMINATION. NAATs will amplify even trace amounts of contaminating nucleic acids. This can lead to false-positive results, which can be a significant limitation of NAATs. There are several ways to minimize contamination.

1. **Using unidirectional workflow:** The reagents and personnel move in only one direction to prevent reagents and reaction tubes from being contaminated by downstream events.
   - A pre-PCR setup area is used to create the master mix (i.e., combine all the components of the NAAT, EXCEPT the template nucleic acid.) DNA and RNA are not brought into this zone.

Closed systems (especially sample to answer systems) are usually preferred over open systems because of their lower risk of contamination.

- A separate location is used for adding template DNA or RNA.
- A separate location is used for amplification.
- PPE is changed between zones.
- Samples and reagents are stored separately.

2. **Environmental controls**
   - Using a dead-air box or other similar space in order to minimize any contamination from air currents (*see* Fig. 17.1)
   - Using positive air pressure in the pre-PCR setup area. This pushes contaminants away from the setup materials.
   - Using negative air pressure in the post-PCR area. This keeps amplified material from spreading.
3. **Exposure to the environment**
   - **Open systems:** The reaction tube is opened after PCR is performed, usually so that further analysis can be done. Because the tube contains amplified material (i.e., huge amounts of template DNA in a small volume), open systems carry a great risk that the work surfaces or other specimens may be contaminated by aerosolized droplets.
   - **Closed systems:** The reaction tube is not opened after PCR so that there is very low risk that aerosols containing amplicons are spread into the environment.
   - **Sample to answer systems** are a subset of closed systems where all steps including lysis, extraction, purification, amplification, and analysis occur within the same cartridge in an automated way. These have very minimal risk of external contamination, do not need separate spaces for setup, and are becoming more popular in clinical laboratories.
4. **Cleaning:** Instruments and surfaces should be cleaned with 10 to 15% sodium hypochlorite (bleach) in order to degrade nucleic acids. Bleach is corrosive, so alcohol is then used to wipe off the bleach and dry quickly. Alcohol alone does not degrade nucleic acids adequately.
5. **Degrade amplicons:** potential carry-over of amplicons from a previous PCR reaction are degraded in the new PCR master mix before the new amplification reaction is allowed to proceed. This is done as follows.
   - dUTP is added as one of the nucleotides in the master mix instead of dTTP. This is so that all amplified material will contain uracil instead of thymine, and will therefore have a different composition than the original template DNA.
   - A DNA repair enzyme called **uracil-*N*-glycosylase (UNG)** is added to the reaction tube.
   - The master mix is heated for 2 minutes at 50°C to allow UNG to degrade any DNA with uracil in it (i.e., any contaminating amplicons that have been carried over from a previous reaction).
   - The master mix is heated to 95°C in order to inactivate UNG.
   - PCR thermocycling is performed as normal.
6. **Wipe tests:** Instruments and surfaces are swabbed and then PCR amplified to see if there is any contaminating nucleic acid in the testing environment. This should be done routinely (weekly or monthly, depending on volume and technical competency).

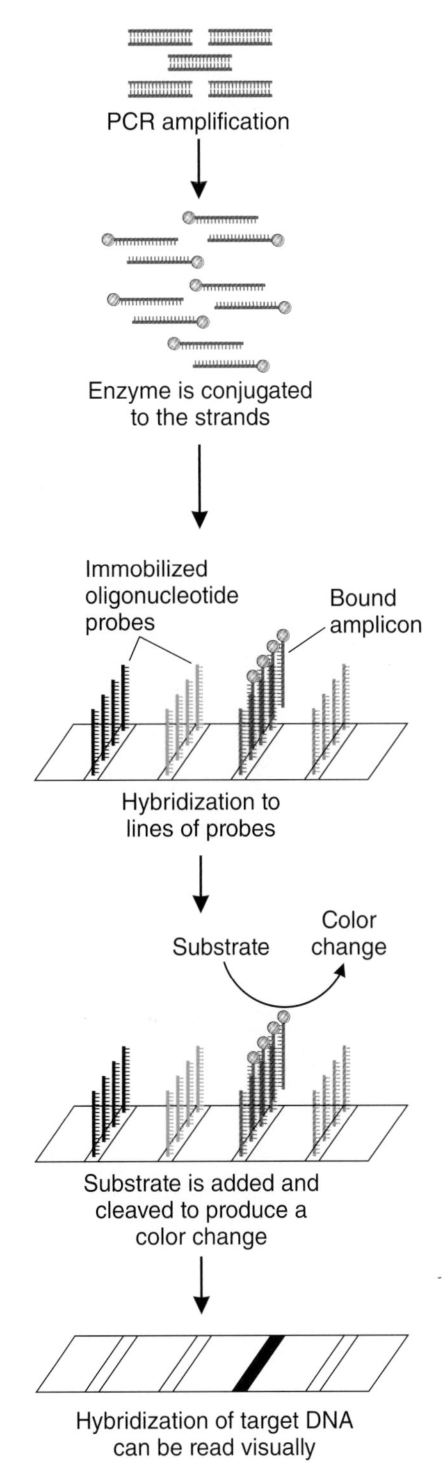

Figure 15.12. Line probe assays. DNA amplicons are tagged with an enzyme label. These are then washed over a membrane that has been attached to probes of known sequences. The enzyme substrate is washed over the strip so that any amplicons that have bound will produce a color change.

Multiple-Choice Questions

1. **Two patients have positive PCR results for HMPV. The first has a C_T value of 12 and the second has a C_T value of 30. Which has a higher viral load?**
   a. Patient with C_T value of 12
   b. Patient with C_T value of 30
   c. Cannot determine without a standard curve
   d. Cannot determine without a calibrator

2. **What effect will lowering the annealing temperature have on PCR?**
   a. Reduce the amount of template DNA that has denatured
   b. Increase nonspecific primer binding
   c. Triple the amount of DNA amplified during each cycle
   d. This is not possible. The annealing temperature is always 56°C.

3. **What is RT-PCR?**
   a. Real-time PCR
   b. Reverse transcription-PCR
   c. Quantitative PCR
   d. Right-ended PCR

4. **Which of the following methods does NOT prevent errors due to carry over contamination?**
   a. Using UNG
   b. Unidirectional workflow
   c. Cleaning surfaces with bleach
   d. Using a calibrator

5. **When is UNG added to the reaction tube?**
   a. Prior to amplification (pre-PCR)
   b. After amplification (post-PCR)
   c. During reverse transcription
   d. During transcription-mediated amplification

6. **How does real-time PCR quantify the amount of amplified DNA?**
   a. Using melt curve analysis
   b. Comparing the results to a standard curve
   c. Using a Poisson distribution of positive droplets
   d. Comparing the C_T values to an internal control

7. **How does droplet digital PCR quantify the amount of amplified DNA?**
   a. Measuring the band intensity on a gel
   b. Comparing the results to a standard curve
   c. Using a Poisson distribution of positive droplets
   d. Comparing the C_T values to an internal control

8. **Which of the following is true regarding melt curve analysis?**
   a. It reveals the presence of nonidentical amplicons.
   b. It is necessary for quantification.

c. It is used to melt primers off template DNA.

d. It can be done on hydrolysis probes.

9. Which of the following viruses needs to be detected by RT-PCR?

a. HBV

b. Adenovirus

c. Rotavirus

d. HSV

True or False

10. Nucleic acids are synthesized in the 3′ → 5′ direction. **T F**

11. The antisense strand is the template for mRNA. **T F**

12. cDNA production is isothermal. **T F**

16

CHAPTER 16

MOLECULAR TECHNIQUES: SEQUENCING

I. OVERVIEW. Sequencing is a technique that is used to determine the exact order of nucleotides (A, T, C, and G) in nucleic acids. This information is incredibly valuable because it can be used for identifying single organisms, populations of organisms, mutations, relationships between outbreak isolates, and even transcription of host genes in response to a pathogen infection. However, sequencing requires significant technical expertise, equipment, and data analytics, so it is not yet common in clinical laboratories. First-generation sequencers can sequence fragments of purified DNA. They are used in some clinical labs for simple applications like identifying organisms. Next-generation sequencing (next gen or NGS) can process tremendous amounts of DNA in parallel, so it can be used for high-throughput, complex applications. However, NGS techniques are also not yet commonly available in clinical labs because they require new technical expertise, specialized instrumentation, and substantial computing power (Table 16.1).

1. **Applications of sequencing**
   - **Whole-genome sequencing:** Used to discover new pathogens and to create new reference sequences. NGS processes much more DNA in parallel and is extremely efficient at sequencing whole genomes (from viruses to human). Sanger sequencing can be used for whole-genome sequencing too, but it is cumbersome and is done in two main ways.
     - **Primer walking**, which is sequencing of consecutive segments of DNA. Primers are created to sequence a fragment of DNA. Once this sequence is known, another primer is created within that region to sequence the next section.
     - **Shotgun method**, which is sequencing of random segments of DNA. The template genome is randomly broken up into hundreds of fragments between 1 and 300 kilobases in length. Each fragment is inserted into a universal plasmid so that it is stable, and the regions on either side of the fragment are known. The fragment can then be sequenced with a universal primer. The shotgun approach is much faster because sequencing of a fragment does not depend on getting the results of the previous fragment first. However, unlike primer walking, it requires significantly more processing power and computational algorithms to assemble all the sequences in the correct order.

- **Organism identification:** A characteristic and conserved gene segment is sequenced and compared to a database of known sequences in order to identify it.
    - Sanger sequencing is used for identification of bacteria, mycobacteria, and fungi but is not performed in clinical labs for viruses.
        - Bacteria and mycobacteria: Sequencing ~500 bp of 16S rRNA can differentiate hundreds of species to the genus and species level.
        - Fungi: Sequencing ~300 bp of 18S, 28S, or internal transcribed spacer (ITS) rRNA can differentiate hundreds of yeasts and molds to the genus and species level.
        - Viruses: not commonly used because there is no single gene region that is highly conserved among all viruses; in fact, viruses even have very different types of nucleic acids (*see* Baltimore classification, chapter 1).
    - NGS can identify bacteria and fungi using the same conserved regions (e.g., 16S, 18S, or ITS rRNA). However, it sequences all DNA so it can also identify pathogens, including viruses, by using the whole patho-

Table 16.1. Comparison between the features and applications of Sanger sequencing and NGS

PARAMETER	SANGER SEQUENCING	NGS
Throughput	Low	High (massively parallel sequencing)
Cost	Moderate	Moderate to high (depends on the number of samples per run)
Requires specialized instrumentation	Yes	Yes
Sample preparation	Complex	Very complex
Amount of data analysis needed	High	Very high
Whole-genome sequencing	Possible but difficult, laborious, and time-consuming	Much simpler, faster, and has higher throughput
Pathogen identification	**Not typically useful for viruses** because there is no single, universally conserved gene across all viruses. **Very useful for bacteria**/most mycobacteria (16S rRNA) and **fungi** (28S or ITS rRNA) and used in some diagnostic labs	Useful for **all organisms**, even without prior knowledge of pathogen. It can utilize both single-gene and multilocus gene-based identification.
Metagenomics	Not useful	Highly useful but requires significant bioinformatical analysis. Can be used to identify the virome/microbiome, the relative abundance of organisms, the transcription of pathogen and human genes, and the correlation of human genes to specific pathogens.
Resistance testing	Identifies mutations in the predominant variant that is present. PCR must be used to amplify select genes first. Currently in use in reference laboratories, most commonly for HIV, HCV, and influenza virus.	Identifies mutations present in both major and minor variant strains, and their relative proportions. In limited use.
Strain relatedness	Possible for a specific virus or pathogen. Discriminatory loci need to be chosen and PCR would be needed to amplify those selected genes.	Useful for identifying the source and progression of outbreaks based on all mutations in the virus.

gen sequence. Sequencing full genomes can provide additional information such as strain typing and resistance genes.

- **Metagenomics:** Sequencing of all nucleic acids in a sample in order to identify all organisms and their relative proportions. This is done to describe microbiological ecosystems. Sanger sequencing cannot be used for this, but NGS can identify all pathogens (including viruses) in a patient specimen ("microbiomes" and "viromes"), all the quasispecies formed in chronic infections, and the relative abundance of each organism.
- **Detection of resistance mutations:** Sanger sequencing is currently used to identify mutations within genes that encode proteins targeted by antivirals. For example, protease, integrase, and reverse transcriptase in HIV and mutations in the neuraminidase gene of influenza A virus. NGS can be used to identify *all* mutations and variations in a virus, although the effects of those mutations may not be known.
- **Strain relatedness:** Sequencing of viruses can identify the source and evolution of viruses over the course of an outbreak, guide selection of vaccine strains, and determine relatedness for taxonomy purposes. Both Sanger sequencing and NGS can be used for identifying similarities between strains, but NGS has greater resolution because it can be used to sequence a larger portion of the pathogen.

2. Important limitations of sequencing

- The assays are labor-intensive, need technical expertise, and require special instruments. They also need robust data analytic tools, especially with NGS.
- Presequencing amplification can introduce artifacts (like only amplifying the most prevalent gene variant or introducing new point mutations).
- The assays will sequence any DNA, even low-level contaminants that are present in reagents, media, materials, or the patient specimen.
- If a single gene is sequenced (e.g., Sanger sequencing), it may not be enough to discriminate between some pathogen species.
- Interpretation of metagenomics data is difficult because we have no real definition of which sequences are part of the normal microbiome/virome. For instance, we currently have a limited understanding of which organisms are commensals versus true pathogens, whether results represent potential contamination, and what the meaning is of relative proportions of organisms.

Amplification prior to sequencing can introduce bias into metagenomics-based analyses because some fragments may be amplified more than others.

II. SEQUENCING BASICS. Sequencing a gene requires significant sample preparation to ensure that the nucleic acids can be read. Then, the input material is placed on specialized instrumentation for sequencing, which records the specific order of bases present in the nucleic acid fragments. Finally, sequence results are analyzed using various bioinformatics tools depending on what application is needed.

1. **Overview of the sequencing process:** There are 4 steps for Sanger sequencing and 5 steps for NGS (Fig. 16.1 and Fig. 16.2).
   - **Extraction and purification:** Nucleic acids are extracted from the specimen. DNA is purified. If RNA needs to be sequenced, it is purified and then converted into cDNA.
   - **(NGS only) Library generation:** DNA is broken up into fragments. Short DNA labels (called adapters) are added onto the ends of the fragments.

A **base call** is the identification of a base in a sequence.

- **PCR:** amplifies target genes so that there is enough material to sequence. Only a few next-generation platforms do not need this step.
- **Sequencing:** Sequencing instruments identify the sequence of bases present in the DNA fragments.
- **Data analysis:** The sequence result is cleaned up and assessed.

2. **Phred quality score** (Q score) measures the probability that a base call in a sequence is correct. A higher Q score indicates greater accuracy (Table 16.2). Sanger sequencing generally has ~Q20 accuracy while the benchmark score for NGS methods is at least Q30.

The terms "bases" and "base pairs" are often used interchangeably, because knowing a base will also define its paired base.

3. **Read length** is the number of base calls a sequencing platform can generate at a stretch. Longer read lengths give greater information, are more likely to be unique, and can produce enough overlap with other sequences to determine their context in a genome. Length is expressed as the number of bases or base pairs.
   - Base pairs (bp)
   - Kilobase pairs (kb or kbp) = 1,000 base pairs
   - Megabase pairs (Mb or Mbp) = 1,000,000 base pairs
   - Gigabase pairs (Gb or Gbp) = 1,000,000,000 base pairs

4. **Depth of coverage** is the average number of times a nucleotide will be sequenced. Greater depth of coverage gives greater confidence in the base call.
   - Greater depth is more important when sequencing a large genome or for genome assembly so that all bases are accounted for. A lower depth of coverage is acceptable when sequencing a small gene for identification purposes.

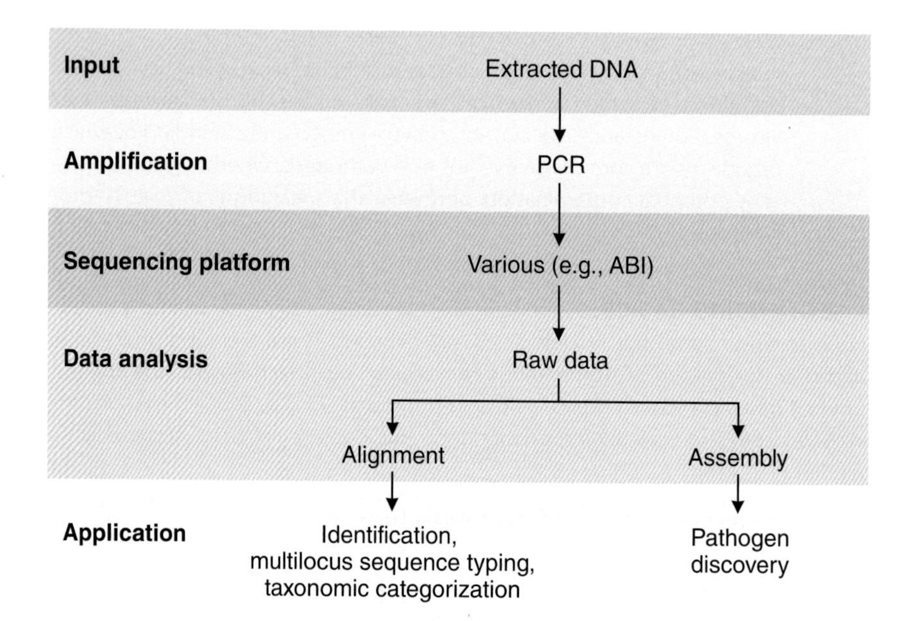

Figure 16.1. Overview of the Sanger sequencing process.

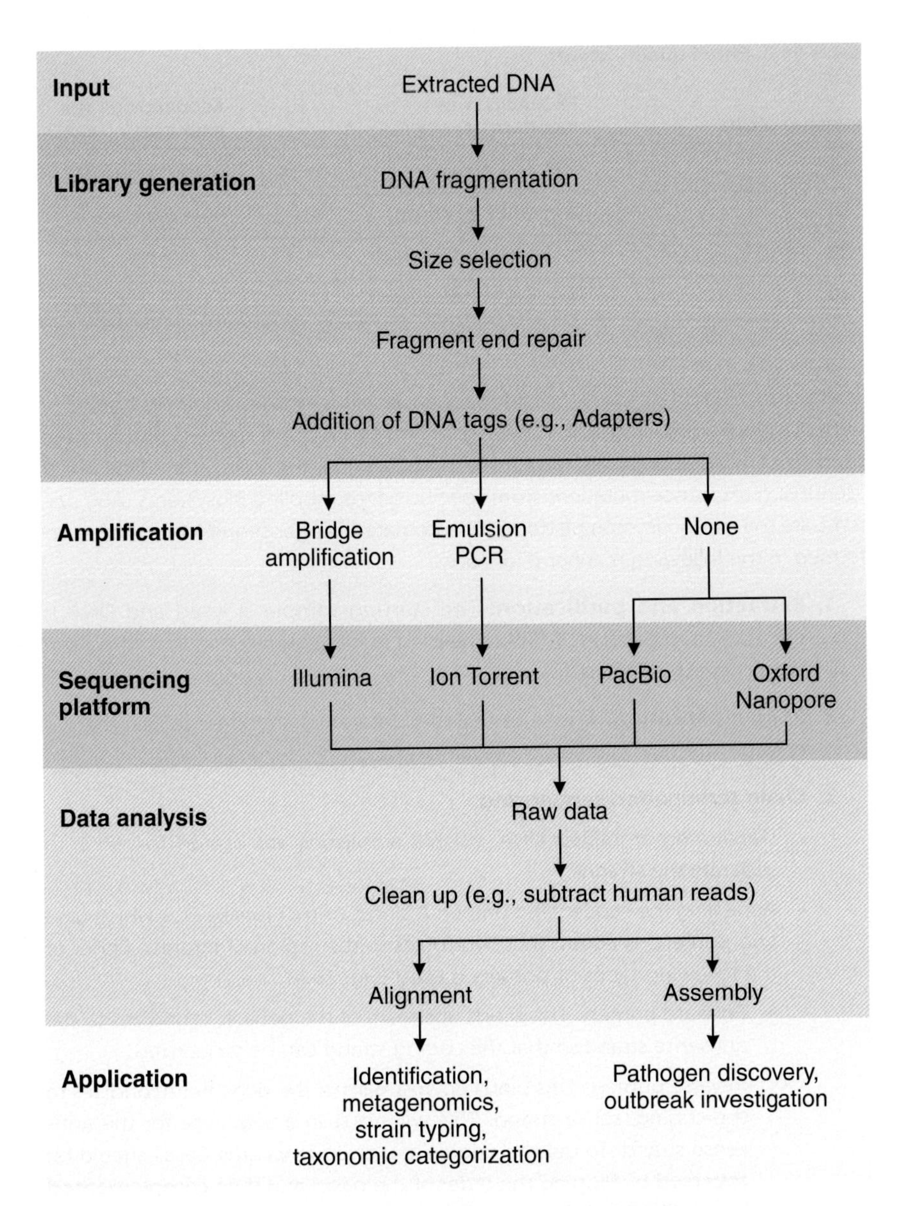

Figure 16.2. Overview of the next-generation sequencing (NGS) process.

- **Deep sequencing** is a nonspecific term, but it typically means >7× coverage. Whole-genome sequencing typically requires ≥30× coverage.
- **Ultradeep sequencing** is also not defined but typically means >1,000× coverage.

III. FIRST-GENERATION SEQUENCING METHODS. First-generation sequencing methods sequence one small section of DNA at a time. **Maxam-Gilbert sequencing** is a chemical cleavage method of sequencing because it identifies the nucleotide pattern by cleaving the template at certain positions. It is highly manual, laborious, and difficult to perform and has been replaced by Sanger sequencing. **Sanger sequencing** was originally developed as an enzymatic chain termination

Sanger sequencing is also known as "dideoxy" or "chain termination" sequencing.

Table 16.2. Phred quality scores

PHRED Q SCORE	PROBABILITY OF INCORRECT BASE CALL	ACCURACY OF THE BASE CALL
10	1 in 10 ("**1** zero")	90% ("**1** nine")
20	1 in 100 ("**2** zeroes")	99% ("**2** nines")
30	1 in 1,000 ("**3** zeroes")	99.9% ("**3** nines")
40	1 in 1,000 ("**4** zeroes")	99.99% ("**4** nines")

method and identifies the nucleotide sequence as it is synthesizing new strands of DNA. Sanger sequencing is the most common sequencing method and is often used in clinical laboratories for simple purposes like organism identification or identifying resistance mutations from specific genes. Applied Biosystems (ABI) systems are the most common platform for automated Sanger sequencing and is performed in the following manner (Fig. 16.3).

1. **Extraction and purification:** The starting sample is lysed and DNA is extracted and purified. If RNA is needed to be sequenced, it is purified and then converted into cDNA.
2. **PCR amplification:** A gene or region of interest is amplified so that there is enough starting material for sequencing.
3. **Chain termination sequencing:**
   - Denaturing amplified DNA: dsDNA amplicons are heated to ~95°C to separate the strands.
   - Annealing the sequencing primer: A single primer (unlike PCR which uses two primers) is added that hybridizes to the region of interest. Either of the following types of primers is used (Fig. 16.4).
     - Forward primers: This binds upstream of the gene. It hybridizes to the antisense strand so that the coding strand can be sequenced.
     - Reverse primers: This binds downstream of the gene but hybridizes to the coding, sense strand. This will result in a sequence for the antisense strand. To get the original gene sequence, the bases should be **reversed** by flipping the order of the bases and then **complemented** by converting each base call to its matched base. For example,

       The sequence result: AATTGC

       After reversing: CGTTAA

       After complementing: GCAATT

ddNTPs are chain terminators.

   - Synthesis: DNA polymerase, dNTPs, and fluorescently labeled **dideoxynucleotides (ddNTPs)** are added together.
     - A ddNTP is a deoxynucleotide that is missing a hydroxyl group. The absence of the 3′-OH group prevents DNA polymerase from adding more nucleotides so incorporation of a ddNTP **terminates** elongation of the chain. There are 4 kinds of ddNTPs that are used, one for each matching dNTP: ddATP, ddCTP, ddGTP, and ddTTP.
     - Each ddNTP is labeled with a different fluorescent dye so that incorporation of each ddNTP can be differentiated by color.
     - During sequencing, DNA polymerase adds dNTPs or ddNTPs as it synthesizes new strands of nucleic acids. The ddNTPs are present in the mixture at a very low concentration (~1:100 compared to the dNTP),

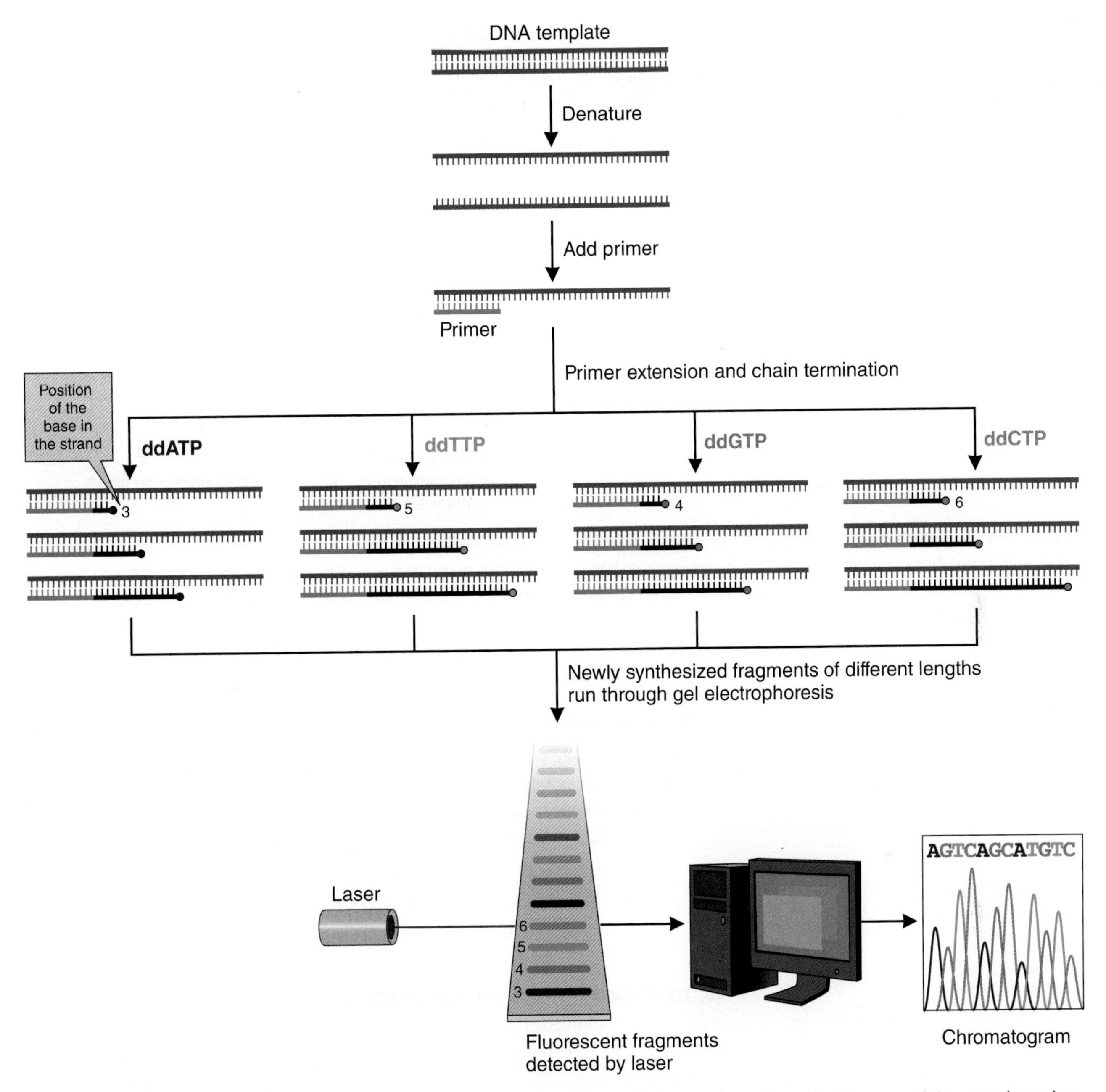

Figure 16.3. Sanger sequencing. A target piece of DNA is denatured. A single sequencing primer binds to one of the strands so that DNA polymerase can synthesize the complementary strand. The polymerase uses mostly dNTPs but will occasionally incorporate a fluorescently tagged ddNTP, which terminates synthesis. All the fragments are then separated by size. A laser then measures the fluorescence at the terminus of each fragment. A computer translates this into a chromatogram, where the colors of fluorescence are converted into specific base calls. The order of fluorescently tagged bases determines the sequence of the target DNA.

so a ddNTP will be introduced only very occasionally. When incorporated, it will terminate elongation.

- After many rounds of synthesis, the resulting fragments are of varying lengths depending on where that ddNTP was incorporated. Each newly synthesized DNA strand ends with a ddNTP.

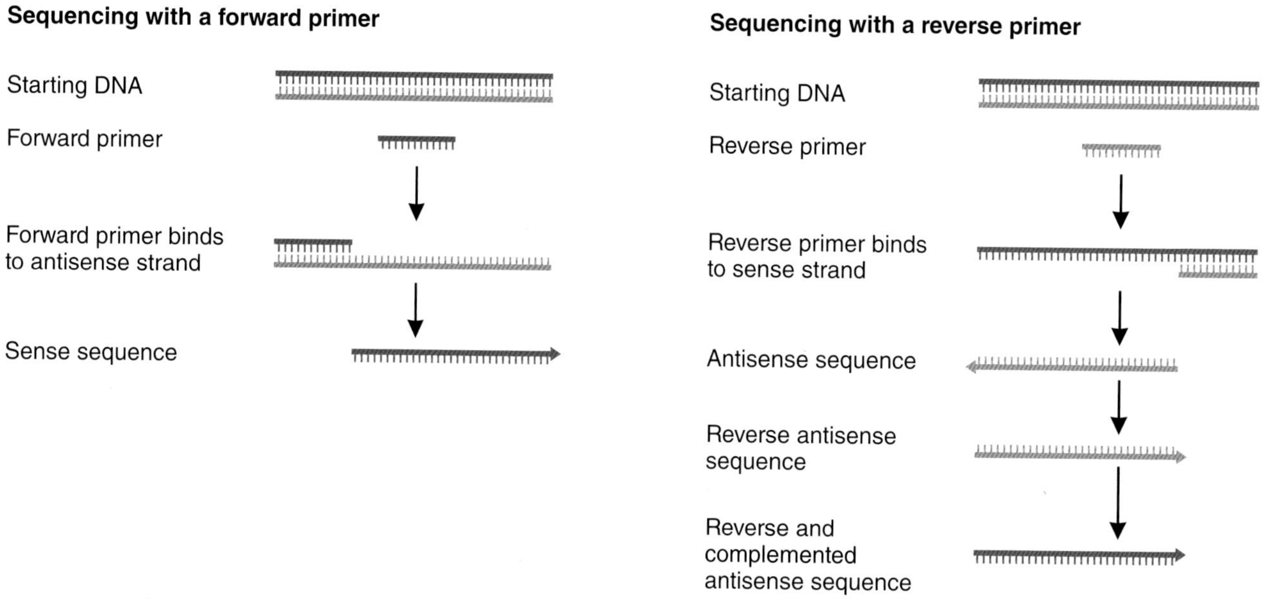

Sequencing with a forward primer

Starting DNA

Forward primer

Forward primer binds to antisense strand

Sense sequence

Sequencing with a reverse primer

Starting DNA

Reverse primer

Reverse primer binds to sense strand

Antisense sequence

Reverse antisense sequence

Reverse and complemented antisense sequence

Figure 16.4. Sequencing with forward versus reverse primers. Left: a forward primer binds to the antisense strand and produces a sense sequence. Right: a reverse primer binds the sense strand and produces an antisense sequence. The sequence will have to be "flipped" (reversed) and then each base will have to be exchanged with the complementary base (complemented) in order to get the final sense sequence.

4. Interpretation

- The newly synthesized fragments are pushed through long, thin, glass capillaries filled with a semiliquid polymer. This is called **capillary gel electrophoresis** and it is used to separate the fragments by size, from smallest (1 base long) to largest (total length of the fragment).
- A laser detects the dye tags associated with each base.
- The fluorescent signals are reported as a **chromatogram.** Software converts each signal peak to the appropriate base call (Fig. 16.3).

Sanger sequencing is accurate for short sections (~500 to 700 bases at a time).

IV. NEXT-GENERATION SEQUENCING. NGS is a nonspecific term to include many different kinds of platforms that are high-throughput and sequence lots of DNA in parallel. This means that NGS systems can identify millions of bases in a single run by sequencing many fragments of DNA simultaneously. Therefore, NGS is useful for sequencing large amounts of DNA, such as the genomes of organisms (whole genomes), actively transcribed genes in response to disease or infection (transcriptomes), or all pathogen genomes in a clinical specimen (microbiomes or viromes). This type of technology has the potential to revolutionize diagnostic testing because a single assay can replace many laborious, slow, insensitive, nonquantitative, and other specialty tests. NGS is typically performed in the following way.

Next-generation sequencing: high-throughput, massively parallel sequencing

1. **Nucleic acid extraction:** For some applications (like metagenomics or for long sequence reads), a significant amount of starting DNA (as much as ~1 to 5 μg) is needed in order to have enough unique sequences and because there is loss of material at each processing step.
2. **Library generation:** This is preprocessing of the input DNA so that it can be read by the sequencer. A "library" is developed by fragmenting the initial DNA and adding tags, called **adapters**, to each fragment (Fig. 16.5).

- **Fragmentation:** Nucleic acids are broken up into many pieces. Depending on the technique used, fragments can be short (100 to 1,000 bp) or very long (5 to 20 kb). Fragmentation can be done in three main ways.
  - Enzymatic: Endonucleases and transposases cut DNA.
  - Mechanical: Sound waves shear the DNA (e.g., acoustic shearing, hydrodynamic shearing, and sonication), or pressure through a tiny hole breaks the DNA into pieces (e.g., nebulization).
  - Chemical
- **Cleanup** or **size selection:** Correctly sized fragments are selected to improve the quality of the results. Other fragments (such as primer or adapter dimers) are excluded.
  - Gel electrophoresis method: The fragments are run on a gel and the bands containing the correctly sized pieces are cut out and purified. This is usually manual.
  - Bead-based method: Solid-phase reversible immobilization (**SPRI**) beads bind to DNA fragments in the presence of polyethylene glycol (PEG). The concentration of PEG controls the size of fragments that are bound. For example, a lower ratio of the bead-PEG solution to DNA will select for larger fragments. The magnetic beads are immobilized with a magnet so that unbound fragments can be washed away. This method can be automated for higher throughput.
- **End repair:** Three enzymes (T4 polynucleotide kinase, T4 DNA polymerase, and Klenow large fragment) are used to blunt the ends of each fragment.
- **Adapter binding:** Adapters are short (20- to 40-bp) nucleotide sequences that are added to the ends of library fragments by an enzyme called **ligase**. Their sequences are platform specific. Adapters are added for the following purposes.
  - Binding to the flow cell during sequencing
  - PCR amplification. Adapter regions can be binding sites for primers so that the sample is enriched for fragments that have bound to adapters.
  - Sorting DNA sequences during data analysis
- **Barcoding:** Barcodes are DNA tags that are used in some libraries to differentiate fragments of DNA. There are two types of barcodes.
  - **Sample barcodes:** a sequence that is inserted into all the fragments of one sample, so that they can be differentiated from fragments in another sample. These DNA tags are often incorporated into the adapters themselves, so they are sometimes called indexed adapters. Sample barcodes allow a sequencing reaction to be multiplexed, whereby several samples with different barcodes can be run simultaneously in a single run, and then sorted by their barcodes during the data processing stage.
  - **Molecular barcodes:** a unique sequence added to every fragment in a library, so that each fragment within a sample can be differentiated from the others. Molecular barcodes allow the user to remove duplicate sequence reads, since fragments with the same molecular barcode all came from the same parent fragment.

3. **DNA amplification:** Many NGS platforms require amplification of the DNA fragments prior to sequencing. PCR amplification increases the amount of

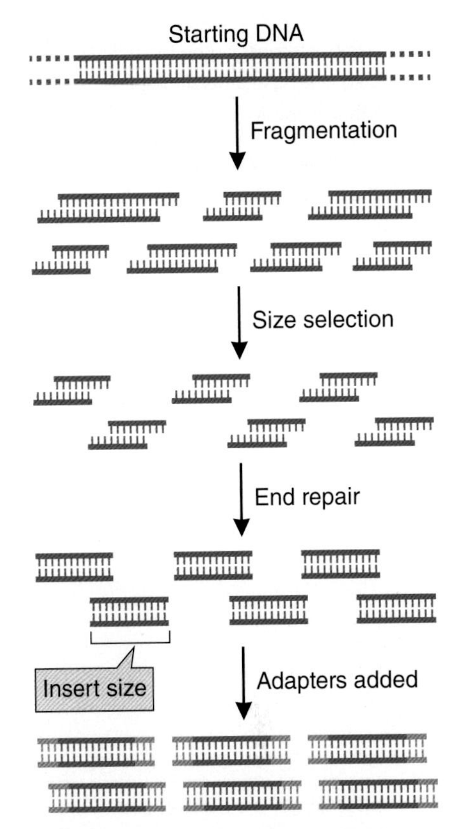

Figure 16.5. Process of generating a library for NGS. Purified DNA is fragmented and enriched for optimally sized fragments. The ends of the fragments are blunted and then attached to adapters.

DNA tags (e.g., adapters or barcodes) are short nucleotide sequences that are incorporated into a DNA fragment and are easily identifiable.

Insert size: size of the DNA fragment
Library fragment size: insert size + adapters and other tags

Sample barcodes: used for multiplexing (sequencing of several samples simultaneously)

Molecular barcodes: used for removing duplicate sequences within a sample

starting material for sequencing and therefore improves specificity of sequencing. However, there are biases that may be introduced because of this step. For instance, low-level point mutations may be lost because of PCR amplification.

4. **Sequencing:** can be performed on several different platforms that have different methodologies, advantages, and limitations (see Table 16.3)

V. NGS PLATFORMS. There are several companies that produce NGS platforms, but they use different processes for sequencing. It is important to understand the chemistry behind each system because it defines whether they are rapid or have a long turnaround time, can read long or short fragments, produce accurate or variable sequences, require preamplification, what amount of DNA they can sequence, and where bias can be introduced (Table 16.3).

1. **Illumina** (MiSeq, HiSeq, NextSeq): This is the most common NGS platform. It produces only short reads (<400-bp fragments) but can sequence huge numbers of fragments simultaneously, so the total amount of DNA sequenced is very large and of good quality. For sequencing to occur, each DNA fragment in the sample is first amplified by bridge amplification. Then, amplicons are sequenced by denaturing them into single strands and synthesizing the complementary strand using tagged nucleotides called reversible dye terminators. Each base that is incorporated into the new strand is recorded to produce the final sequence. The complete process is as follows.

A DNA library is a group of DNA fragments that has adapters.

Bridge amplification forms clusters of each fragment on the surface of the flow cell. Clusters are useful because they magnify the presence of any fluorescence.

- A fragment library is created from the starting template material.
- The library fragments is captured on a flow cell. This is a solid surface containing oligonucleotide fragments that hybridize to the DNA adapters.
- Each of the library fragments is amplified simultaneously by **bridge amplification** (Fig. 16.6). Like classical PCR, the DNA template strands are denatured, annealed to primers, and then the complementary strands are elongated using polymerases. Unlike classical PCR, this type of PCR occurs right on the flow cell, with one end of the DNA fragments being bound to the surface but the free ends being able to bend over

Table 16.3. Comparison of NGS methods

SEQUENCING PLATFORM	GENERATION	METHOD OF AMPLIFICATION	SEQUENCING TECHNOLOGY	LENGTH OF READ[a]	QUALITY SCORE	TOTAL NUMBER OF NUCLEOTIDES SEQUENCED[b]	RUN TIME[c]
Applied Biosystems	Sanger	PCR	Chain termination	Moderate	Q20	500–1,000 b	Very fast
Illumina	NGS	Bridge amplification	Reversible dye terminators	Short	Q30	<15 Gb or 500 Gb (depending on instrument)	Moderate
Ion Torrent	NGS	Emulsion PCR	Voltage change	Short	Q20	~1 Gb	Very fast
PacBio	NGS	None	Single-molecule sequencing	Long	Q10–Q50	<0.1 Gb	Very fast
Oxford Nanopore	NGS	None	Direct single-molecule sequencing	Long	Q10	<1 Gb	Very fast

[a]Short, 100 to 400 bases; moderate, 700 to 1,000 bases; long, >4,000 bases.

[b]b, base; Gb, gigabase.

[c]Moderate, 1 to 3 days; very fast, <6 hours.

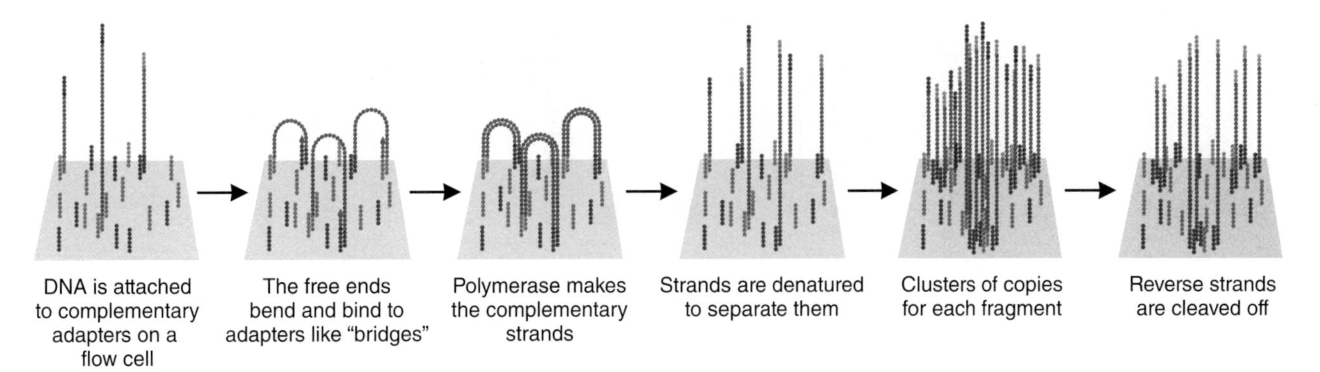

Figure 16.6. Bridge amplification. DNA library fragments are bound to a solid phase. They bend over to form "bridges" during amplification. Groups of amplicons form clusters on the flow cell.

and attach to any free oligonucleotide primers nearby (i.e., the strands form "bridges"). Because of this, clusters of amplicons form around the original template after many rounds of amplification.

- Sequencing occurs by synthesis using **reversible dye terminators**. These are special kinds of nucleotides that are attached to a dye and temporarily terminate any further synthesis in a cycle. This pause allows the sequencer to record the base when it is incorporated. The dye is then removed so that the process can continue with the next reversible dye terminating nucleotide (Fig. 16.7).
    - A primer is added that binds to the top of the fragment.
    - The surface of the flow cell is flooded with just one kind of nucleotide reversible dye terminator. This will be incorporated in any cluster where there is a complementary base. Any further synthesis is terminated temporarily.
    - The excess, unincorporated reversible dye terminators are washed off.
    - A laser activates the dye and a camera takes a picture of the flow cell. If the nucleotide was incorporated into a cluster, the camera detects the signal at that location. The fluorescent tag is chemically removed so that the cycle can be repeated for the next type of nucleotide.
    - A computer converts the optical signals captured by the camera into the appropriate base calls for each cluster.
- With Illumina systems, sequencing can occur from either direction of the DNA fragment (depending on which direction a fragment binds to the flow cell), so it is called **paired-end sequencing**.

Illumina performs **sequencing by synthesis** using reversible terminators.

2. **Ion Torrent** (Personal Genome Machine, Proton): This platform sequences a small amount of total DNA with relatively poor accuracy for an NGS system, but it is very fast and the instrument has a small footprint. It works by measuring tiny changes in pH that occur when new nucleotides are incorporated into a growing DNA strand.
    - A fragment library is created from the starting template material.
    - The fragments are amplified by **emulsion PCR** (Fig. 16.8). For this method, PCR usually occurs on beads that are individually suspended in water droplets. These droplets are like microscopic reaction chambers, and they

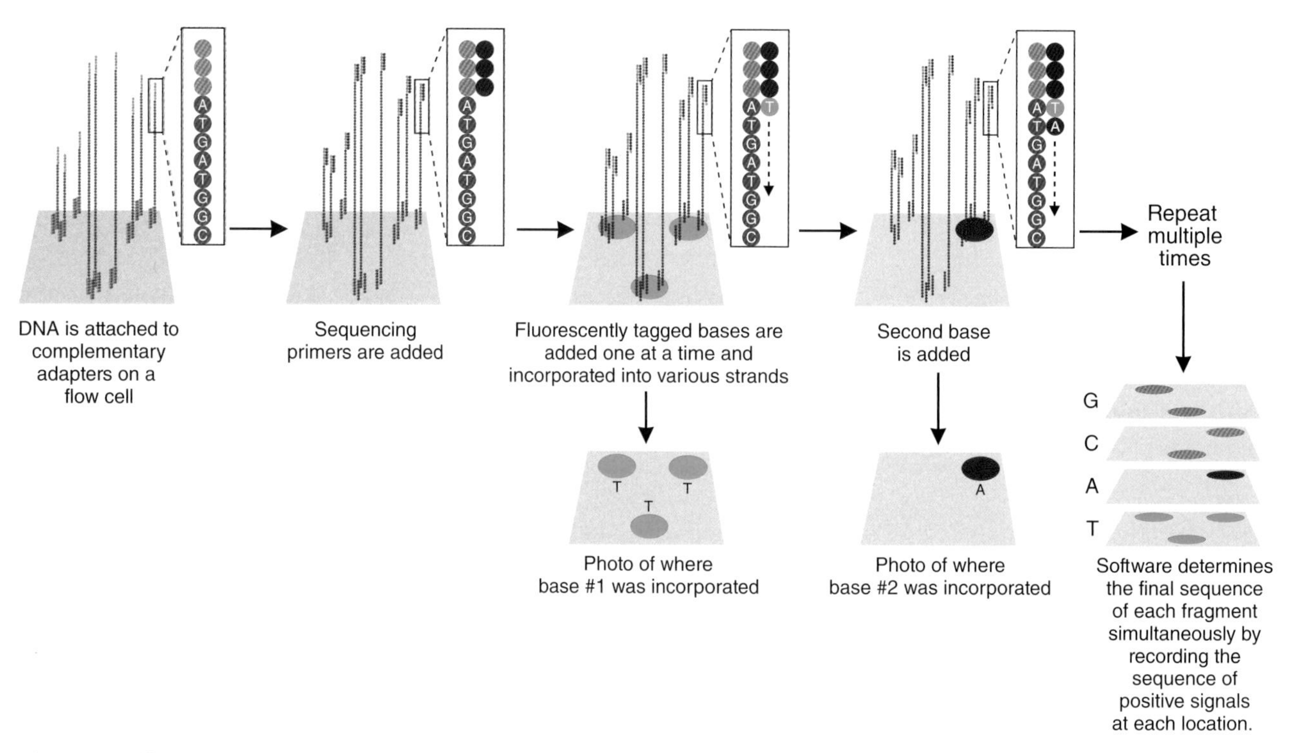

Figure 16.7. Illumina sequencing by synthesis. Reversible terminators are added to DNA clusters and a camera takes pictures to capture the location of any fluorescence that has been incorporated.

allow thousands of PCRs to occur simultaneously in a single tube. Emulsion PCR on beads is performed as follows.

- Tiny capture beads (~1 μm in diameter) are vortexed into a water-oil emulsion so that each one is individually suspended in a water droplet.
- The beads have primers attached to them so that DNA fragments will anneal and elongation can occur.
- The bead is covered with amplified DNA at the end of PCR. This occurs simultaneously in each water droplet and PCR is done on thousands of fragments in parallel.

Emulsion PCR: Thousands of different PCR reactions are performed simultaneously by separating each reaction into its own water droplet within a water-oil emulsion.

- Sequencing is then performed on a microchip using a **pyrosequencing**-based technique (Fig. 16.9). This methodology is named after the release of **pyrophosphate** (two phosphates), which is cleaved off from dNTPs every time one is added to a growing DNA chain. A proton (H^+) is also released with every pyrophosphate. These charged protons cause tiny changes in the pH, which can be measured and represent incorporation of a new dNTP.
  - The beads are washed into tiny wells of the microchip so that there is only one bead per well.
  - DNA polymerases in each well of the chip copy the DNA template that is present on each bead.
  - Only one type of nucleotide is flooded across the chip at a time. As each nucleotide is incorporated, a pyrophosphate and H^+ are produced.

Ion Torrent methods recognize when the next known nucleotide has been incorporated by measuring tiny changes in pH.

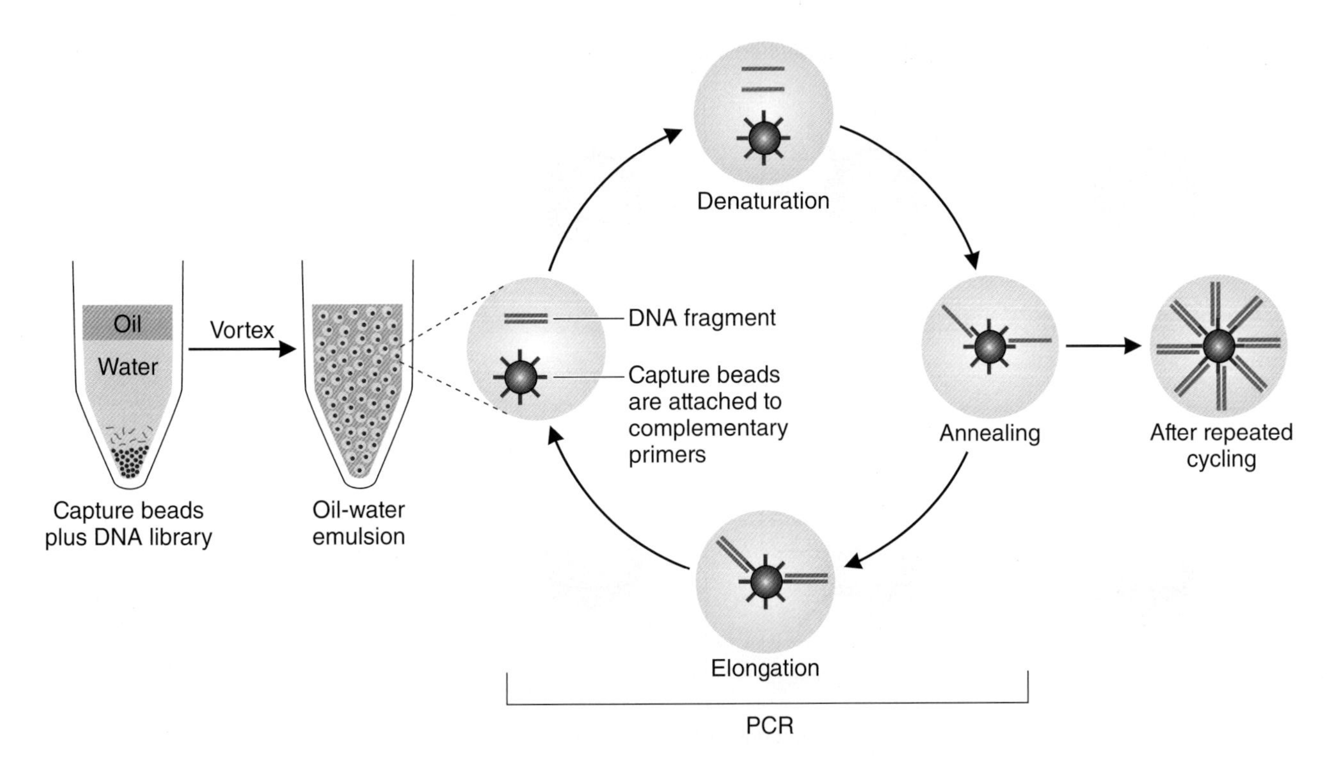

Figure 16.8. Emulsion PCR. Capture beads are emulsified into tiny water droplets within a water-oil emulsion. PCR is performed on each bead.

Each well acts as an individual voltmeter to measure any voltage changes that occur.

3. **PacBio** (SMRT, single-molecule real-time sequencing): This technology uses direct observation of a DNA polymerase as it adds nucleotides to a growing chain (Fig. 16.10). Unlike other common sequencing platforms, the template does not have to be amplified first and the sequence can be determined in real time. In this system, DNA polymerase can produce long reads but the total amount of DNA sequenced is small.
   - DNA is fragmented. A library is created by adding bell-shaped, or hairpin, adapters (called SMRTbell adapters) to each end of the fragments. This makes the whole DNA fragment circular.
   - The genomic library is added to a chip containing thousands of tiny reaction wells called **zero-mode waveguides** so that there is no more than a single circularized fragment in each well. Each of these wells contains a single DNA polymerase.
   - DNA polymerase synthesizes the complementary strand of the circular fragment using fluorescently tagged nucleotides. The fluorescent dyes are attached on the terminal phosphate of each nucleotide so that every time a nucleotide is incorporated, the fluorescent tag is cleaved off and produces a signal. Therefore, unlike with other sequencing methods, all four nucleotides can be added to the chip simultaneously.
   - The zero-mode waveguides are carefully designed to act like high-powered microscopes so that as DNA synthesis occurs, the fluorescent dye signals can be observed directly and interpreted into base calls.

DNA is circularized for sequencing.

Amplification is not needed for PacBio systems.

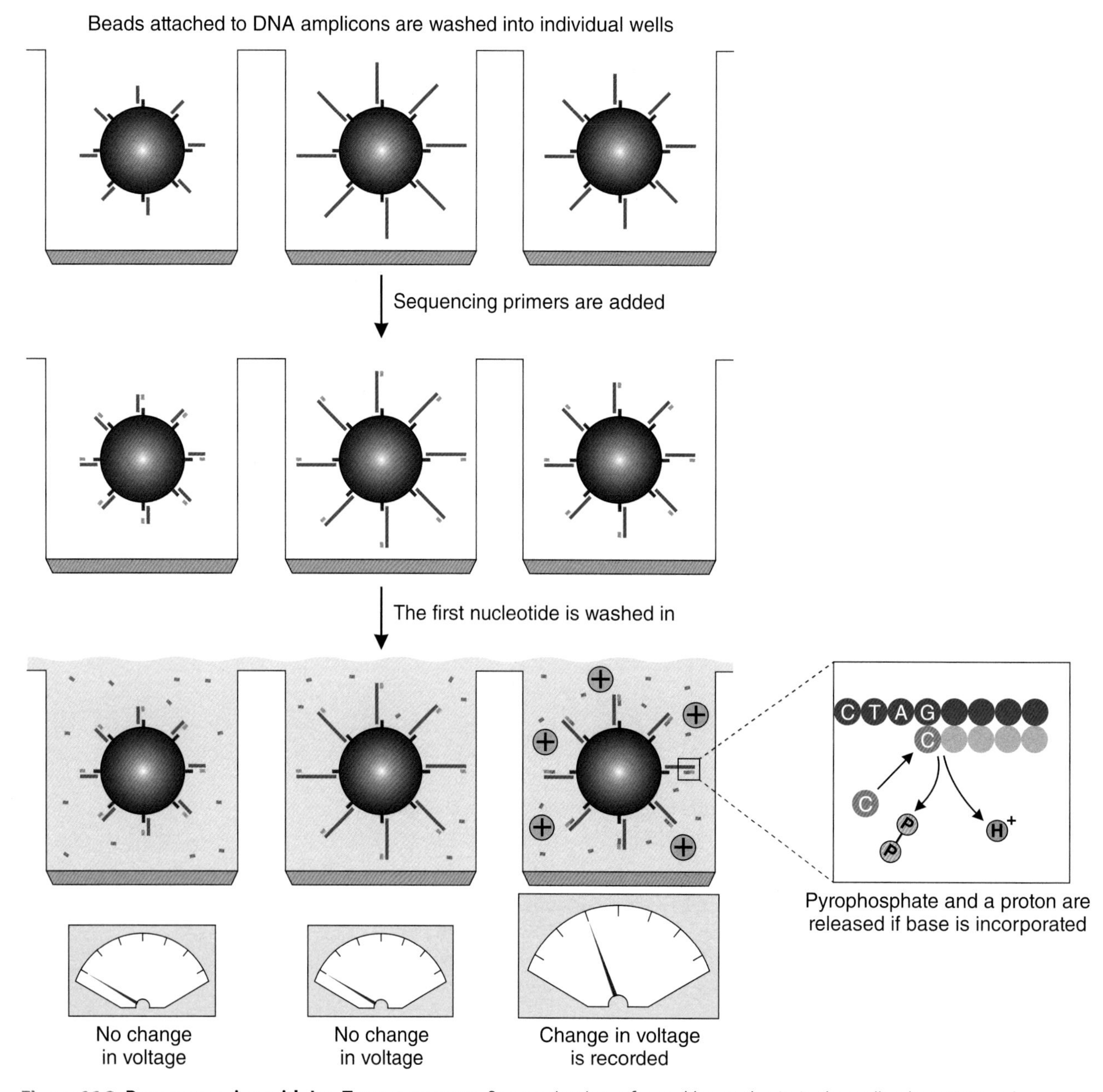

Figure 16.9. Pyrosequencing with Ion Torrent systems. Sequencing is performed by synthesis. In the wells where DNA polymerase incorporates the nucleotide, a pyrophosphate and an H^+ are produced. This causes a slight change in pH, which is measured using a highly sensitive voltmeter.

Single-pass SMRT sequencing has a very high error rate. Multiple-pass sequencing through circular consensus sequencing significantly improves the error rate.

- The error rate of sequencing individual copies is high (>10%). To improve this, **circular consensus sequencing** is used, where circularized fragments between 0.5 and 2 kb in length will cycle through sequencing at least 3 to 5 times. This allows the system to generate a consensus sequence that has good overall accuracy.

4. **Oxford Nanopore** (MinION): a miniaturized, portable sequencing system. Like with the PacBio technology, DNA does not have to be amplified and the individual strands are sequenced and observed in real time. How-

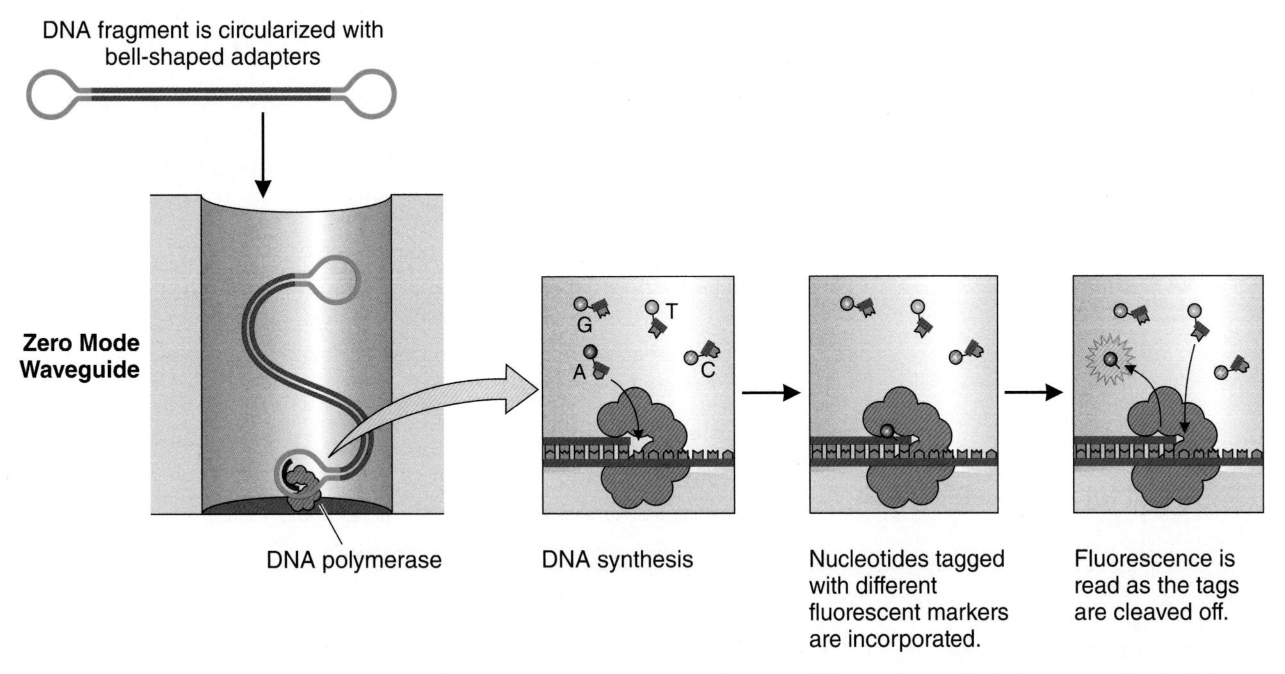

Figure 16.10. PacBio sequencing systems. DNA fragments are circularized using SMRTbell adapters. Each fragment is washed into a well called a zero mode waveguide, which contains a DNA polymerase and all four fluorescently tagged bases. As the polymerase synthesizes the complementary strand, it cleaves off the fluorescent tag, which can be directly detected and converted into a base call.

ever, unlike the other platforms, DNA does not need to be synthesized or cleaved. Rather, the nucleotide sequences are read directly from the strand using changes in electrical current (Fig. 16.11). Because of this, nanopores can produce long reads very fast, but the accuracy is low.

- A library is created by adding one hairpin adapter to the end of the fragments. This connects both the template and the complement strands so that the whole strand can be read without interruption. Another adapter is added to the other free end so that the strand can be captured at the nanopore.
- The DNA library is added to a chip. This chip is a scaffold for hundreds of tiny channels called **nanopores**. Nanopores may be protein, synthetic, or a hybrid of the two.
- A **processive enzyme** is attached to the entry of each pore. This will separate the double-stranded DNA template and feed one of the strands through the pore (the strand of nucleotides will "process" through).
- Each nucleotide that passes through the pore produces a characteristic change in the electric current across the nanopore's surface. These characteristic changes are converted into base calls.

Nanopore sequencing is different from the other methods because there is **no synthesis**. It "reads" the nucleotides directly off the strand, as opposed to recording which nucleotides are added as a complementary strand is being synthesized.

VI. COMPUTATIONAL DATA ANALYSIS. Both Sanger sequencing and NGS data need to be analyzed and interpreted. Analysis for NGS is considerably more complex because these techniques produce enormous amounts (i.e., terabytes) of data.

1. **Input:** Raw sequence data are usually stored in 2 main formats (Fig. 16.12).
   - **FASTA** is a text file that contains a header/description line and begins with ">." The nucleotides are displayed in sequence as single letters

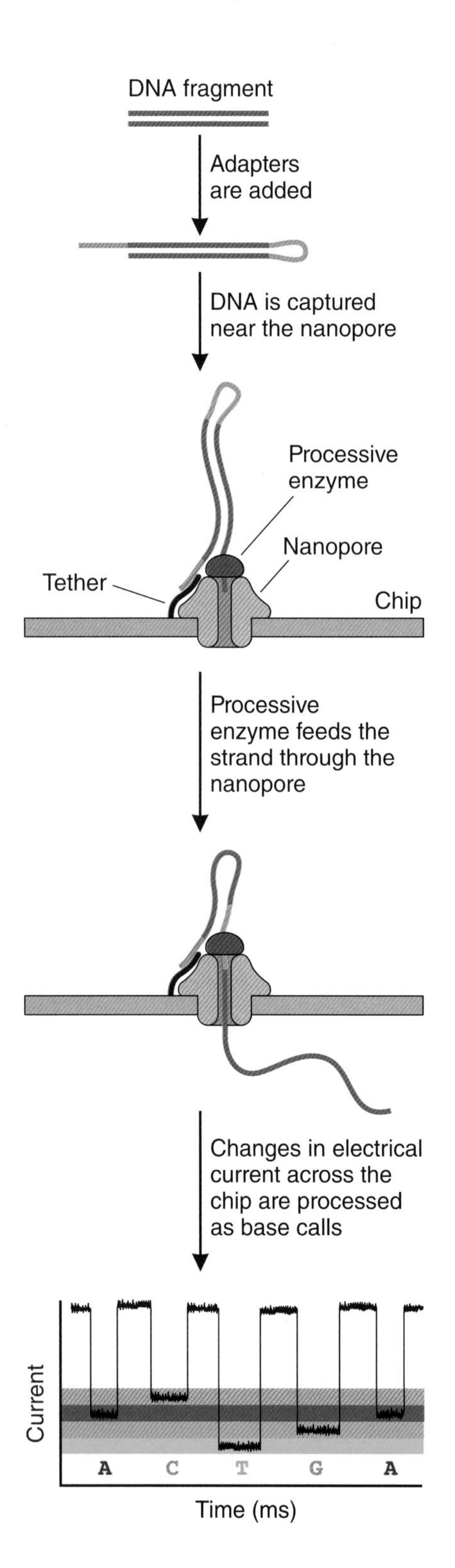

starting on the next line. This format is very simple for raw sequences and contains no formatting (not even spaces or line numbers).

- **FASTQ** is a compact text format that has a total of four lines. These four lines contain sequence data and their Phred quality.
    - Line 1: the header/description line, which begins with "@"
    - Line 2: nucleotide sequence
    - Line 3: another descriptive line that begins with a "+"
    - Line 4: a string of characters, one for each nucleotide. Each character is a code for the quality of the corresponding base call. Characters from the American Standard Code for Information Interchange (ASCII) +33 are typically used.

2. Data preprocessing or **cleanup**

- Fragment sequences are sorted by barcodes, if applicable.
- Adapter sequences are trimmed.
- Low-quality sequence reads are removed.
- Nontarget reads are eliminated. For example, if DNA from a whole specimen was used, then reads that map to the human genome are subtracted.

3. Data processing

- **Alignment** (or read mapping): when a sequence is compared (or mapped) to reference sequences. This is often used to identify the organism or identify variants (e.g., point mutants, deletions, insertions, and rearrangements).
    - Public databases: new sequence data are usually compared to known sequences in huge databases, such as **GenBank** (from the National Center for Biotechnology Information [NCBI]), the European Molecular Biology Laboratory (EMBL), and the DNA Databank of Japan (DDBJ).
    - Common software that performs alignments include the following.
        - **BLAST** (Basic Local Alignment Search Tool)
        - SAMtools
        - BWA (Burrows-Wheeler Aligner)
        - Bowtie2
        - SOAP (Short Oligonucleotide Analysis Package)
    - "Dark matter": sequences that cannot be taxonomically assigned because they do not match anything in the database. This is a significant problem for viruses because they are relatively diverse and unknown.
- **Assembly:** Shorter sequences are assembled in the correct order to create a larger sequence. This is usually done to build genomes that have not been sequenced before. Assembly can be done in several ways.
    - **Reference sequence:** Shorter sequences are stitched together based on homology to a reference sequence (for example, against the genome of a similar species).

Figure 16.11. Nanopore sequencing. Adapters are added to linearize the complementary strands. Another adapter tethers the template DNA near the nanopore. A processive enzyme feeds the DNA fragment into the channel and characteristic changes in current across the pore are registered as unique base calls.

	FASTA	FASTQ
Line 1	>GENESEQ	@GENESEQ
Line 2	AGTCCTTGGAATACCCGGAA	AGTCCTTCCAATACCCGGAA
Line 3		+
Line 4		;!;!;;;9;7;~.7;393D3

Figure 16.12. Example of FASTA and FASTQ formats for sequence data.

- ***De novo* assembly:** when a novel sequence of DNA is stitched together strictly using overlapping segments, without the benefit of a reference sequence. This requires significant time and computational power and is usually used when sequencing totally novel viruses.
- **K-mers:** often used for assembling short reads. Algorithms will break up a sequence into shorter sequences of length "k," which are then aligned together based on their overlap of k-1 substrings (using a De Bruijn graph).
- Common software used for assembly
    - Velvet (e.g., Illumina)
    - SPAdes (e.g., Illumina and PacBio)
    - MIRA (e.g., Ion Torrent)
    - CLC Genomic Workbench

4. **Postprocessing:** Sequence data can be stored in 2 main formats.
    - **SAM:** stores a human-readable version of the processed sequence data
    - **BAM:** the binary version of the SAM format
5. **Data visualization:** Specific programs are used to visualize the data. They can be used for interpretation or identification, to identify issues, to identify low-quality base calls, and to see where reads align to other sequences.

Multiple-Choice Questions

1. Which of the following results in increased throughput on NGS platforms compared to Sanger sequencing?

a. NGS uses capillary gel electrophoresis.

b. NGS is significantly faster.

c. NGS processes much more DNA in parallel.

d. NGS requires more computational analysis.

2. What is sequence alignment?

a. Assembling *de novo* genomes

b. Cleanup of sequence data by removing extraneous sequences

c. Matching a newly generated sequence with a reference sequence

d. Choosing the correct sequencing platform

3. What is emulsion PCR?

a. PCR that occurs in an oil-water emulsion

b. Sequencing that occurs in water and then oil

c. PCR that occurs using bridge amplification

d. All of the above

4. Bridge amplification is used by which sequencing platform?

a. Illumina
b. Ion Torrent
c. Oxford Nanopore
d. Sanger sequencing

5. Which method is used to improve the accuracy (Phred quality score) of sequencing results?

a. Reversible dye terminators
b. Chain termination
c. Sequencing by synthesis
d. Circular consensus sequencing

6. Which step must occur with NGS that does NOT need to occur in Sanger sequencing?

a. DNA extraction
b. Library generation
c. Data analysis
d. None of the above

7. Once the library has been generated, which of the following systems is portable?

a. Illumina
b. Applied Biosystems
c. Oxford Nanopore
d. Ion Torrent

8. Which of the following systems uses pyrosequencing?

a. Illumina
b. Oxford Nanopore
c. PacBio
d. Ion Torrent

9. PacBio and Oxford Nanopore are both single-molecule systems that do not require amplification prior to sequencing. How are these systems different?

a. DNA polymerase is attached to the pore in nanopore systems and to the zero-mode waveguides in PacBio systems.
b. DNA polymerase is absent in nanopore systems.
c. Nanopore systems typically have higher Phred scores.
d. None of the above; they are the same except that nanopore systems produce longer reads.

10. What is a sequence chromatogram?

a. A gel showing sequence bands
b. The location of "clustered" sequences in Illumina platforms
c. The output from Applied Biosystems sequencers showing peaks of fluorescence
d. The sequence of colors from zero-mode waveguides

True or False

11. "Dark matter" is DNA that cannot be physically sequenced. **T F**

12. NGS can be used for 16S rRNA sequencing. **T F**

13. NGS is not a very useful tool for metagenomics studies. **T F**

14. Alignments are more computationally challenging than *de novo* assembly. **T F**

15. The universal gene used for Sanger sequencing of all viruses is the long terminal repeat region. *(May require reading outside of this chapter.)* **T F**

SECTION IV

PREVENTION AND MANAGEMENT OF VIRAL INFECTIONS

17

CHAPTER 17

BIOSAFETY

I. OVERVIEW. Pathogens may cause a range of effects on host populations, from local outbreaks to global pandemics. Health care workers have an occupational risk of exposure to infectious viral pathogens. There are several measures that can be taken to protect hospital and laboratory staff, such as personal protective equipment (PPE) and specialized laboratory handling procedures. Infection control measures, such as isolation and decontamination, prevent viral pathogens from spreading to health care workers and hospital patients.

II. BIOSAFETY CATEGORIZATION. Viral pathogens are categorized in several different ways depending on their level of infectivity, whether they are potential agents of bioterrorism, or if they need to be monitored for public health purposes (29).

1. **Biosafety levels (BSLs):** Viral and bacterial pathogens are classified into BSLs 1 through 4 based on how pathogenic, easily spread, and treatable they are (Table 17.1). Hospitals and laboratories can use this classification to develop policies for pathogens based on the level of risk to employees and patients. For example, most clinical laboratories maintain BSL2 practices, which means that they are equipped to handle moderate-risk pathogens. Because of this, clinicians should notify laboratory staff if specimens or isolates are higher risk, so that labs can consider quarantining the specimen, sending it to a high-risk laboratory, and/or performing minimal testing.

Box 17.1. Important terms describing the distribution of disease

1. Sporadic: rare occurrences of a disease in an area
2. Endemic: the normal prevalence of a disease in an area
3. Outbreak: a sudden rise (greater than expected) in disease cases in a limited area
4. Epidemic: a rise (greater than expected) in the number of cases in a larger area
5. Pandemic: a widely distributed epidemic

Table 17.1. Description of Biosafety Levels (BSLs) (29)

BSL	HAZARDOUS MATERIAL	LABORATORY PRACTICES	EXAMPLES[a]
BSL1	Agents that have very **low risk** of causing disease	Standard laboratory practices and equipment are used. Work can be done on an open bench. Access to the lab is limited.	Free nucleic acids or nonpathogenic viruses
BSL2	**Moderate-risk** agents that are capable of causing moderate or serious disease but are **not** readily transmitted by inhalation	Most work can be done on an open bench, but PPE and hoods are used for specimens when aerosols are generated. Equipment is routinely decontaminated, lab personnel have specific training in handling pathogens, and access to the lab is controlled.	Most viruses, including dengue virus, HBV, HCV, HIV, and viruses in the *Herpesviridae* family
BSL3	Pathogens with **high risk** of serious disease	All work is performed in hoods, and workers wear additional PPE. All waste must be decontaminated when it leaves the lab. The facility is specially designed: it is under negative pressure, exhaust is HEPA filtered, there is an anteroom, and access is restricted.	EEEV, WEEV, VEEV, Japanese encephalitis virus, yellow fever virus, Rift Valley fever virus, St. Louis encephalitis virus, and hantaviruses
BSL4	Dangerous agents with a **high risk of fatal disease,** are readily transmissible (e.g., inhalation), and have no treatment or vaccines	Personnel are highly trained and access is tightly restricted. All work is done within a class III biosafety cabinet or with extensive PPE (i.e., positive-pressure suits). A shower is taken on exit, with a change of clothes.	Ebola virus, CCHFV, Lassa virus, and Marburg virus

[a]See abbreviations list at the front of this book for virus name expansions.

2. **Reportable pathogens** are agents that must be reported to public health agencies if detected in a patient. These pathogens are unusual, are high risk, or can cause significant outbreaks, so their prevalence is monitored on a statewide and/or national scale.

3. **Select agents** are pathogens or toxins that can have a major impact on national security, public health, and safety. They are typically a subset of reportable pathogens and may be used as potential agents of bioterrorism. They are under strict regulations regarding possession and transport. Once identified, diagnostic laboratories are required to either transfer or destroy the pathogen within 7 days (http://www.selectagents.gov).

4. **Laboratory response network:** a network of public health and other (e.g., military) laboratories across the country that are equipped to provide an effective response to biological and chemical threats.

5. **Sentinel laboratories:** any laboratory that is in a position to recognize potential threats (e.g., clinical laboratories). The role of sentinel laboratories is to recognize and escalate potential threat agents to the Laboratory Response Network. These labs must be able to comply with requirements regarding shipping and destruction of the pathogen.

Reportable agent: a pathogen that is monitored by public health facilities

Select agent: a potential agent of bioterrorism

III. ISOLATION AND QUARANTINE. Infection control can place patients with transmissible infections under isolation precautions and PPE restrictions in order to prevent transmission of infectious agents to other patients and staff. The most common types of isolation precautions are **standard, contact, droplet,** and **airborne** isolation (Table 17.2). Different levels of isolation are used depending on how the pathogen is transmitted and the type of exposures that can occur (Table 17.3).

Table 17.2. **PPE for different isolation precautions**

PRECAUTION STATUS	PPE	NOTES
Standard precautions	Gloves	Includes hand hygiene for basic protection against transmission. Gloves are used when needed; for example, when the area is visibly soiled.
Contact precautions	Gloves, gown	Used for organisms that are transmitted by direct or indirect (e.g., fomites) contact
Droplet precautions	Gloves, gown, mask	Used for organisms transmitted over small distances through droplets. Patients should have single rooms or separation of ≥3 feet from other patients.
Airborne precautions	Gloves, gown, N95 respirator	Used for organisms that may be transmitted over a long range through fine, aerosolized particles that stay suspended in the air

Table 17.3. **CDC-based recommendations for isolation precautions of patients with viral infections (31)**

VIRUS TYPE OR VIRUS-INDUCED SYNDROME	TYPICAL/RECOMMENDED ISOLATION PRECAUTIONS[a]	EXAMPLES OF VIRUSES
Arboviruses	S	EEEV, WEEV, VEEV, St. Louis encephalitis virus, California encephalitis virus, West Nile virus, dengue virus, yellow fever virus, Colorado tick fever virus
Zoonotic viruses	S	Hantavirus, rabies virus
Gastrointestinal viruses	S or C, if patient is diapered or incontinent	Adenovirus, norovirus, HAV, enteroviruses, rotavirus
Respiratory viruses	C and D, or D alone	Adenovirus, rhinovirus, influenza virus, mumps virus
	C	HMPV, RSV, parainfluenza virus
	A, D, and C	SARS virus, MERS virus
Hepatitis viruses	S	HAV to HEV
Viruses that cause exanthems	S	Enteroviruses and parechoviruses
	S; add **A** and C if patient is immunocompromised or has disseminated disease	VZV (chickenpox and shingles)
	A	Measles virus
	D	Rubella virus (exception: neonatal infection)
	D	Parvovirus B19
	A and C	Variola (smallpox) virus
	C, until lesions are crusted over	Vaccinia virus, monkeypox virus, disseminated or severe HSV mucocutaneous disease
	S	Molluscum contagiosum
Viruses that cause conjunctivitis	C	Adenovirus, enterovirus 70, coxsackievirus A24
Viruses that cause neonatal/congenital disease	C	HSV, rubella virus
Viruses that cause paralysis	C	Enteroviruses (poliovirus, enterovirus D68)
Viruses that cause latent/persistent infection	S	HIV, CMV, EBV, most cases of HSV
Viruses that cause hemorrhagic fevers	S, D, and C	Lassa virus, Ebola virus, Marburg virus, CCHFV

[a]A, airborne; C, contact; D, droplet; S, standard precautions.

Table 17.4. Description and function of biosafety cabinets (BSCs) (29)

TYPE	DESIGN	PROVIDES PROTECTION FOR
Dead-air box (also known as a PCR workstation)	• Does not have airflow in order to prevent contamination of materials via air currents	Materials
BSC[a] class I	• Air is drawn in through the open front. • Air is **exhausted** through a HEPA filter.	Personnel
BSC class II	• There are 4 distinct types, with different amounts of air being vented and recirculated. • Air is drawn in through a **front grill** and filtered through a HEPA filter, and clean air is pushed down onto the materials. • Air is exhausted through a HEPA filter down onto the specimen.	Personnel and materials
BSC class III (also known as a glove box)	• Is **totally sealed** from the room. Specimens are handled through permanently attached gloves. • Air is taken in through a HEPA filter. • Air is exhausted through **two** HEPA filters.	Personnel and materials (maximum protection)

[a]BSC, biosafety cabinet.

IV. PERSONAL PROTECTIVE EQUIPMENT. PPE includes personal barriers like gloves, gowns, masks, or respirators and is used to protect health care personnel or lab staff from patients or lab materials (30) (see Table 17.2). The sequence of putting PPE on and taking it off is very important, because an incorrect sequence can lead to exposure and transmission of hazardous organisms.

1. **Donning:** the process of putting PPE on (use "bottom up" rule): gown, mask/respirator, goggles, and gloves last
2. **Doffing:** the process of taking PPE off. In general, the most contaminated piece (e.g., gloves) is taken off first.
   - Technique 1 (use alphabetical rule): gloves, goggles, gown, and mask/respirator. Wash hands.
   - Technique 2 (use "dirtiest first" rule): gloves and gown together, goggles, and mask/respirator. Wash hands.

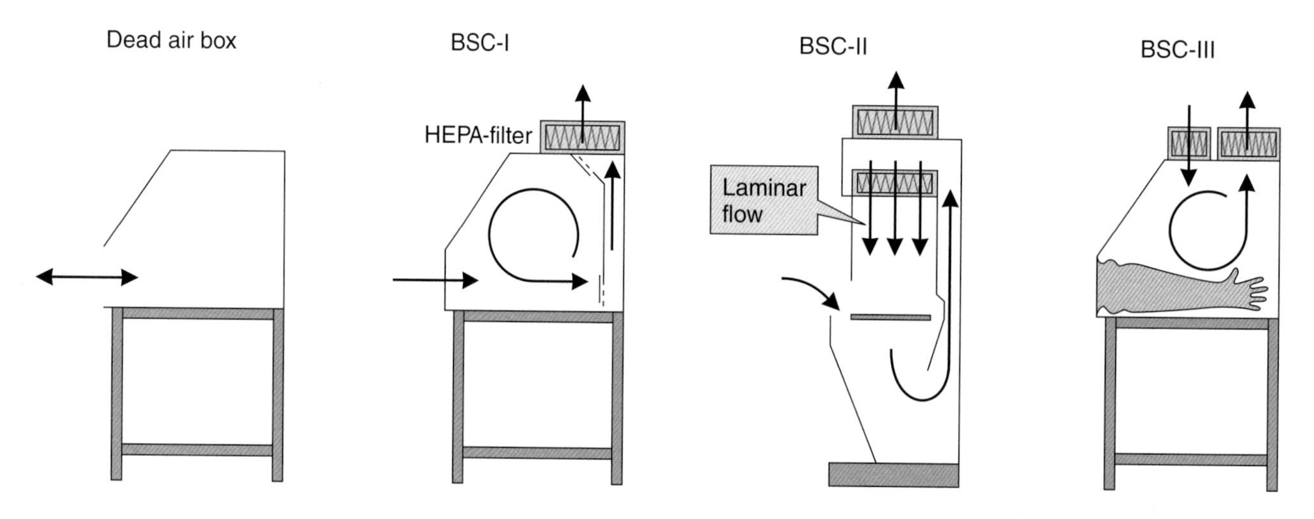

Figure 17.1. Comparison of biosafety cabinets (BSCs). Arrows indicate direction of airflow.

V. BIOSAFETY CABINETS. BSCs, or "hoods," are pieces of lab equipment that provide controlled and protective environments to handle reagents, assays, and samples. There are 3 main types, with differences in airflow and amount of protection provided (Table 17.4 and Fig. 17.1). Most human specimens are handled in a biosafety cabinet under at least BSL2 conditions because of their potential to harbor infectious agents.

Multiple-Choice Questions

1. **What is the best description of a select agent?**
   a. Reportable agents
   b. Potential agents of bioterrorism
   c. BSL4 agents
   d. Agents without any cure
2. **Yellow fever virus can cause fever, viremia, diarrhea, and "black vomit" and can even be fatal. However, it requires only standard isolation precautions in a hospital. Why?** *(May require reading outside of this chapter.)*
   a. It rarely causes severe disease.
   b. It is treatable.
   c. It is not transmitted person-to-person.
   d. All of the above
3. **Most viruses that are worked with in diagnostic laboratories are categorized within which BSL?**
   a. 1
   b. 2
   c. 3
   d. 4
4. **Why isn't HIV handled under BSL4 conditions?**
   a. There is absolutely no risk of exposure to lab personnel.
   b. There is a vaccine.
   c. It does not cause serious or fatal disease.
   d. It is not readily transmitted (e.g., not transmitted by the respiratory route).
5. **Inpatients infected with most respiratory viruses require droplet isolation in addition to other precautions. Which of the following respiratory viruses requires only contact precautions?**
   a. Influenza virus
   b. Mumps virus
   c. Rhinovirus
   d. RSV
6. **Which of the following viruses requires airborne isolation?**
   a. Influenza virus
   b. Parainfluenza virus
   c. Measles virus
   d. Parvovirus B19

True or False

7. Inpatient isolation precautions are used to protect the patient. **T F**

8. Airborne and droplet isolation precautions are the same. **T F**

9. All patient specimens should be treated as potentially infectious. **T F**

10. During doffing, gloves are the first article of PPE that should be removed. **T F**

18

CHAPTER 18

VACCINES

I. TYPES OF PROTECTIVE IMMUNE RESPONSES. To date, vaccines are the best defense against viruses (in fact, most vaccines are designed to be used against viruses). Most vaccines work by inducing an active immune response. This is because the resulting antibody response is highly specific to a pathogen and provides long-lived protection. Some viruses can be neutralized by passive transfer of preformed antibodies. This is a less robust form of protection but is useful in cases where individuals cannot produce their own antibody response rapidly enough. Finally, a few vaccines utilize cell-based immunity in order to stimulate both the B-cell and T-cell arms of the immune response.

1. **Natural infection:** typically elicits the most robust immune response because it results in maximum antigen exposure and activates both humoral and cellular responses. However, it is a potentially dangerous way of acquiring immunity because of the morbidity and mortality associated with viral infections.
2. **Active immunity:** production of an active antibody response against an antigen. This is sometimes called the **humoral response**.
   - **Naive individuals** have not been exposed to a particular pathogen and do not have antibodies against it.
   - In the humoral response, IgM antibodies are typically generated first, between 1 and 3 weeks after exposure. However, these do not persist long term.
   - IgG antibodies are typically generated between ~2 and 4 weeks after exposure and are highly specific and long-lived. Vaccines rely on production of IgG to be effective.
   - Antibodies can inactivate pathogens by the following methods.
     - **Neutralization:** Neutralizing antibodies are highly protective because they coat the virus and prevent it from infecting cells (viruses are "neutralized").
     - **Complement fixation:** Antibodies bind the virus and trigger proteins called "complement" to lyse infected cells and recruit phagocytic cells.
     - **Opsonization:** Antibodies bind the virus to mark them for phagocytosis.

"Chickenpox parties" (and other similar virus "parties") were used to intentionally expose children to contagious viruses. These are not recommended because of the potential severity and risk of disease.

Antigens are substances (usually proteins) that elicit an immune response.

Antibodies are also called immunoglobulins (Ig).

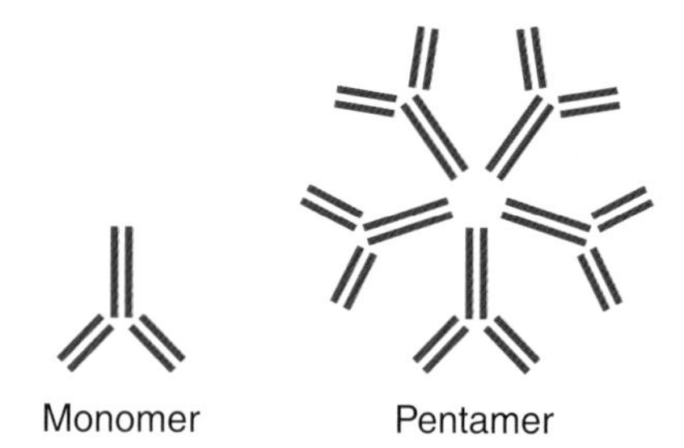

Figure 18.1. Arrangement of antibodies in monomers (IgG) or pentamers (IgM).

Infants are protected by transplacental IgG antibodies for ~6 months after birth.

- Active immunity results in long-lived **memory responses**. Memory B cells are highly persistent and produce neutralizing antibodies rapidly upon any subsequent exposure to the pathogen.

3. **Passive immunity:** the process of transferring preformed antibodies into a naive host. These antibodies can neutralize pathogens, but protection is transient because it lasts only as long as the antibodies are present.
   - **Maternal antibodies:** antibodies produced during pregnancy that provide protection to the fetus and neonate.
     - Maternal IgA is secreted into breast milk and provides mucosal protection against gastrointestinal pathogens (32).
     - IgM antibodies are **pentamers** and are able to bind up to 10 epitopes (Fig. 18.1). As such, they are too large to cross the placenta.
     - IgG is a **monomer** that can bind up to two epitopes (Fig. 18.1). It is small enough to pass through the placenta and can provide passive immunity to the fetus.
     - Maternal IgG antibodies made in the second and third trimesters are transported across the placenta and circulate in the neonate's blood. They protect the fetus through gestation and up to ~6 to 8 months after birth (33, 34). For this reason, immunization against several viruses (inactivated vaccines only) is recommended during pregnancy.
   - **Intravenous immunoglobulin (IVIG):** an intravenous solution of immunoglobulins that are pooled from the plasma from several thousand healthy donors
     - Plasma from healthy individuals contains circulating antibodies against many viruses and can be used to protect people who are unable to fight an infection (e.g., transplant patients).
     - Examples: human rabies immunoglobulin (anti-rabies) and hepatitis B immunoglobulin (anti-HBV)

4. **Cellular immune response:** activation of CD4$^+$ (helper) and CD8$^+$ cytotoxic (killer) T cells. Vaccines that use this cellular response attempt to trigger a broadly activating response because helper T cells stimulate B cells, other T cells, and macrophages. Killer T cells selectively target and destroy infected cells.

II. VACCINE TYPES. A vaccine should be broad enough to cover multiple strains of a viral pathogen and specific enough that it does not cross-react with any self-antigens (35). There are six main types of vaccines, with various levels of immunogenicity and risk (Fig. 18.2 and Table 18.1).

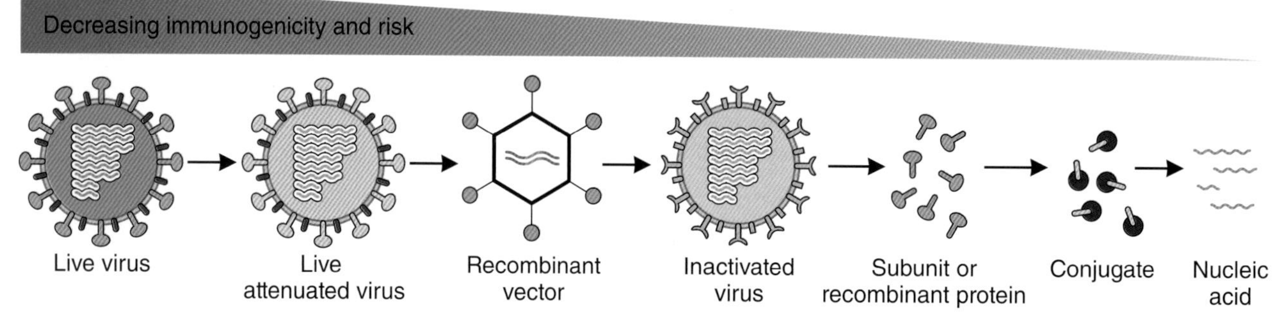

Figure 18.2. Vaccine types in order of decreasing immunogenicity and risk of causing infection.

1. **Live attenuated vaccines** contain live virions that have been modified to be less pathogenic.
   - Virus strains can be attenuated by passage in nonhuman cells. After multiple rounds of replication, viruses select for mutations that allow them to grow better in the new cell type and poorly in human cells. Strains that are stable and have decreased virulence in humans can be used for immunization.
   - Live attenuated viruses simulate a true infection and therefore tend to elicit highly robust, multifactorial immune responses.
   - Importantly, replicating viruses, even attenuated ones, carry a risk of actual infection.
     - Immunization with attenuated vaccine strains is usually asymptomatic in individuals with healthy immune systems. They may rarely cause a mild infection.
     - In immunosuppressed individuals, an attenuated virus may cause disease that can result in significant morbidity and mortality.
     - Replicating virions, even if they are attenuated, can be shed from healthy recipients and can spread to at-risk individuals.

Immunosuppressed, pregnant, elderly (>65 years), and very young (<6 months) persons should not get live attenuated vaccines.

2. **Inactivated vaccines** contain whole virions that have been inactivated by chemicals, heat, or radiation. These vaccines are highly antigenic because they are made from live virions, but they do not replicate. As a result, they typically induce only antibody (and not cellular) responses and cannot cause infection.

For inactivated "killed" vaccines, remember: "**P**op and **I killed A R**abid **J**apanese **T**ick" (polio, influenza, hepatitis A, rabies, Japanese encephalitis, tick-borne encephalitis)

3. **Subunit vaccines** contain only the antigenic portion of the pathogen. This minimizes the risk of nonspecific immune activation but decreases the overall breadth of the immune response.
   - A pathogen gene encoding an antigenic protein is recombined into another organism, like a bacterial or yeast cell, so that high levels of the protein can be churned out. So these are sometimes called **recombinant** or recombinant subunit vaccines.
   - These proteins can also be purified from viral preparations.

Recombinant (subunit) vaccines are different from recombinant vector vaccines.

4. **Live recombinant vector vaccines:** Genes from the target virus are recombined into a viral vector (e.g., a carrier virus). These hybrid viruses are highly antigenic because they simulate a real viral infection. However, they do not cause the disease because the vector and target viruses are not complete viruses.
5. **Conjugate vaccines** are used when the pathogen contains only poorly antigenic targets (such as carbohydrates). These weakly antigenic targets are conjugated to other antigens that are known to stimulate the immune system vigorously.
6. **DNA vaccines:** a newer, uncommon type of vaccine that is designed to stimulate cell-mediated immunity instead of humoral immunity. These vaccines comprise nucleic acids (DNA) instead of proteins or conjugated carbohydrates. When this DNA is delivered into a cell, new proteins are translated from it; these are then processed and displayed on the surface of the cell. This display activates T cells, so the host develops a response to kill the infected cells at the start of viral replication (unlike humoral immunity, which neutralizes the viruses once they are already produced).

Table 18.1. Vaccines against viruses (36–39)

AGENT(S)	COMMON VACCINE NAME	INDICATION	VACCINE TYPE	NUMBER OF DOSES	DELIVERY	COMMENT(S)
Adenovirus (40)		Military personnel	Live attenuated	2 tablets, each containing a different serotype	**Oral**	Vaccinates against serotypes 4 and 7
Dengue virus (41)		For use only in areas of endemicity. Not licensed in the USA.	Live recombinant vector	3	Subcutaneous	Contains all 4 serotypes. Can cause **ADE**.
Hantaviruses (42)		Available only outside the USA (China, Korea)	Inactivated	3; first two 1 month apart, 3rd after 1 year	Intramuscular	Contains an adjuvant
HAV	HepA	Childhood, routine; unvaccinated travelers to a region of endemicity	Inactivated	2	Intramuscular	
HBV	HepB	Childhood, routine	Subunit	3 (new 2-dose series exists for >18-year-olds)	Intramuscular	Vaccine target is **sAg**. Vaccination also prevents HDV infection.
HEV (43)		Only licensed in **China** for people >16 years old	Subunit	2	Intramuscular	
HPV	HPV2, -4, or -9	Late childhood, routine; unvaccinated sexually active adults	Subunit	2 or 3, depending on brand	Intramuscular	Vaccine target is the **L1** protein. Vaccinates against two, four, or nine different serotypes. All formulations include strains HPV16 and 18.
Influenza virus	Trivalent, inactivated	Yearly	Inactivated	1; high-dose formulations available for ≥65 years old	Intramuscular	Vaccinates against two A strains and one B strain
	Quadrivalent, inactivated	Yearly	Inactivated	1	Intramuscular or intradermal for those 18–64 years old	Vaccinates against two A and two B strains
	Recombinant (trivalent)	Yearly	Inactivated	1	Intramuscular	Vaccinates against two A strains and one B strain
	LAIV (quadrivalent)	Yearly	Live attenuated	1	**Intranasal**	Vaccinates against two A and two B strains
Japanese encephalitis virus (44)		Travel to Asia for >1 mo; routine in areas of endemicity	Inactivated; live attenuated vaccines available outside the USA	2, 1 mo apart	Intramuscular or subcutaneous	

Measles, mumps, and rubella viruses	MMR	Childhood, routine	Live attenuated	2	Subcutaneous	Rare: can cause congenital rubella syndrome if administered during pregnancy
Poliovirus	Inactivated poliovirus (e.g., Salk)	Childhood, routine (used in areas of low endemicity)	Inactivated	4	Intramuscular or subcutaneous	Vaccinates against all 3 serotypes
	Oral poliovirus (e.g., Sabin)	Childhood, routine (used **only in areas of endemicity**)	Live attenuated	4	**Oral**	Vaccinates against all 3 serotypes. May cause **vaccine-derived polio**.
Rabies virus		People with occupational risk (e.g., **veterinarians**); international travelers to high-risk areas.	Inactivated	3, for preexposure 2, for postexposure with prior vaccination 4, for postexposure and no previous vaccination	Intramuscular or intradermal	Subunit vaccines available for animals
Rotavirus	RV1 or RV5	Childhood, routine	Live attenuated	2 or 3, depending on brand	**Oral**	Vaccinates against one or five serotypes. Risk of **intussusception**.
Smallpox virus	Vaccinia	People with occupational risk (e.g., lab workers, some military personnel)	Live attenuated	1	**Scarification**	
Tick-borne encephalitis virus (45)		Not available in the USA. Consider for travelers to Europe on a long trip and extensive outdoor exposure.	Inactivated	2 or 3, over the course of >6 months	Intramuscular	
Yellow fever virus (46)		Travel to countries where it is endemic	Live attenuated	1, at least 10 days before travel	Intramuscular or subcutaneous	Rare complication: **vaccine-associated** viscerotropic or neurologic disease
VZV (chickenpox)	VAR, varicella	Childhood, routine	Live attenuated	2	Subcutaneous	
VZV (shingles)	ZOS, herpes zoster	Patients >60 years old	Live attenuated	1	Subcutaneous	Contains a high dose of the varicella vaccine
		Patients >50 years old. Can be used in **immunocompromised** individuals	Subunit	2	Intramuscular	Contains an adjuvant. This is the preferred formulation.

III. VACCINE DELIVERY.

1. **Timing of administration**
   - Most vaccines are **prophylactic**, which means that they are designed to be administered prior to exposure in order to prevent disease.
   - **Therapeutic vaccines** are administered after exposure in order to prevent the infection from progressing. Examples: rabies and anthrax vaccines
2. **Strain coverage:** Several viruses or virus strains may be delivered in a single injection. A **monovalent** vaccine contains a single virus or antigen, while a **polyvalent** vaccine contains multiple viruses or antigens.
3. **Adjuvants:** substances that are highly antigenic and help to activate the immune response. These are often chemicals, or other antigenic proteins, that are delivered with the target substance if it is not antigenic enough on its own. The most commonly used chemical adjuvants contain aluminum.
4. **Common routes of administration:** intramuscular, intradermal, subcutaneous, and intranasal (Fig. 18.3)
5. **Contraindications:** Vaccines should not be delivered in the following situations.
   - Anaphylactic allergy to components in the vaccine (egg, neomycin, or latex). Nonanaphylactic allergies are not a contraindication.
   - Immunosuppressive conditions, including pregnancy, for live vaccines
   - Current moderate to severe illness. Vaccination should typically be delayed until after the existing, acute illness has subsided.
6. **Vaccine schedules**
   - A **prime** immunization is the first, or primary, vaccination dose that is used to expose the immune system, generate antibodies against a target antigen, and develop immunological memory.
   - A **boost** is any repeat immunizations against the same target. A "booster shot" is used to recall and augment the original immune response.
   - Testing for antibody response can be done by serology. This is not necessary unless a physician is concerned that the patient is not producing robust antibody responses (i.e., they are immunosuppressed). It can also be used to confirm prior vaccination if there is no documentation.

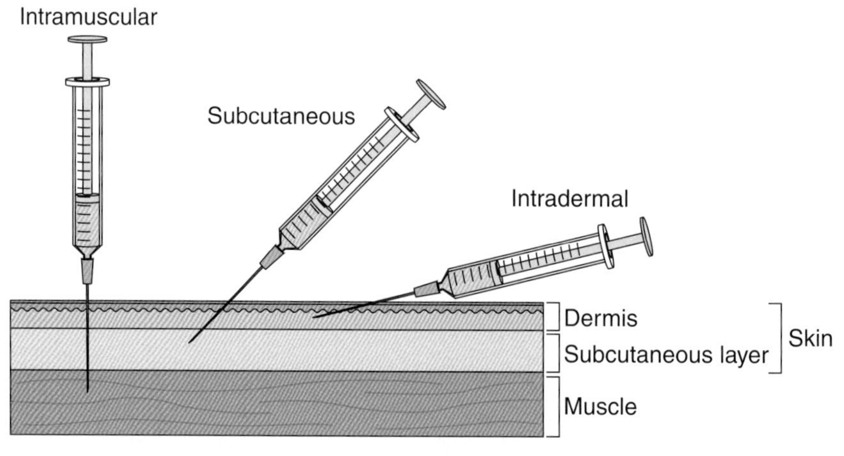

Figure 18.3. Routes of vaccine administration.

		Months									Years										
Vaccine	**Birth**	**1**	**2**	**4**	**6**	**9**	**12**	**15**	**18**	**19-23**	**2-3**	**4-6**	**7-8**	**9-10**	**11-12**	**13-15**	**16-18**	**19-21**	**21-26**	**27-49**	**50-60**	**>60**
Hepatitis B	HBV 1	HBV 2			HBV 3																	
Rotavirus			Rota 1	Rota 2	Rota*																	
Diphtheria, pertussis, and tetanus			DTaP 1	DTaP 2	DTaP 3			DTaP 4				DTaP 5				Td#						
Haemophilus influenzae, type b			H.flu 1	H.flu 2	H.flu*		H.flu 3															
Streptococcus pneumoniae, 13-valent			S.pne 1	S.pne 2	S.pne 3		S.pne 4															
Poliovirus, inactivated			Polio 1	Polio 2	Polio 3							Polio 4										
Measles, mumps, and rubella							MMR 1					MMR 2										
Chickenpox							VZV 1					VZV 2										
Shingles																					VZV 1, 2	
Hepatitis A							HAV 1, 2															
Neisseria meningitidis															N.men 1		N.men 2					
Human papillomavirus														HPV 1, 2, 3								
Influenza					Flu, yearly																	

Figure 18.4. Typical schedule for routine vaccinations in the United States. *, extra dose, if applicable; #, every 10 years.

- Many countries recommend vaccines for routine administration depending on which viruses are endemic in the area, vaccine formulation, and the age of the recipient (e.g., children or elderly patients). A typical vaccination routine in the United States is described in Fig. 18.4.

7. **Vaccine combinations:** Some vaccine formulations combine several traditional vaccines together (e.g., MMR plus VZV or hepatitis A plus hepatitis B). This helps reduce the total number of injections.

IV. OTHER CONSEQUENCES OF VACCINATION

1. **Complications**
   - Side effects are rare and minor. The most common is mild redness and swelling at the injection site.
   - Live attenuated vaccines can rarely cause disease, but the others cannot.
   - **Imitation of infection**: infection-like symptoms after vaccination (e.g., fever) that occur because the body is mounting an immune response
   - Adverse events are reported voluntarily to the CDC and FDA through the Vaccine Adverse Event Reporting System.
2. **Herd immunity:** when enough people have been vaccinated such that an infectious agent cannot spread efficiently in a community. This results in the following.
   - The remainder of the population is also protected. This includes at-risk individuals who cannot get vaccines, like very young infants, pregnant women, immunosuppressed or elderly individuals, etc.
   - The risk of outbreaks in the community is lower.
3. **Original antigenic sin:** a peculiar phenomenon that occurs when a robust immune response is generated against an initial infecting virus strain, but the recalled response is non-neutralizing against similar strains of the virus. This can lead to antibody-dependent enhancement (ADE), which is when the non-neutralizing antibodies enhance the infectivity of the virus so that disease from the secondary infection is more severe than primary infection. ADE occurs with a few viruses like dengue and influenza viruses. For example, a second infection with a different dengue virus serotype will cause worse infection than the first time the patient was infected. In fact, the patient is more likely to have the hemorrhagic form of disease upon secondary infection. Therefore, development of a dengue vaccine is complicated because it can cause ADE.

ADE is an unusual phenomenon where secondary infection with a slightly different virus strain can produce worse, instead of milder, disease. It can be seen with influenza and dengue viruses.

Multiple-Choice Questions

1. **Which of the following groups should routinely receive the MMR vaccine?**
   a. One-year-old infants
   b. Transplant patients
   c. Pregnant women
   d. People >65 years old
2. **Which of the following vaccines protects against HCV?**
   a. Hepatitis A vaccine
   b. Hepatitis B vaccine

c. Hepatitis D vaccine

d. No vaccine is available that protects against HCV.

3. Which of the following vaccines protects against HDV?

a. Hepatitis A vaccine

b. Hepatitis B vaccine

c. Hepatitis D vaccine

d. No vaccine is available that protects against HDV.

4. What is the difference between the chickenpox and shingles vaccines?

a. They are exactly the same.

b. They contain different strains of VZV.

c. Shingles vaccine contains a subunit or different dosage of VZV.

d. Chickenpox vaccine is inactivated, while shingles vaccine is live attenuated.

5. A 36-year-old male receives the intramuscular seasonal influenza vaccine in November. After 2 days he presents with fever, malaise, and myalgias. Which of the following is most likely?

a. He received a live attenuated vaccine, which can cause mild disease.

b. This is an imitation effect.

c. He contracted parainfluenza virus, which also circulates at this time.

d. This is an egg allergy.

6. Vaccination with inactivated vaccines during pregnancy

a. Is contraindicated due to potential transmission of virus to the infant

b. Is contraindicated because the mother cannot produce antibodies

c. Is necessary because the mother's immune system is suppressed

d. Is necessary for increased transmission of maternal antibodies to the fetus

7. Vaccination with live attenuated viral vaccines during pregnancy

a. Is contraindicated due to potential transmission of virus to the infant

b. Is contraindicated because the mother cannot produce antibodies

c. Is necessary because the mother's immune system is suppressed

d. Is necessary for increased transmission of maternal antibodies to the fetus

8. A vaccine exists for only one of the following human herpesviruses. Which one? *(May require reading outside of this chapter.)*

a. HHV-1

b. HHV-2

c. HHV-3

d. HHV-4

e. HHV-5

f. HHV-6

g. HHV-7

h. HHV-8

9. Which of the following vaccines is no longer administered routinely in the United States?

a. Smallpox

b. Polio

c. Hepatitis A

d. Rotavirus

10. Antibody-dependent enhancement (ADE) is a phenomenon in which

a. Antibody levels increase 4-fold after secondary exposure

b. Disease is more severe upon secondary exposure

c. Increased virus levels are toxic

d. Increased antibody levels increase potential for transmission

11. Vaccines help prevent outbreaks among humans even if everyone is not immunized. What is this an example of?

a. Transmissibility of vaccine strains

b. The anti-vaxxer phenomenon

c. Herd immunity

d. None of the above. Vaccines only protect individuals and do not prevent outbreaks.

True or False

12. Some formulations of the polio vaccine can cause polio. **T F**

13. The rabies vaccine should never be given postexposure. **T F**

14. An influenza vaccine should not be administered to children under the age of one. **T F**

15. The live attenuated influenza vaccine (inhalational) causes intussusception in pediatric patients. **T F**

19

CHAPTER 19

ANTIVIRALS

I. OVERVIEW. Most viral infections are self-limited and do not need antivirals. Some common infections (such as those caused by influenza virus, herpesvirus, or RSV) can become severe and may warrant the use of antivirals, especially in high-risk groups, like neonates and immunocompromised hosts. There are some viral infections that should be monitored for potential treatment (e.g., HBV and HCV) and only a few viral infections that should always be treated (e.g., HIV). Antivirals are difficult to develop because viruses are highly mutable and are able to develop resistance. Also, viruses vary widely (*see* chapter 1) and do not share common features that can be used as universal drug targets. However, antivirals have been developed against some groups of similar viruses. The most common mechanisms of action for antivirals include interference with essential steps in the virus life cycle, such as the following (*see* Table 19.11).

1. **Viral attachment and entry:** Antivirals can block receptors that the virus uses to bind or fuse to the host cell in order to prevent viral attachment and entry
2. **Integration** (if applicable): Antivirals prevent integration (e.g., HIV) in order to inhibit a persistent infection (*see* Fig. 19.5).
3. **Viral replication:** There are three main ways antivirals interfere with viral replication (Fig. 19.1).
   - Direct binding to replicative enzymes, like polymerases. This can be a highly specific drug target because many viruses require virus-specific enzymes for replication.
   - Nucleotide or nucleoside analogs: Nucleotides are substrates for polymerases, so mimics are used to block or disrupt nucleic acid synthesis and prevent further nucleotides from being added.
   - Introducing mutations: A large number of errors incorporated during viral replication can cause the virus population to degenerate.
4. **Viral proteins and assembly:** Antivirals that interfere with viral assembly increase the proportion of nonviable (noninfectious) virions. For example, **viral protease** is an important drug target because it is used to cleave polyproteins into individual proteins for structure and replication. **Boosters** (e.g.,

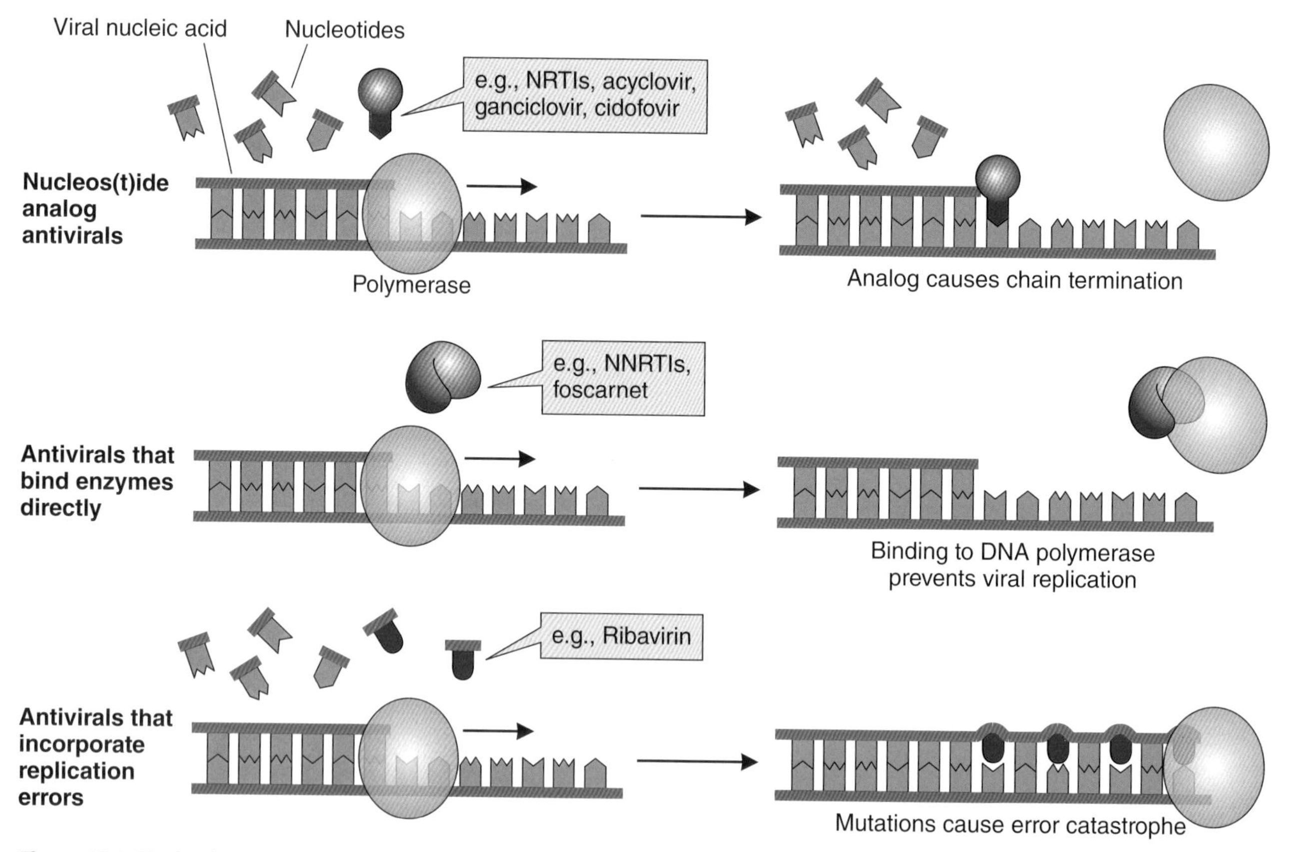

Figure 19.1. Mechanisms of antivirals against viral replication.

ritonavir) can be used with HIV protease inhibitors to decrease clearance of protease inhibitors and prolong their half-lives.

5. **Release of viral particles:** Viruses that are prevented from being released from cells cannot go on to infect new cells (*see* Fig. 19.4).

6. **Activating the immune response (immunomodulators).** Some therapies (e.g., interferons) trigger the immune system's natural defenses against viruses.

II. ANTIVIRALS THAT CAN COVER SEVERAL VIRUS TYPES. (Table 19.1; *see also* Table 19.11)

1. **Interferons:** These are cytokines that are naturally produced during the innate immune response and interfere with virus replication.

Cytokines: small proteins used for intracellular signaling

   - **Mechanism:** Interferons have antiviral, antiproliferative, and immunomodulatory effects and trigger signaling cascades to activate other immune system components (Fig. 19.2).
   - Recombinant type I interferons are used as antiviral therapy against some DNA and RNA viruses.
     - They are available in two forms: standard or attached to **p**oly**e**thylene **g**lycol (PEG). PEGylated interferons are more common because they have reduced clearance and longer half-lives.
     - Interferon alpha-2a is used in the United States, but interferon alpha-2b is also used in other countries (these forms have different dosing protocols).

Table 19.1. **Antivirals that can cover several virus types**

DRUG CLASS	DRUGS	MECHANISM OF ACTION	ROUTE OF ADMINISTRATION	ADVERSE EFFECT(S)	NOTES
Nucleos(t)ide analogs	Cidofovir	Cause chain termination when incorporated	Intravenous	Renal toxicity, neutropenia	Mutations in DNA polymerase genes (like UL54) can cause resistance.
	Ribavirin	Cause mutagenesis when incorporated	Oral, inhalation	Fetal defects, hemolytic anemia	RNA viruses (used in combination with other drugs for HCV). Can also work against DNA viruses.
Immune modulators	Interferon alpha-2a and -2b, with or without PEGylation	Trigger innate antiviral immune responses	Subcutaneous	Flu-like symptoms, bone marrow suppression, autoimmune and neuropsychiatric disorders	

- Side effects: many, because it triggers an immune response instead of acting on a specific viral component.
  - Flu-like symptoms (e.g., headache, fever, and aches)
  - Bone marrow suppression
  - Neuropsychiatric disorders
  - Autoimmune disorders

2. **Ribavirin** can be used against DNA and RNA viruses.
   - Mechanism
     - RNA viruses: acts as a nucleoside analog. It is incorporated into a growing chain, but instead of causing chain termination, it introduces a large number of mutations that are lethal to the virus. This is called "**error catastrophe**."
     - DNA viruses: unclear
   - Used mostly for RSV, hemorrhagic fever viruses, and HCV
     - RSV: aerosolized formulation is used for pediatric patients.
     - HCV: oral formulation is used. It is ineffective as monotherapy and should always be used in combination with other drugs.
     - Hemorrhagic fever viruses: intravenous or oral formulations are effective against many types of hemorrhagic fever viruses, except Ebola virus.
   - Can cause **hemolytic anemia** and myelosuppression. It is also teratogenic and can cause **birth defects** in the infant if taken by either parent (it is contraindicated in pregnant women and male partners of pregnant women). The aerosolized form may be a risk to health care workers.

3. **Cidofovir** has activity against several DNA viruses.
   - Mechanism: Acts as a nucleotide analog and terminates DNA synthesis if it is incorporated into a growing chain.
   - Used mostly for CMV (especially retinitis in AIDS patients). It can also work against acyclovir-resistant herpesviruses and against adenoviruses.
   - Resistance is due to mutations in viral DNA polymerase genes (UL54 in CMV).
   - Causes neutropenia and renal toxicity. To reduce the risk, intravenous prehydration and concomitant **probenecid** are used.

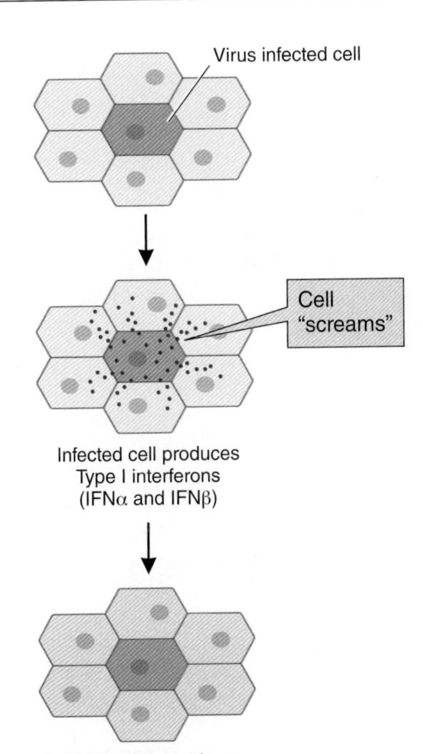

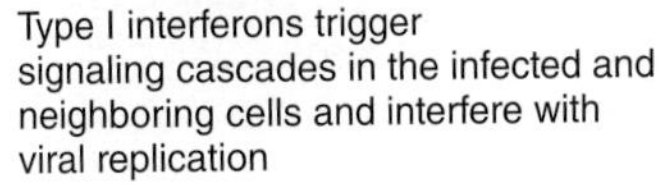

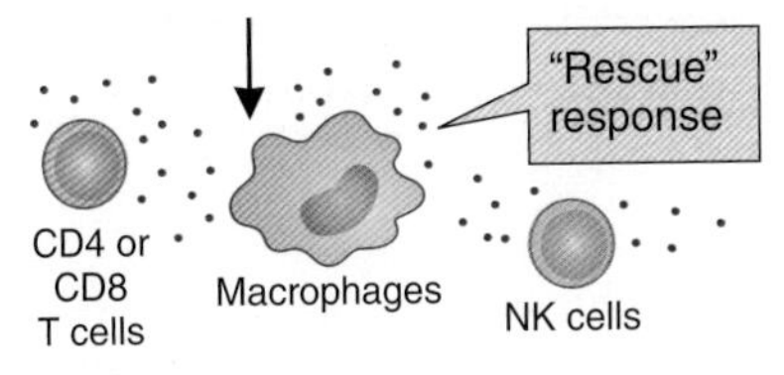

Figure 19.2. **Interferon response against viral pathogens.**

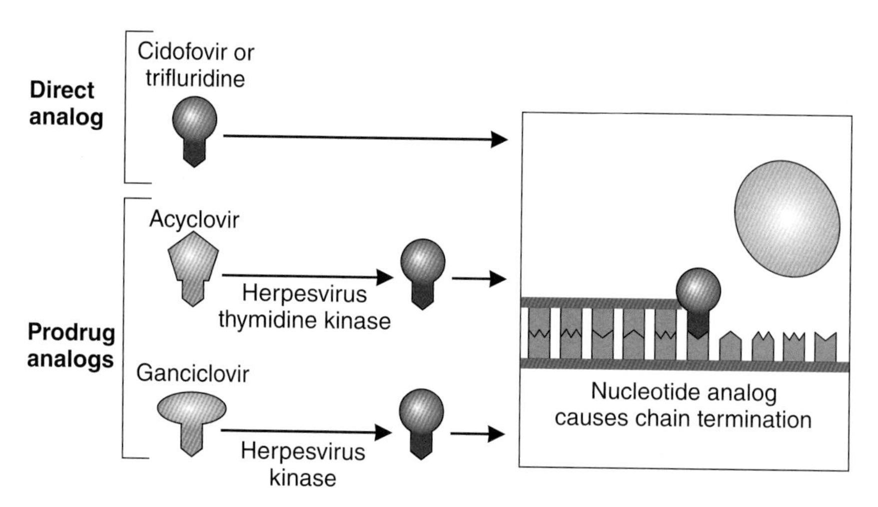

Figure 19.3. **Mechanism of action of nucleotide analogs used for treatment of herpesviruses.**

III. ANTIVIRALS AGAINST HERPESVIRUSES. (Fig. 19.3 and Table 19.2; *see also* Table 19.11)

1. **Acyclovir:** active against HSV, VZV, and EBV but not CMV.
   - Mechanism: is activated when a viral thymidine kinase phosphorylates it. It then acts like a guanosine analog that is incorporated into a growing DNA chain, but it terminates any subsequent elongation.
   - Is not active against CMV because CMV does not have thymidine kinase.
   - Typically administered intravenously due to poor bioavailability. The oral prodrug version is valacyclovir.
   - Penciclovir has a similar mechanism and is topical. Its oral prodrug version is famciclovir.
   - Resistance is rare and is due to mutations in UL23 and UL30, the genes encoding viral thymidine kinase and DNA polymerase.

Acyclovir is activated by herpesviral **thymidine kinase**.

Ganciclovir is activated by herpesviral viral **kinase**.

2. **Ganciclovir:** active against herpesviruses, including CMV.
   - Mechanism: is activated when a viral kinase phosphorylates it. Once activated, it also functions as a guanosine analog that terminates elongation.
   - Typically administered intravenously due to poor bioavailability. The oral prodrug version is valganciclovir.
   - In CMV, the viral kinase is encoded by the gene UL97 and the viral DNA polymerase is encoded by UL54. Mutations in these two genes are associated with resistance.

Acyclovir does NOT work against CMV (CMV does not have thymidine kinase).

Ganciclovir DOES work against CMV (CMV does have a viral kinase).

3. **Letermovir:** a new antiviral that is used for prophylaxis against CMV
   - Mechanism: inhibits the viral terminase complex, which is normally involved in cleaving CMV polyproteins and packaging viral DNA into the capsid
   - It does not exhibit cross-resistance against other antivirals. Mutations in UL56 and UL89 can cause resistance.
4. **Foscarnet:** primarily used for herpesviruses, especially CMV
   - Mechanism: is a pyrophosphate analog. It binds to a site on DNA polymerase and reverse transcriptase and prevents them from cleaving

Table 19.2. Antivirals against herpesviruses

DRUG CLASS	DRUG	MECHANISM OF ACTION	ROUTE(S) OF ADMINISTRATION	ADVERSE EFFECTS	NOTE(S)
Nucleoside analogs (guanosine)	Acyclovir	Are activated by **viral thymidine kinase.** Cause chain termination when incorporated into DNA.	Intravenous	Has very few side effects because it needs activation by the virus. Intravenously: renal toxicity Myelosuppression, neutropenia, thrombocytopenia	Mutations in **UL23** and **UL30** cause resistance. **CMV is intrinsically resistant.**
	Valacyclovir		Oral		
	Penciclovir		Topical		
	Famciclovir		Oral		
	Ganciclovir	Are activated by **viral kinase**. Cause chain termination when incorporated into DNA	Intravenous		Mutations in CMV **UL97** and **UL54** cause resistance.
	Valganciclovir		Oral		
Nucleoside analogs (other)	Trifluridine	Cause chain termination when incorporated into DNA	Topical	Higher toxicity than antiherpes guanosine analogs	Used in ocular infections (e.g., keratitis)
Binds directly to viral polymerases	Foscarnet	Binds to **DNA polymerase** and reverse transcriptase and prevents them from cleaving the pyrophosphate off nucleotides	Intravenous	Renal toxicity, seizures	May be used for acyclovir-resistant HSV. However, mutations in the CMV UL54 gene may also result in cross-resistance to ganciclovir, cidofovir, and foscarnet.
Inhibits assembly	Letermovir	Binds the viral **terminase complex** and prevents it from cleaving polyproteins and packaging viral DNA	Oral, intravenous		Used for **CMV prophylaxis.** Mutations in UL56 and UL89 can cause resistance.

the pyrophosphates from nucleotides. As a result, addition of nucleotides stops.

- Resistance due to mutations in viral DNA polymerase genes (UL54)

  It is not activated by viral kinases, so it is used in infections with acyclovir/ganciclovir-resistant viruses.
- Causes renal toxicity and can cause seizures due to alterations in minerals and electrolytes.

5. **Cidofovir:** active against all herpesviruses, including CMV. It is used against acyclovir-resistant strains (mutations in UL97).

IV. ANTIVIRALS AGAINST HUMAN PAPILLOMAVIRUS. (Table 19.3; *see also* Table 19.11)

1. **Imiquimod**
   - Mechanism: triggers the innate immune response
     - Binds toll-like receptor 7 and triggers proinflammatory signaling cascades
     - Also triggers apoptosis
   - Is a topical antiviral that is used to treat condyloma acuminata (genital warts)

Table 19.3. **Antivirals against HPVs**

DRUG CLASS	DRUGS	MECHANISM OF ACTION	ROUTE OF ADMINISTRATION	ADVERSE EFFECTS	NOTES
Immune modulators	Imiquimod	Triggers inflammation and apoptosis	Topical		
	Interferons		Intralesional, topical, systemic		Systemic administration may not be effective.
Cytotoxic agents	Podophyllin, podophyllotoxin	Inhibits cell division by binding to tubulin	Topical		

2. **Podophyllotoxin (podofilox), and podophyllin**
   - Mechanism: inhibits cell division and growth of proliferating cells by binding to tubulin
   - Applied topically to genital warts for several weeks
3. **Interferons:** administered topically or intravenously or injected into the lesion. It may take several weeks to produce an effect.

V. ANTIVIRALS AGAINST INFLUENZA VIRUS. (Table 19.4; *see also* Table 19.11)

1. **Oseltamivir, zanamivir, and peramivir**
   - Mechanism: bind to neuraminidase and prevent it from cleaving the virus off the infected host cell. As a result, the virus cannot spread to other cells (Fig. 19.4).

Table 19.4. **Antivirals against influenza**

DRUG CLASS	DRUGS	MECHANISM OF ACTION	ROUTE OF ADMINISTRATION	ADVERSE EFFECTS	NOTES
Inhibits assembly	Amantadine, rimantadine	Inhibit viral assembly. Also block M2 ion channel and prevent viral RNA from entering the host cell.	Oral		Influenza **A virus only**. **Not used** anymore due to high levels of resistance.
Neuraminidase inhibitors	**Oseltamivir**	Bind to neuraminidase and prevent virions from being released	Oral		**No activity >48 hours** after symptoms
	Zanamivir		**Inhalation**	Bronchospasms (in patients with asthma or COPD)	
	Peramivir		Intravenous		
Inhibitors of viral transcription	Baloxavir marboxil	Binds to cap-dependent endonucleases to prevent the virus from "snatching" the cap from host mRNA and using it for its own RNA transcription.	Oral	Headache and diarrhea	Administer within 48 hours of symptoms. Only **one** dose is needed.

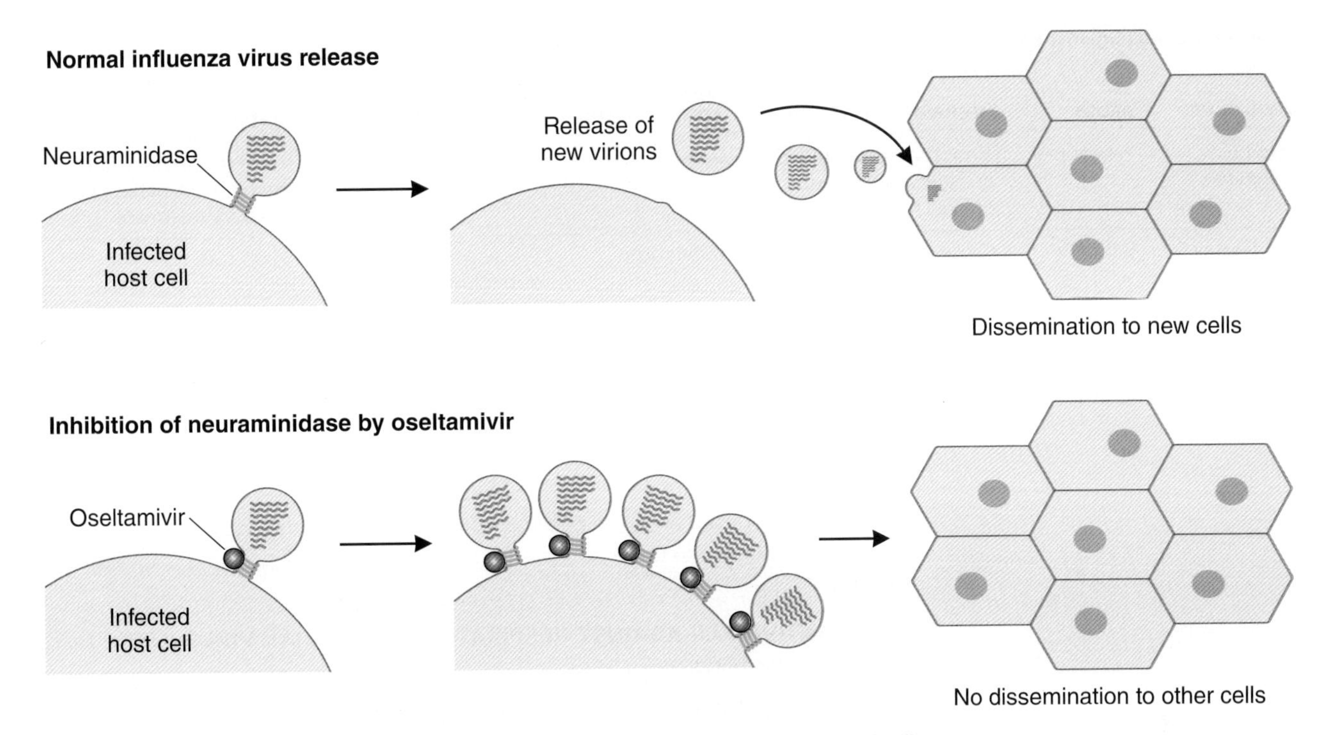

Figure 19.4. Mechanism of action of neuraminidase inhibitors used for treatment of influenza.

- Work against both influenza A and B
- Must be administered within 48 hours of symptoms.
- Indicated for people with severe, progressive disease or people at risk of complications
- Oseltamivir is oral, zanamivir is an oral inhalation, and peramivir is intravenous.
- Resistance is rare and mostly against oseltamivir. Resistance can occur when there are mutations in the neuraminidase gene. The most frequent mutations seen are:
  - **H275Y** mutation in seasonal H1N1 strains and E119V, R292K, and N294S mutations in H3N2 strains
  - Point mutations at E117, D197, H273, and R374 in influenza B virus

Oseltamivir = **o**ral
Zanamivir = no**z**e
Peramivir = **p**arenteral

2. Baloxavir marboxil

- Mechanism: Inhibits viral transcription by binding to cap-dependent endonucleases. This prevents influenza virus from using these endonucleases to "snatch" the transcription primer (called the cap) from host mRNA so that its viral genomes can be transcribed.
- Administer a single oral dose within 48 hours of infection.
- A very recently approved antiviral that is currently indicated for people >12 years old.

3. Amantadine and rimantadine

- Mechanism: interfere with viral assembly. They also inhibit the M2 ion channel of influenza A virus and prevent viral RNA from being released into the host cell.

Table 19.5. Drugs against RSV

DRUG CLASS	DRUGS	MECHANISM OF ACTION	ROUTE OF ADMINISTRATION	ADVERSE EFFECTS	NOTES
Fusion inhibitors	Palivizumab	Bind RSV **F protein** and prevent virus-cell and cell-cell fusion	Intramuscular		Used **prophylactically** for high-risk babies; not used for treatment
Nucleoside analogs	Ribavirin	Cause **mutagenesis** when incorporated	**Inhalation**	Sudden deterioration of respiratory function, hemolytic anemia	High cost, potential toxicity to health care workers

- Are only for influenza A virus; there is no activity against influenza B virus because the latter does not have the M2 protein
- Not used anymore because influenza virus has developed significant resistance to them

VI. ANTIVIRALS AGAINST RESPIRATORY SYNCYTIAL VIRUS. (Table 19.5; *see also* Table 19.11)

1. **Palivizumab:** an antibody that prevents RSV infection
   - Mechanism: a monoclonal antibody that binds to the **F protein** on RSV's envelope and inhibits fusion between the virus and the cell. It also prevents cells from fusing to each other (syncytia formation is normally a hallmark of RSV infection).
   - Should be used prophylactically because it prevents infection. It does not treat the infection after it has occurred.
   - Only used on babies at high risk of hospitalization with RSV

Palivizumab is like a vaccine because it is used prophylactically to prevent infection; it does not treat infections once they have occurred.

2. **Ribavirin:** typically used only in extremely severe RSV infections (e.g., young children sick enough to be hospitalized). It is administered through inhalation, so there may be occupational hazards to health care workers.

VII. ANTIRETROVIRALS AGAINST HIV. There are 6 classes of drugs with different mechanisms of action (Table 19.6 and Table 19.7)

1. **CCR5 antagonists** (entry inhibitors)
   - Prevent HIV from binding to cells
   - The antagonists attach to CCR5 on T cells and macrophages, which most HIV strains use as a coreceptor for binding to cells.
   - In about 10% of infections, HIV uses the CXCR4 coreceptor instead. These strains do not respond to CCR5 antagonists.
2. **Fusion inhibitors**
   - Prevent HIV from fusing to the cell's envelope by binding to gp41 on the virus envelope
   - Fusion inhibitors have "-fu-" in the middle of the name.
3. **Nucleoside reverse transcriptase inhibitors (NRTIs)**
   - These are nucleoside (or nucleotide) analogs that reverse transcriptase incorporates into the growing cDNA and cause chain termination.
   - NRTIs are prodrugs and must be phosphorylated to become active.

Table 19.6. Antiretrovirals against HIV

DRUG CLASS	DRUG(S) (ABBREVIATION[S])	MECHANISM OF ACTION	ROUTE(S) OF ADMINISTRATION	ADVERSE EFFECTS	NOTES
NRTIs	Abacavir (ABC), didanosine (DDI), emtricitabine (FTC)[a], stavudine (D4T)	Cause chain termination when incorporated into cDNA	Oral	Lactic acidosis and severe hepatomegaly with steatosis, mitochondrial toxicity	Most end in **"-ine."** Abacavir cannot be used in patients with HLA-B*5701 due to high risk for fatal hypersensitivity reaction.
	Zidovudine (AZT)		Oral, **intravenous**	Lactic acidosis, hematologic toxicities (neutropenia, anemia, myopathy)	
	Lamivudine (3TC)[a]		Oral	Lactic acidosis, hepatomegaly	
	Tenofovir disoproxil fumarate (TDF)[a]		Oral	Renal and bone toxicity	Nucleo**t**ide analog
	Tenofovir alafen-amide (TAF)		Oral	Reduced renal and bone toxicity	
NNRTIs	Delavirdine, doravirine, efavirenz, etravirine, nevira-pine, rilpivirine	Directly bind and inactivate reverse transcriptase	Oral	Hepatotoxicity rash, including Stevens-Johnsons syndrome and toxic epidermal necrolysis	Have "**-vir-**" in the middle
Protease inhibitors	Atazanavir, daruna-vir, fosamprenavir, indinavir, nelfinavir, saquinavir, tipranavir, lopinavir	Inhibit HIV proteases	Oral	Hyperglycemia, new-onset diabetes, lipohypertrophy, hyperlipidemia	End in "-**navir.**" All require boosting. High genetic barrier to resistance
Integrase strand transfer inhibitors	Raltegravir, dolute-gravir, elvitegravir	Inhibit HIV integrase	Oral	Insomnia	Have "**-tegra-**" in the middle
Entry inhibitors	Maraviroc	CCR5 antagonist	Oral	Upper respiratory tract infections	
Fusion inhibitors	Enfuvirtide	Inhibits fusion of the virion with the host cell	**Subcutaneous**	Injection site reactions	Have "**-fu-**" in the middle
Boosters (pharmacokinetic enhancers)	Cobicistat	Prolong half-life of protease inhibitors	Oral		Used with atazanavir or darunavir
	Ritonavir[b]	Prolong half-life of other protease inhibitors. Have some protease inhibitor activity.	Oral		

[a]Can also be used for HBV, but with different dosing.

[b]Can also be used for HCV.

- Most are cleared renally (exception: abacavir).
- The names of most end in "-ine" (exceptions: abacavir and tenofovir).

4. Non-nucleoside reverse transcriptase inhibitors (NNRTIs)

- These bind directly to reverse transcriptase and prevent it from making cDNA.
- The names of most have "-vir-" in the middle.
- All are metabolized via CYP enzymes.

Table 19.7. **Combination antiretrovirals against HIV**

COMBINATION DRUG BRAND NAME	COMPONENTS[a]
Atripla	Efavirenz, emtricitabine, tenofovir
Biktarvy	Bictegravir, emtricitabine, tenofovir
Cimduo	Lamivudine, tenofovir
Combivir	Lamivudine, zidovudine
Complera	Emtricitabine, rilpivirine, tenofovir
Delstrigo	Doravirine, lamivudine, tenofovir
Epzicom	Abacavir, lamivudine
Evotaz	Atazanavir, cobicistat
Genvoya	Cobicistat, elvitegravir, emtricitabine, tenofovir
Juluca	Dolutegravir, rilpivirine
Odefsey	Emtricitabine, rilpivirine, tenofovir
Prezcobix	Cobicistat, darunavir
Stribild	Cobicistat, elvitegravir, emtricitabine, tenofovir
Symfi or Symfi Lo	Efavirenz, lamivudine, tenofovir
Symtuza	Darunavir, cobicistat, emtricitabine, tenofovir
Triumeq	Abacavir, dolutegravir, lamivudine
Trizivir	Abacavir, lamivudine, zidovudine
Truvada	Emtricitabine, tenofovir

[a]Older combinations with tenofovir may contain TDF while newer combination may use TAF.

5. **Integrase strand transfer inhibitors** (Fig. 19.5)
    - Bind to HIV integrase and prevent the virus from integrating
    - The names of integrase inhibitors have "-integra-" in the middle.

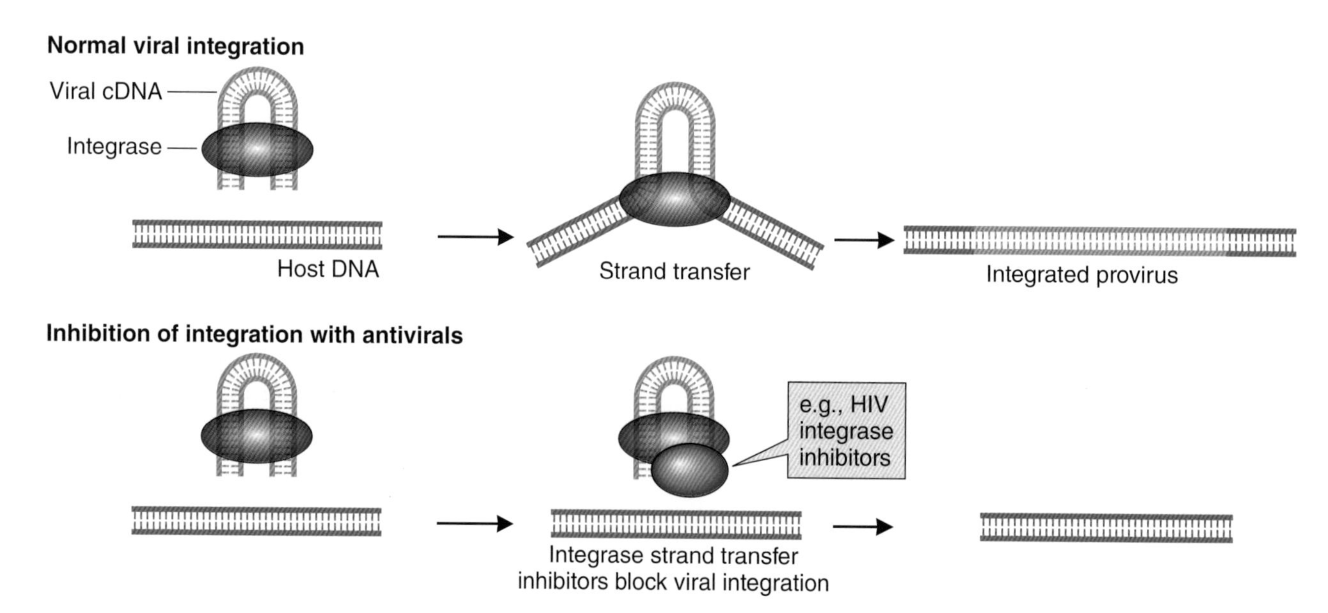

Figure 19.5. **Mechanisms of action of integrase inhibitors used for treatment of HIV.**

Table 19.8. **Antivirals against HBV**

DRUG CLASS	DRUG(S) (ABBREVIATION[S])	MECHANISM OF ACTION	ROUTE OF ADMINISTRATION	ADVERSE EFFECT(S)	NOTES
NRTIs	**Lamivudine (3TC)**[a]	Cause chain termination when incorporated into cDNA	Oral	Lactic acidosis, hepatomegaly	
	Tenofovir disoproxil fumarate (TDF), Tenofovir alafenamide (TAF)[a]		Oral	Renal and bone toxicity	Nucleotide analog
	Adefovir		Oral	Nephrotoxicity	Nucleotide analog
	Telbivudine, entecavir		Oral	Increase in creatine phosphokinase, peripheral edema	
Immune modulators	PEGylated interferon alpha-2a or -2b	Modify and degrade viral DNA, interfere with protein synthesis	**Subcutaneous**	Flu-like symptoms, bone marrow suppression, autoimmune and neuropsychiatric disorders	Usually **combined with ribavirin**

[a]Higher doses are used against HIV.

6. **Protease inhibitors (PIs)**
   - Bind to HIV protease. This protein is necessary for cleaving the long HIV polyprotein into individual proteins so the virus will not be assembled (Fig. 19.6).
   - Are administered with a boosting agent
   - The names end in "-navir."
   - Resistance develops only if HIV develops many mutations (high genetic barrier to resistance).

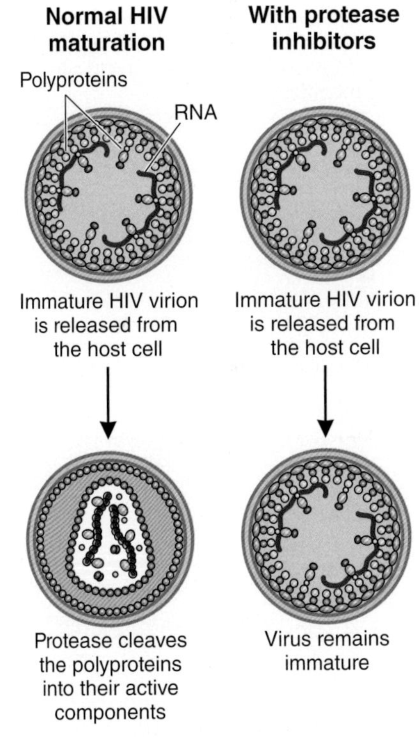

Figure 19.6. **Mechanism of action of protease inhibitors used for treatment of HIV.**

VIII. ANTIVIRALS AGAINST HEPATITIS B VIRUS. (Table 19.8; *see also* Table 19.11)

1. **NRTIs**: Despite being very different viruses, HBV and HIV both have reverse transcriptase, so some NRTIs that work against HBV also **overlap** as antiretrovirals.
2. **Interferon alpha:** Interferons modify and degrade HBV DNA. PEGylated interferon is usually preferred over conventional interferon. It is used for chronic HBV.

Some NRTIs against HIV also work against HBV because they both have reverse transcriptase.

IX. ANTIVIRALS AGAINST HEPATITIS C VIRUS. The choice of antivirals used for the treatment of HCV is dependent on patient factors as well as the virus genotype (Tables 19.9 to Table 19.11; *see also* Table 8.3).

1. **Direct-acting antivirals (DAA):** These inhibit HCV directly. There are four main classes that are efficacious against HCV, although they may result in reactivation of HBV.
   - **Protease inhibitors (PIs):** bind to the NS3/4A protease, which is used to cleave HCV's polyprotein. Without protease, the virus cannot assemble new virions and cannot make proteins that are needed for viral replication.

Table 19.9. Antivirals against HCV

DRUG CLASS	DRUG(S)	MECHANISM OF ACTION	ROUTE OF ADMINISTRATION	ADVERSE EFFECTS	NOTES
NPI	**Sofosbuvir**	Causes chain termination when incorporated	Oral		Always **used in combination** with other HCV drugs (like ledipasvir or ribavirin and PEGylated interferon alpha)
NNPI	Dasabuvir	Binds RNA polymerase **NS5B** and prevents it from cleaving the HCV polyprotein	Oral	Reversible nephrotoxicity	
Protease inhibitors	Telaprevir, boceprevir, simeprevir, paritaprevir, grazoprevir, asunaprevir	Inhibit HCV protease, NS3/4A	Oral	Serious skin reactions, including rash, pruritus, Stevens-Johnson syndrome and DRESS syndrome	Have "-**pre**-" in the middle
Inhibits assembly	**Daclatasvir**, elbasvir, ledipasvir, ombitasvir, velpatasvir	Bind **NS5A**, which interferes with replication and assembly	Oral		End in "-**asvir**"
Immune response modifiers	PEGylated interferons alpha-2a and -2b	Modify and degrade viral DNA, interfere with protein synthesis	**Subcutaneous**	Flu-like symptoms, bone marrow suppression, autoimmune and neuropsychiatric disorders	Usually **combined with ribavirin**
Booster	Ritonavir[a]	Prolongs half-life of other protease inhibitors. Has some protease inhibitor activity.	Oral		

[a] Can also be used against HIV.

Table 19.10. Combination antivirals against HCV

COMBINATION DRUG BRAND NAME	COMPONENTS	HCV GENOTYPE TARGET(S)
Epclusa	Sofosbuvir, velpatasvir	1, 2, 3, 4, 5, and 6
Harvoni	Ledipasvir, sofosbuvir	1, 4, 5, and 6
Mavyret	Glecaprevir, pibrentasvir	1, 2, 3, 4, 5, and 6
Technivie	Ombitasvir, paritaprevir, ritonavir	4
Viekira Pak or Viekira XR	Dasabuvir, ombitasvir, paritaprevir, ritonavir	1
Vosevi	Sofosbuvir, velpatasvir, voxilaprevir	1, 2, 3, 4, 5, and 6
Zepatier	Elbasvir, grazoprevir	1, 4

Table 19.11. Types of antivirals used for different groups of viruses by mechanism of action

TYPE OF ANTIVIRAL	ANTI-HCV	ANTI-RETROVIRALS	ANTI-HBV	ANTI-HERPESVIRUS	ANTI-PAPILLOMAVIRUS	ANTI-INFLUENZA VIRUS	ANTI-RSV	ANTI-SMALLPOX VIRUS	OTHER VIRUSES
Protease inhibitors	NS3/4A PI (e.g., simeprevir)	PI (e.g., atazanavir)							
Nucleos(t)ide analogs	NPI (e.g., sofosbuvir), ribavirin	NRTI (e.g., tenofovir)	NRTI (e.g., tenofovir)	Guanosine analogs (e.g., acyclovir, ganciclovir), cidofovir, foscarnet, trifluridine					Cidofovir, ribavirin
Agents that bind directly to viral polymerases	NNPI (e.g., dasabuvir)	NNRTI (e.g., efavirenz)							
Inhibitors of viral transcription						Baloxavir marboxil			
Inhibitors of integration		Integrase inhibitors (e.g., raltegravir)							
Inhibitors of fusion		Fusion inhibitors (e.g., enfuvirtide)					Monoclonal antibody (e.g., palivizumab)		
Binding inhibitors		CCR5 antagonists (e.g., maraviroc)							
Inhibitors of release						Neuraminidase inhibitors (e.g., oseltamivir)		Tecovirimat	
Boosters	E.g., ritonavir	E.g., ritonavir							
Inhibitors of viral assembly	NS5A inhibitors (e.g., ledipasvir)			Letermovir		E.g., amantadine			
Agents that stimulate antiviral immunity	Interferon alpha-2a with ribavirin		Interferon alpha-2a		E.g., imiquimod, interferon				Interferon alpha-2a
Cytotoxic agents					E.g., podophyllotoxin				

- **NS5A inhibitors:** interfere with the NS5A viral protein, which is normally involved in replication and assembly.
- **Nucleos(t)ide polymerase inhibitors (NPI):** indirectly inhibit the action of the NS5B viral RNA polymerase. Nucleos(t)ide analogs are incorporated into new RNA strands and cause chain termination. A new NPI called **sofosbuvir** is curative for as many as 90% of cases. It is very expensive.
- **Nonnucleoside polymerase inhibitors (NNPI):** bind directly to NS5B RNA polymerase and prevent it from replicating the genome.

2. **PEGylated interferon alpha-2a** indirectly fights HCV infection by stimulating the innate antiviral immune response. It can be used with or without **ribavirin**. This is an older treatment regimen that was widely used but is being replaced with direct-acting antivirals. Contraindications: alcohol, liver cirrhosis, and organ transplant.

Multiple-Choice Questions

1. **Which drugs can be used to treat both HIV and HBV?**
   a. Lamivudine and tenofovir
   b. Lopinavir and ritonavir
   c. Entecavir and tenofovir
   d. Entecavir and PEGylated interferon

2. **Which of the following is most likely to cause rash (including Stevens-Johnson syndrome)?**
   a. Ritonavir
   b. Atazanavir
   c. Nevirapine
   d. Acyclovir

3. **Which of the following drugs act on the influenza A virus M2 protein?**
   a. Oseltamivir
   b. Zanamivir
   c. Acyclovir
   d. Amantadine

4. **Which of the following drugs still have low rates of resistance in influenza virus?**
   a. Acyclovir
   b. Amantadine
   c. Oseltamivir
   d. Rimantadine

5. **Which of the following CANNOT be used against CMV?**
   a. Acyclovir
   b. Cidofovir
   c. Ganciclovir
   d. Foscarnet

6. Which of the following antivirals used for CMV should be coadministered with probenecid to reduce the risk of renal toxicity?

a. Cidofovir

b. Oseltamivir

c. Efavirenz

d. Ribavirin

7. Which of the following is not a direct-acting antiviral(s) against HCV?

a. NS3/4A inhibitors

b. NS5A inhibitors

c. NS5B inhibitors

d. PEGylated interferon

8. Which of the following is true about acyclovir?

a. It is not a prodrug, so it is active immediately.

b. It is used against HPV.

c. It needs to be activated by viral kinase.

d. It needs to be activated by viral thymidine kinase.

9. A high genetic barrier to resistance is an advantage of which class of antivirals?

a. NNRTIs

b. NRTIs

c. Integrase inhibitors

d. Protease inhibitors

10. Which antiviral should NOT be used for patients who are HLA-B*5701 positive due to high risk for fatal hypersensitivity reaction?

a. Abacavir

b. Efavirenz

c. Tenofovir

d. Delavirdine

11. What is the difference between NNRTIs and NRTIs?

a. NRTIs bind directly to reverse transcriptase, while NNRTIs mimic nucleos(t)ides.

b. NRTIs bind to CCR5, while NNRTIs bind to reverse transcriptase.

c. NRTIs mimic substrates of polymerases, while NNRTIs bind directly to polymerases.

d. NRTIs bind to CD4, while NNRTIs mimic nucleos(t)ides.

True or False

12. Oseltamivir should be administered within 48 hours of symptoms. **T F**

13. Intramuscular injections of palivizumab should be given to a 3-year-old with severe pneumonia. **T F**

14. Imiquimod does not act directly against HPV. **T F**

15. Oseltamivir cannot be used against influenza B virus. **T F**

SECTION V

THE REGULATORY ENVIRONMENT FOR LABORATORY TESTING

20

CHAPTER 20

REGULATORY REQUIREMENTS

I. OVERVIEW. Diagnostic laboratories and assays are regulated to make sure that they produce accurate, reliable, and timely patient results. In the United States, the Food and Drug Administration (FDA), the Centers for Medicare and Medicaid Services (CMS), and the Centers for Disease Control and Prevention (CDC) are all involved in laboratory regulation (47).

II. CLASSIFICATION OF DIAGNOSTIC ASSAYS. The FDA classifies all assays that are performed on human samples and used for the diagnosis of disease. There are several categories of ***in vitro*** **diagnostic (IVD)** assays, and they have different regulatory requirements to show reliability, efficacy, and safety (48).

1. **FDA classification of assays**
   - **Research use only (RUO) assays:** These assays are technically still in the research phase of development.
     - They are not approved for clinical use and cannot be used for a clinical diagnosis. But they also do not have to follow the same good manufacturing practices. Also, adverse events (like misleading test results) do not have to be reported.
     - Testing must be clearly designated. For example, label results "For Research Use Only."
   - **Investigational use only (IUO) assays:** This is typically an interim label that is used for assays in the product testing phase of development.
     - Like RUOs, these assays do not have the same stringent manufacturing and reporting requirements as fully approved assays, are not approved for clinical use, and cannot be used for a clinical diagnosis.
     - Testing must be clearly designated. For example, label results "For Investigational Use Only."
   - **FDA-approved assay:** a commercially manufactured test that is intended to be used for diagnosis. For the assay to be FDA approved, it must pass through the FDA's **premarket approval (PMA) process**, which is highly stringent and involves a large trial using clinical samples to collect data on safety and efficacy.

- **FDA-cleared assay:** a commercially manufactured test that is intended for diagnosis, but there is already a similar test that has been FDA approved. In this case, the manufacturer can submit a (slightly less stringent) **510K submission** to the FDA, which also includes trials with clinical samples.
- **Analyte specific reagents (ASRs)** are not comprehensive diagnostic assays but are test components that are used to detect a target or ligand (like antibodies). These materials do have to follow good manufacturing practices but do not have established performance characteristics. As a result, they can be used as part of clinical diagnostic tests but cannot be marketed as a "kit."
- **Laboratory-developed tests (LDTs)** are assays typically developed at an institution to address a specific need for which there is relatively small national demand (they are sometimes known as "home brews" or "in-house-developed tests").
  - LDTs have not received any FDA review, approval, or clearance.
  - Results from these tests must be reported with a disclaimer indicating that the test is not approved or cleared by the FDA but has been validated internally for clinical diagnostic purposes.
  - Because LDTs have traditionally been used on a small scale, the FDA has not regulated them very tightly, although there have been recent steps to change that. Instead, the quality and reliability of these tests have been regulated by CMS (see below).
- **Modified FDA-approved/cleared tests** (**ModTs**): These are FDA-approved or -cleared tests that have been modified and used off-label. Like LDTs, results must also be reported with a disclaimer. Off-label usage includes changes to the following.
  - The specimen source or type
  - The original procedure
  - The reagents that are used
  - How the test is interpreted

CE marking: This is NOT an FDA category but is a designation for tests that meet European performance requirements. This is similar to FDA clearance or FDA approval, but these tests CANNOT be marketed in the United States as is.

2. **Test complexity:** The FDA assigns complexity levels for FDA-approved or -cleared tests.
   - **Waived complexity:** tests that require very minimal training and laboratory knowledge but still have almost no risk of being performed incorrectly. If they are performed incorrectly, they have negligible risk of doing harm to the patient. Many (but not all) point-of-care tests are waived because they are simple and rapid.
   - **Moderate and high complexity:** are nonwaived tests that require more laboratory training. Labs performing these tests must do proficiency testing (see below) to ensure that they are being performed correctly. LDTs and ModTs are automatically considered high complexity.

III. CENTERS FOR MEDICARE AND MEDICAID SERVICES. CMS is an agency that is responsible for regulating clinical testing on humans in the United States. In 1988, it developed laws called the Clinical Laboratory Improvements Amendments (**CLIA '88**). These establish the minimum quality standards for laboratories depending on the complexity of testing being performed.

CLIA is a set of laws to ensure quality in lab testing. The CMS is a regulatory agency responsible for enforcing them.

1. **CLIA** applies to all laboratories that provide information to diagnose, prevent, or treat disease. It does not include research labs.
    - Whether a lab is required to follow CLIA depends on the kind of testing it does, not on if it accepts Medicare/Medicaid payments.
    - **CLIA exempt:** a lab in a state that has laws that are the same as or more stringent than CLIA. States included are Washington and New York.
2. **Laboratories are inspected** to make sure that they are complying with their quality standards. CMS can inspect laboratories itself, or it can outsource inspections to certain organizations with **deemed status**, such as the following.
    - American Association of Blood Banks
    - American Association for Laboratory Accreditation
    - American Osteopathic Association
    - American Society for Histocompatibility and Immunogenetics
    - College of American Pathologists (CAP)
    - COLA
    - The Joint Commission
3. **CLIA has different requirements** depending on the complexity of testing being performed. A laboratory is required to purchase one of the following certificates to reflect the complexity of testing being performed.
    - Certificate of Waiver: Only waived tests can be performed.
    - Certificate for Provider Performed Microscopy Procedures: Certain microscopy (moderate-complexity) tests can be performed by clinicians.
    - Certificate of Compliance: Labs can perform moderate- and high-complexity testing and must be inspected every 2 years by CMS or a CMS agent.
    - Certificate of Accreditation: Labs can perform moderate- and high-complexity testing and must be inspected every 2 years by a deemed accreditation agency.
    - Certificate of Registration: Moderate- and high-complexity testing can be performed until the certificates of compliance or accreditation are issued (i.e., it is an interim certificate).

The FDA categorizes tests and test complexity. CMS regulates test quality.

4. **Verification or Validation** (*see* chapter 21): CLIA requires that laboratories verify or validate their laboratory tests to make sure that they work as intended. LDTs and ModTs require a greater level of assessment than FDA-approved/cleared tests.
5. **Proficiency testing (PT):** Laboratories are "tested" with standardized CMS-approved samples to prove that they can perform tests correctly and produce accurate patient results. These challenges must be treated like routine patient samples (49).
    - PT is absolutely required for all tests except for waived tests.
    - **Formal PT:** tests for which a formalized system of PT exists. Challenges can be purchased and usually consist of 3 events per year, with 5 challenge samples in each event. Formal PT is available for only a certain number of tests.
    - **Alternative PT:** must be done for the remaining tests that do not have formal PT. There should be at least 2 events per year, each with at least 1 challenge sample.

Table 20.1.

DESCRIPTION OF PT FAILURE	OCCURRENCE	CONSEQUENCE
Unsatisfactory performance	1 failed PT	Failures must be investigated.
Unsuccessful performance	2 consecutive or 2 out of 3 failed PTs	Failures must be investigated and submitted with corrective actions to the lab's accrediting agency.
Repeat unsuccessful performance	Another unsuccessful performance, or 3 out of 4 failed PTs within 6 events	Lab must cease testing of that analyte on patient samples for 6 months. Failures and corrective action plans should be documented and submitted. The lab can resume testing of the analyte if it can complete two additional PT events successfully.

- Records need to be maintained for at least 2 years.
- Each CLIA certificate in a health care system is required to perform PT, not each site. So if an assay is performed at multiple sites within a health system but they are all under the same CLIA certificate, then the health system only needs to enroll in one PT program.
- PT samples should never be sent to or discussed with another laboratory until after the end reporting date. This includes samples that a laboratory would have sent out for testing had they been patient specimens. This is to make sure that laboratories are not "checking" their results against another laboratory.
  - Similarly, labs must not compare, or "check," their results against other methods even in their own laboratory (unless that is how a patient specimen would be treated).
  - They must not be tested more or fewer times than a regular patient sample.
  - They must be tested by personnel who test regular patient samples (i.e., not performed just by the "best techs").
- Consequences of failing PT: Failing PT suggests technical or operational issues regarding testing of an analyte. It therefore has significant implications for the lab (Table 20.1).
  - A score of <80% on a testing event is considered a failed PT.
  - Failed PT may occur because of incorrect testing, but also because of incorrect participation in the PT program, late submissions, and even clerical errors.
  - PT failures are not bound by the calendar year.

The passing score for PT is 80%.

PT samples must be treated in the same way as regular patient samples (except that PT samples must never be referred out).

IV. CDC SUPPORT OF CLIA. The CDC supports CLIA by providing information and **scientific and technical expertise**. It distributes educational resources and conducts studies on laboratory quality improvement. It also develops and monitors best practices for laboratories, including the PT program.

V. CLINICAL AND LABORATORY STANDARDS INSTITUTE. The Clinical and Laboratory Standards Institute (CLSI) is a volunteer organization that develops **consensus guidelines** for operating laboratory assays and developing tests. The guidelines are created with input from representatives of government, industry, and health care professions. So these documents are often considered the standard for meeting regulatory requirements. CLSI documents are reviewed every 3 to 5 years (or before, if needed).

VI. BILLING AND CODING. Both the cost of performing a test and how much a payer is willing to pay for it affect the kind of assays and test algorithms a laboratory can offer. Health care facilities are reimbursed for medical and laboratory services depending on a complex set of rules.

1. The following parties reimburse health care facilities for medical services.
   - The patient, directly
   - Private insurers
   - Public insurers: primarily Medicare and Medicaid. Because these parties are responsible for reimbursing a huge portion of medical services, they can drive reimbursement rates.
     - Medicaid: for people with low incomes
     - Medicare: for people ≥65 years old and for people <65 years old with certain chronic diseases.
2. Medical services are reimbursed in two main ways.
   - **Capitation** (or prospective payment): when a payer pays a predetermined amount for a certain clinical circumstance (a "lump sum")
     - These clinical circumstances are categorized into "diagnosis-related groups," or **DRGs**. For example, common DRGs are sepsis, heart failure, joint replacement, and urinary tract infections.
     - In this system, it does not matter how expensive or how cheap the laboratory testing is because the cost the care provider recoups will be the same.
     - This makes it difficult for labs to provide expensive testing even though it might provide a better diagnosis. On the other hand, it incentivizes labs and hospitals to minimize unnecessary testing.
   - **Fee for service** (or retrospective payment): when a payer pays a fee for each procedure that is performed
     - This means that care providers that do more lab testing can charge more, which can result in unnecessary testing.
     - On the other hand, it also means that care providers can afford to offer tests that are better than others, even if they are more expensive.
3. **Standardized codes** are used to standardize medical services and make them comparable across institutions. This helps to document and bill for services. There are several types of coding systems.
   - **ICD-10** translates a diagnosis into a system of 7-digit alphanumeric codes. This is an internationally recognized system.
   - **CPT** translates services like medical, surgical, and laboratory procedures into 5-digit numeric codes.
     - This is a U.S.-based system and is administered by the American Medical Association (private). It is the most commonly used system for coding services.

ICD-10 encodes the patient's disease.

CPT or HCPCS level I encodes medical services provided for the patient.

- When a cutting-edge new test is developed, it takes more than 1 year for the CPT editorial panel to create a relevant CPT code. This can cause delays in billing and therefore implementation of the test.

HCPCS level II encodes medical services that are not included in CPT or HCPCS level I.

- **HCPCS** (pronounced "hicks-picks")
  - **Level I:** identical to CPT codes but administered by CMS (public) and used for Medicare, Medicaid, and other third-party payers
  - **Level II** translates services and supplies that are not included in the CPT code system into 5-digit alphanumeric codes. This is further broken down into categories from A to V codes.
- **LOINC** translates clinical or laboratory test observations into 3- to 7-digit numbers. It is more granular than a CPT code, so a single CPT code can map to multiple LOINC codes. It encodes 6 specific pieces of information about a test:
  - Specimen type (e.g., CSF or blood)
  - What the analyte was (e.g., HBV antigen or HSV DNA)
  - Method used (e.g., PCR, physical signs, or fever)
  - Scale (e.g., quantitative versus qualitative)
  - Timing (e.g., instantaneous or a 24-hour urine)
  - Property that was measured (e.g., concentration or catalytic rate)

LOINC encodes granular test information.

- **SNOMED CT** is similar to the ICD system because it also is an international coding system for diagnoses and other clinical information, but it is more comprehensive. Also, it is built for electronic systems and computer algorithms and so it can be read by nonhuman systems.

SNOMED CT encodes the patient's disease and is software readable.

Multiple-Choice Questions

1. Proficiency samples should be treated like clinical samples. How should the laboratory proceed with a test request on a proficiency sample that is normally sent to a reference laboratory?

a. Send out to that reference laboratory

b. Send back to the provider of the proficiency

c. Do not send out and report as "referred to a reference laboratory"

d. Process in-house anyway, by whatever means possible

2. Which of the following coding systems uses a numerical, computer-readable set of medical terminology?

a. HCPCS level I

b. CPT

c. LOINC

d. SNOMED CT

3. How often are laboratories with a Certificate of Accreditation typically inspected?

a. Every year

b. Every 2 years

c. Every 5 years

d. Twice a year

4. Waived complexity tests

a. Are always point-of-care instruments

b. Are never reimbursed by Medicare or Medicaid

c. Do not need proficiency testing

d. Always require follow-up testing

5. Which of the following test categories requires a premarket approval?

a. FDA approval

b. FDA clearance

c. Research use only

d. Investigational use only

True or False

6. The FDA regulates laboratory quality. **T F**

7. The FDA classifies all diagnostic assays. **T F**

8. CLIA is a set of laws governing laboratory quality. **T F**

9. LDTs are not approved or cleared by the FDA. **T F**

10. Running an FDA-approved assay on a nonapproved source requires additional validation for that source. **T F**

21

CHAPTER 21

ASSAY PERFORMANCE AND INTERPRETATION

I. OVERVIEW. To make sure that diagnostic tests produce accurate and reliable results, laboratories are required to **verify or validate** their laboratory tests by monitoring their performance characteristics. The **performance characteristics** of a particular diagnostic assay describe how accurately and reliably it detects an intended target. Some of the common metrics used to define performance characteristics are precision, accuracy, reportable range, reference range, limit of detection, stability of reagents and specimens, and cross-reaction because of interfering substances.

1. **Ways to assess test performance:** The terms verification and validation have had several varying definitions, which can be confusing. Currently they are used by the CAP and CLSI (*see* chapter 20) as one-time testing that is done on an assay prior to diagnostic use to ensure that it performs as intended (50, 51). Once the assay is in use, quality assurance processes are used to ensure that the test continues to perform as expected.
   - **Verification:** confirmation of the performance characteristics of an FDA-approved or -cleared assay. This involves relatively basic testing to show precision, accuracy, reportable range, and reference range (PARR).
   - **Validation:** establishment and confirmation of the performance characteristics of lab-developed tests (LDTs), including modified FDA-approved/cleared tests (ModTs). These assays need to undergo a more extensive process to prove that they are reliable and accurate, such as precision, accuracy, reportable range, reference range, analytic sensitivity, and analytic specificity (PARR+AS+AS).
   - **Quality assurance:** subsequent, ongoing processes that are followed to confirm that the test continues to perform correctly under diagnostic testing conditions.
2. **When to perform a verification/validation**
   - Prior to implementation of a new nonwaived test or instrument
   - If the test system is moved to a substantially new location, since the change in environment (e.g., temperature, personnel, or humidity) can affect performance
   - If substantial modifications are made to an FDA-approved/cleared test (a modified test). The modifications should be validated.

P: Precision

A: Accuracy

R: Reportable range

R: Reference range

AS: Analytic sensitivity

AS: Analytic specificity

Verification/validation: performed **once** to establish testing
Quality assurance: performed on an **ongoing** basis to monitor and ensure correct testing

Verification: confirmation of FDA-approved/cleared performance characteristics. Required to perform PARR.

Validation: establishment and confirmation of non-FDA-approved/cleared tests (LDTs and modified tests). Required to perform PARR+AS+AS.

3. The difference between "clinical" and "diagnostic" performance characteristics

- Clinical sensitivity and specificity: the ability of a test to detect and correctly diagnose a *clinical* infection
- Diagnostic sensitivity and specificity test the technical performance of the test and its ability to correctly detect a *target*. Laboratories are only able to measure and control the diagnostic factors of their assays, so unless specified, the numbers released by laboratories are usually diagnostic measurements.
- Example 1: PCR on blood has very high diagnostic sensitivity for HSV-1. That means that if HSV-1 is present in the blood, PCR will be able to detect it. However, HSV-1 PCR from blood will not do a good job detecting an actual clinical infection because HSV-1 resides in neuronal cells and usually is not present in the blood. As a result, PCR on blood has a very low clinical sensitivity.
- Example 2: PCR for JC virus on a urine sample has a high diagnostic specificity. So if JC virus is detected, it is highly likely that JC virus is present. However, PCR for JC virus on urine has a very low clinical specificity since JC virus can be present in the urine of both asymptomatic and symptomatic individuals. So its presence alone should not be used to diagnose progressive multifocal leukoencephalopathy.

Usually diagnostic and clinical sensitivity/specificity will correlate because detecting a target usually means detecting a clinical disease. However, this is not always true.

4. Samples used for verification/validation studies

- Residual clinical samples: use these wherever possible, so that the assay is challenged with the same specimen that will be used during active clinical testing.
- Well-characterized samples: these include proficiency samples, quality control material, commercially produced verification panels, material from public health labs, etc. These are useful because the true result is well-established.
- Samples should cover the full range of analytes and their detection levels. For instance, testing should include analytes, negatives, and high and low positives.
- The number of samples needed to confirm each performance characteristic is not well defined. See Table 21.1 for general guidelines.

Precision = reproducibility

II. PRECISION. Precision describes how reproducible the test is (Fig. 21.1). In other words, if the assay is repeated many times on the same specimen, will it produce the same result? When verifying or validating a test for the first time, both intra-assay and interassay precision should be measured.

1. **Intra-assay precision** measures how repeatable an assay is if it is repeated multiple times on the **same** instrument, on the same day, or by the same operator. This tests whether there is there some issue with the test itself that makes it imprecise, or if there are some internal factors (like subjectivity of interpretation) that reduce reproducibility.
2. **Interassay precision** measures how repeatable an assay is if the test is repeated over **different** days, instruments, and operators. This tests whether there is some external factor (like humidity) or the ability of operators to perform the assay that can affect precision.

Accuracy = trueness

III. ACCURACY. Accuracy is a measure of how likely the test results are to be "true" or "actually correct" (Fig. 21.1). In other words, if an assay gives a result of

Table 21.1. Guidelines and common approaches for verification or validation of a new assay[a]

PARAMETER	UNMODIFIED FDA-APPROVED/CLEARED TEST		MODIFIED FDA-APPROVED/CLEARED TESTS AND LDTS		ACCEPTABILITY CRITERIA
	QUALITATIVE	QUANTITATIVE	QUALITATIVE	QUANTITATIVE	
Precision (reproducibility) (53–55)	1 or 2 controls over 3 days OR 20 days OR 2 negatives and 2 positives run in duplicate over two runs	2 negatives and 2 positives at different concentrations run in duplicate over two runs OR 2 positive samples at different concentrations in triplicate over 3–5 days OR use reportable range data	3 positive samples at different concentrations, run in triplicate on one day/run and over 3 days	Same as qualitative assay OR 3 samples at above and below the LOD, in duplicate, over 20 days	≥95% agreement
Accuracy (bias) (51, 53, 55)	≥20 samples total, including positives and negatives	≥20 samples total, including positives and negatives	≥20 samples total OR 30 positives and 10 negatives OR 50 positive and 100 negatives	Same as qualitative, but include positives across analytical measurement range	FDA approved: ≥90% agreement Non-FDA approved: ≥95% agreement
Reportable range (analytical measurement range) (53)	Test weakly and strongly positive samples	Positive samples at 3 concentrations of the range OR in duplicate at 5–7 concentrations across the range	Not applicable	Positive samples at 3 concentrations of the range OR in duplicate or triplicate, at 7–9 concentrations across the expected range	Within 0.5 log of expected value for quantitative tests
Reference range (53)	Use manufacturer's range OR test 20 prospective samples.	Use manufacturer's range OR test 20 prospective samples.	40–120 prospective samples	40–120 prospective samples	
Analytic sensitivity	Not applicable	20 replicates at LOD	Serial dilutions to establish range. Then 20 replicates at LOD.	Serial dilutions to establish range (usually >60 samples). Then 20 replicates at LOD.	≥95% detected at LOD
Analytic specificity (cross-reactivity)	Not applicable	Not applicable	No specific number	No specific number	≥90% agreement

[a]FDA-approved/cleared assays must test PARR and LDTs must test PARR+AS+AS. However, there are currently no specific requirements for testing precision, accuracy, reportable range, reference range, analytic sensitivity, and analytic specificity.

HIV positive, how confident are we that the result is accurate? To measure this, several positive and negative samples are tested by the assay and its results are compared to those from a reference standard method.

1. There are two main measures of accuracy: sensitivity and specificity (*see* Fig. 21.3).
   - **Sensitivity** is the ability of an assay to detect the target of interest, if it is present. In other words, if a target of interest is present in a sample, then a highly sensitive assay will detect it more often than an assay with low sensitivity.
   - **Specificity** is a measure of how likely it is that the assay detects only the target of interest and not something else. In other words, highly specific assays will only detect the target of interest, whereas a less specific assay will cross-react with other substances and result in false positives.

Sensitivity is the ability of an assay to detect a target of interest, if present.

Specificity is how often the assay will identify the **correct** target of interest, and not something else.

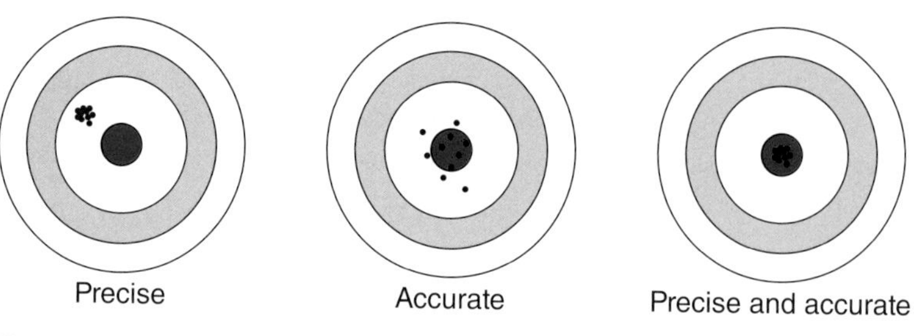

Figure 21.1. **Principle of precision and accuracy.**

The new test is compared to a reference standard to determine if it is detecting the true, "actually correct" results. The comparator may be wrong, so it is important to choose the most accurate comparator method.

2. **Comparator methods:** Accuracy of testing can be determined only by comparing results of the new test with another method of diagnosis. The **reference standard** (or **comparator method**) is the method that is assumed to produce the true, "actually correct" results (*see* Fig. 21.3). There are several types of reference standards.
    - **Clinical diagnosis:** clinical signs that are highly pathognomonic or diagnostic of a pathogen identification. Most viral pathogens have overlapping symptoms, so it is difficult to find a clinical symptom that defines a viral infection. One example is the presence of Koplik's spots, which is highly correlated with measles. On the other hand, not all measles cases have Koplik's spots.
    - **Gold standard test:** another assay that has been shown to produce the most accurate results in the past. It is important to note that these tests are not always perfectly accurate, and this can affect your perception of the new test. For example, a gold standard that does a bad job of detecting a pathogen (has poor sensitivity) will make it seem like the new test is detecting many false positives. This might occur when culture is used as the gold standard for PCR.
    - **Consensus standard:** Several tests are used together to produce the most accurate result. For example, culture, histopathology, and clinical symptoms for adenovirus can be taken together to verify a PCR result.
    - **Spiking:** A known pathogen is put into a sample before testing it. This is the most accurate way to ensure that a pathogen is actually present, but it is also the most artificial method.
3. **Agreement:** In addition to sensitivity and specificity, the new method and the comparator methods are compared to determine how closely they agree.
    - The **percent agreement** is the percent of specimens that give the same answer (positive or negative) between the new test and the reference standard.
    - **Discrepants** are results that do not match between the new and comparator methods.
    - Methods to compare agreement of two quantitative assays:
        - **Correlation:** plot results of one assay against results of the second assay. The Pearson's correlation coefficient (***R* value**) is an indication of how well the two assays agree.
        - **Agreement: A Bland-Altman** plot can be used to visualize the differences between the results of the assays. The average value for each sample is plotted on the *x* axis and the difference is plotted on

the *y* axis. Two assays agree very well if the differences are small (i.e., the points will lie on the mean line if the assays produce exactly the same results).

4. **Receiver-operator curve (ROC) analysis:** There is usually a tradeoff between sensitivity and specificity. If you decrease the stringency (i.e., decrease the cutoff value) of defining a positive result (i.e., the assay becomes more sensitive), it often starts to cross-react with other things too (i.e., it becomes less specific). An ROC curve can be used to chart the relationship between sensitivity and specificity so that the best cutoff value is chosen to maximize both parameters and therefore the overall accuracy of the test (Fig. 21.2).
    - Area under the curve = 1 (the test has a 100% probability of detecting a pathogen and gets the answer right every single time).
    - Area under the curve = 0.5 (the test has a 50% probability of detecting the pathogen and getting the answer right).

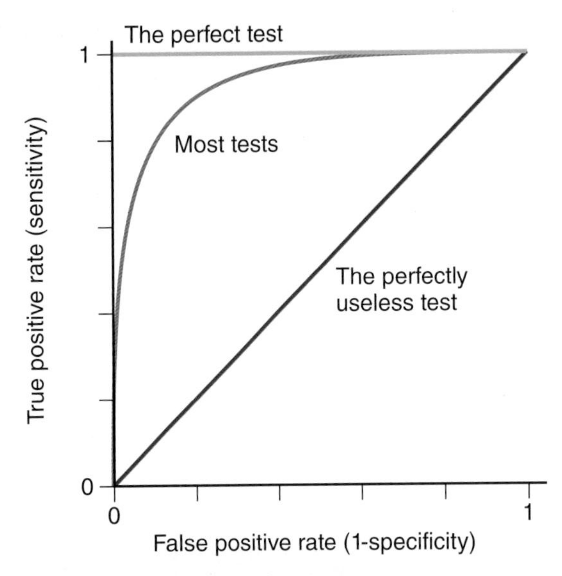

Figure 21.2. Understanding a ROC.

ROC curves: The perfect test has 100% sensitivity and 100% specificity no matter what cutoff value is used. The perfectly *useless* test is as useful as a toss of a coin—regardless of what cutoff is used, the sensitivity and specificity are both ~50%. Most diagnostic tests fall somewhere in the middle.

5. **Types of samples used**
    - **Retrospective** patient samples are samples previously characterized by another assay. This kind of sample is useful if the disease condition is rare, since positives can be stored over days to years.
    - **Prospective** patient samples are samples that are tested by the new method and the comparator method as they come in, so the results are not previously known. These samples are useful to show the actual prevalence and positivity rate of the disease at the testing location. Note that prevalence can be seasonal.
    - **Spiked samples:** Known material that has been purchased or characterized is "spiked" into the testing matrix. This kind of sample is contrived, but it can be used for quantification or if the disease condition is extremely rare.

IV. REPORTABLE RANGE. The reportable range (or analytical measurement range) is the range of target that the assay can detect.

1. **Qualitative assays:** the result is either positive or negative. Assess the range by testing negatives, weak positives, and strong positives.
2. **Quantitative assays:** the reportable range may be a wide range; testing should cover the entire range of the assay.

V. REFERENCE RANGE. The range that is considered "normal," or the baseline, is the reference range. Note that sometimes "normal" is not easy to define. For example, most of the population has antibodies against HSV-1, so this may or may not be considered "normal."

1. **Qualitative assays:** The result can be either positive or negative. For many viruses, the "normal" result is negative. For example, all uninfected individuals should have a negative HIV serology and a negative HIV viral load.
2. **Quantitative assays:** The "normal" value is often a range.

VI. ANALYTIC SENSITIVITY. Analytic sensitivity is a measure of the quantitative range that can detected by an assay.

1. **Limit of detection (LOD):** the smallest amount of target that can be **detected** reliably.
   - The LOD needs to be detected at least 95% of the time (e.g., 19 out of 20 times).
   - The LOD is affected by the specimen matrix, analyte, and method.
2. **Lower limit of quantification (LLOQ):** the smallest amount of target that can be **quantified** reliably. This value is usually higher than the LOD. For example, an HIV PCR might be able to quantify HIV down to 30 copies/ml and detect HIV down to 10 copies/ml.
   - The LLOQ is the lowest value that can be quantified 95% of the time.
   - The **linear range** is the range of the assay that is quantifiable. Values above and below this range may be detectable but not accurately quantifiable.

VII. ANALYTIC SPECIFICITY. Analytic specificity (or **cross-reactivity**) identifies substances or pathogens that can cause false-positive or false-negative results.

1. It can be more useful to call analytic specificity (PARR+AS+**AS**) "cross-reactivity" or "interferences" because it can get confused with specificity that is a component of accuracy (P**A**RR+AS+AS) (52).
2. This is tested by running samples that may contain interfering substances or isolates that are very closely related.

Positive and negative predictive values (PPV and NPV) are not static. They change depending on how prevalent a disease is in a population.

Low prevalence = decreased PPV and increased NPV

High prevalence = increased PPV and decreased NPV

VIII. PREDICTIVE VALUE. Predictive value is extremely valuable to clinicians because it summarizes the probability that a test result truly represents the presence or absence of disease. Unlike sensitivity and specificity, the predictive value of an assay can change depending on the prevalence of a disease (*see* Fig. 21.3 for an example).

1. **Prevalence** is the percentage of the population that has the disease as determined by the reference method. This metric can be determined only if the sample set tested by the reference method is representative of the population (for example, you cannot determine prevalence if only positive samples are tested). Low prevalence of a disease will decrease the PPV and increase the NPV and vice versa.
2. **Positive predictive value (PPV)** is the probability that a positive result given by an assay is real; in other words, the percentage of true positives detected by an assay to the number of positive calls it makes (so, this highlights true positives and **false positives**). Note that this is not the same as sensitivity, which is the probability that a positive case will be detected (a percentage of true positives detected by an assay to the number of reference standard positives; this highlights true positives and **false negatives**).
3. **Negative predictive value (NPV)** is the probability that a negative result given by an assay is real. In other words, the percentage of true negatives to the number of negative calls the assay makes (so this highlights true negatives and **false negatives**). Note that this is not the same as specificity, which is the probability that a negative case will be detected (the number of true negatives detected to the number of reference standard negatives; this highlights true negatives and **false positives**).

Important: An assay with excellent sensitivity and specificity can still perform poorly (more false positives and false negatives) if used in population with very low prevalence of disease.

A

		Reference Standard			
		+	−	Total	Predictive value
New Test Platform	+	True positive (TP)	False positive (FP)	New test positives (NtP)	Positive predictive value = TP/NtP
	−	False negative (FN)	True negative (TN)	New test negatives (NtN)	Negative predictive value = TN/NtN
	Total	Reference positives (RP)	Reference negatives (RN)	Total samples = RP + RN = NtP + NtN	Prevalence = RP/Total samples
	Accuracy	Sensitivity = TP/RP	Specificity = TN/RN		Percent agreement = TP+TN/Total samples

B

		Reference Standard			
		+	−	Total	Predictive value
New Test Platform	+	95	99	194	PPV = 49.0%
	−	5	9,801	9,806	NPV = 99.9%
	Total	100	9,900	10,000	Prevalence = 1%
	Accuracy	Sensitivity = 95%	Specificity = 99%		Percent agreement = 99%

Figure 21.3. (top) Performance characteristics of a new test compared to a reference standard and (bottom) example of performance characteristics for a new test when tested on 10,000 prospective specimens. Note that despite having excellent agreement, the PPV of the new test is low if prevalence is low.

IX. ASSAY INTERPRETATION. The diagnostic performance of a test can be quantified by measuring how well it works compared to another method. The following key variables are used to describe its performance characteristics (Fig. 21.3).

1. **True positives:** the number of cases detected by the new assay that were also positive by the reference method

2. **True negatives:** the number of negative cases by the new assay that were also negative by the reference method
3. **False positives:** the number of positives detected by the new assay that were negative by the reference method
4. **False negatives:** the number of negatives detected by the new assay that were positive by the reference method
5. **Accuracy**
   - **Sensitivity** = true positives/all reference standard positives
   - **Specificity** = true negatives/all reference standard negatives
6. **Predictive value**
   - **Positive predictive value** = true positives/all positives by the new assay
   - **Negative predictive value** = true negatives/all negatives by the new assay
7. **Total samples:** all positives and negatives tested by the new assay. This should be the same as all the positives and negatives tested by the reference standard.
8. **Prevalence:** the total number of actual positives (as determined by the reference method)/total number of samples
9. **Percent agreement:** the percentage of samples that have the same results (i.e., agree) by both methods.

Multiple-Choice Questions

1. **Verification of laboratory tests is a term used for**
   a. FDA-approved or -cleared tests
   b. Laboratory-developed assays
   c. Modifications of FDA-approved or -cleared tests
   d. All of the above
2. **What is the "sensitivity" of an assay?**
   a. Its ability to predict how prevalent a target of interest will be in the population
   b. Its ability to identify only the correct target of interest
   c. Its ability to detect the target of interest if it is there
   d. None of the above
3. **What is assay precision?**
   a. The ability to reproduce the same result repeatedly
   b. The ability to detect a target with high sensitivity
   c. The ability to detect the correct result
   d. All of the above
4. **Which of the following occurs when the prevalence of a disease decreases?**
   a. PPV increases.
   b. False positives increase.
   c. Sensitivity increases.
   d. Specificity decreases.

5. The specificity of an assay is calculated by dividing the number of ______ detected by the assay by the total number of ______ detected by the comparator assay.

a. True positives, positives

b. True negatives, negatives

c. False positives, positives

d. False negative, negatives

True or False

6.	The limit of detection is the smallest amount of the target that can be quantified.	**T F**
7.	Diagnostic sensitivity and clinical sensitivity are the same thing.	**T F**
8.	There are specific requirements for the number of samples that should be tested for accuracy and precision.	**T F**
9.	The reference standard defines the "true" or "correct" result.	**T F**
10.	An assay verification needs to be performed once a year.	**T F**

REFERENCES

1. **Drews SJ.** 2016. The taxonomy, classification, and characterization of medically important viruses, p 5–6. *In* Loeffelholz MJ, Hodinka RL, Young SA, Pinsky BA (ed), *Clinical Virology Manual*, 5th ed. ASM Press, Washington, DC.
2. **Bean B, Moore BM, Sterner B, Peterson LR, Gerding DN, Balfour HH Jr.** 1982. Survival of influenza viruses on environmental surfaces. *J Infect Dis* **146:**47–51.
3. **Tamerius J, Nelson MI, Zhou SZ, Viboud C, Miller MA, Alonso WJ.** 2011. Global influenza seasonality: reconciling patterns across temperate and tropical regions. *Environ Health Perspect* **119:**439–445.
4. **Robinson CC, Loeffelholz MJ, Pinsky BA.** 2016. Respiratory viruses, p 262. *In* Loeffelholz MJ, Hodinka RL, Young SA, Pinsky BA (ed), *Clinical Virology Manual*, 5th ed. ASM Press, Washington, DC.
5. **Centers for Disease Control and Prevention, National Center for Immunization and Respiratory Diseases (NCIRD).** 2017. Types of influenza viruses. https://www.cdc.gov/flu/about/viruses/types.htm. Accessed 9 February 2018.
6. **Kilbourne ED.** 2006. Influenza pandemics of the 20th century. *Emerg Infect Dis* **12:**9–14.
7. **Centers for Disease Control and Prevention.** 2009. 2009 H1N1 early outbreak and disease characteristics. https://www.cdc.gov/h1n1flu/surveillanceqa.htm. Accessed 9 February 2018.
8. **Peiris JSM, de Jong MD, Guan Y.** 2007. Avian influenza virus (H5N1): a threat to human health. *Clin Microbiol Rev* **20:**243–267.
9. **Centers for Disease Control and Prevention, National Center for Immunization and Respiratory Diseases (NCIRD).** 2017. Rapid influenza diagnostic tests. https://www.cdc.gov/flu/professionals/diagnosis/clinician_guidance_ridt.htm. Accessed 9 February 2018.
10. **Jafri HS, Wu X, Makari D, Henrickson KJ.** 2013. Distribution of respiratory syncytial virus subtypes A and B among infants presenting to the emergency department with lower respiratory tract infection or apnea. *Pediatr Infect Dis J* **32:**335–340.
11. **Jacobs SE, Lamson DM, St George K, Walsh TJ.** 2013. Human rhinoviruses. *Clin Microbiol Rev* **26:**135–162.
12. **Richardson SE, Tellier R, Mahony J.** 2004. The laboratory diagnosis of severe acute respiratory syndrome: emerging laboratory tests for an emerging pathogen. *Clin Biochem Rev* **25:**133–141.
13. **Wald A, Corey L.** 2007. Persistence in the population: epidemiology, transmission. *In* Arvin A, Campadelli-Fiume G, Mocarski E, Moore PS, Roizman B, Whitley R, Yamanishi K (ed), *Human Herpesviruses: Biology, Therapy, and Immunoprophylaxis*. Cambridge University Press, Cambridge, United Kingdom.
14. **World Health Organization.** 2017. Herpes simplex virus. http://www.who.int/mediacentre/factsheets/fs400/en/. Accessed 9 February 2018.
15. **Centers for Disease Control and Prevention, National Center for Immunization and Respiratory Diseases, Division of Viral Diseases.** 2016. Preventing varicella-zoster virus (VZV) Transmission from zoster in healthcare settings. https://www.cdc.gov/shingles/hcp/hc-settings.html. Accessed 9 February 2018.
16. **Centers for Disease Control and Prevention.** 2016. Smallpox: prevention and treatment. https://www.cdc.gov/smallpox/prevention-treatment/index.html. Accessed 9 February 2018.
17. **Centers for Disease Control and Prevention.** 2015. Hepatitis E FAQs for health professionals. https://www.cdc.gov/hepatitis/hev/hevfaq.htm. Accessed 9 February 2018.
18. **Terrault NA, Bzowej NH, Chang K-M, Hwang JP, Jonas MM, Murad MH, American Association for the Study of Liver Diseases.** 2016. AASLD guidelines for treatment of chronic hepatitis B. *Hepatology* **63:**261–283.
19. **Gaardbo JC, Hartling HJ, Gerstoft J, Nielsen SD.** 2012. Thirty years with HIV infection-nonprogression is still puzzling: lessons to be learned from controllers and long-term nonprogressors. *AIDS Res Treat* **2012:**161584.
20. **Kapler R.** 2016. Understanding the CDC's updated HIV test protocol. *MLO Med Lab Obs* **48:**8, 10, 12–13, quiz 14.
21. **McQuillan G, Kruszon-Moran D, Markowitz LE, Unger ER, Paulose-Ram R.** 2017. Prevalence of HPV in adults aged 18–69: United States, 2011–2014. *NCHS Data Brief* **280:**1–8.
22. **Dow DE, Cunningham CK, Buchanan AM.** 2014. A review of human herpesvirus 8, the Kaposi's sarcoma-associated herpesvirus, in the pediatric population. *J Pediatric Infect Dis Soc* **3:**66–76.

23. **World Health Organization.** 2014. What we know about transmission of the Ebola virus among humans. http://www.who.int/mediacentre/news/ebola/06-october-2014/en/. Accessed 9 February 2018.
24. **Jaax N, Jahrling P, Geisbert T, Geisbert J, Steele K, McKee K, Nagley D, Johnson E, Jaax G, Peters C.** 1995. Transmission of Ebola virus (Zaire strain) to uninfected control monkeys in a biocontainment laboratory. *Lancet* **346:**1669–1671.
25. **World Health Organization.** First antigen rapid test for Ebola through emergency assessment and eligible for procurement. http://www.who.int/medicines/ebola-treatment/1st_antigen_RT_Ebola/en/. Accessed 9 February 2018.
26. **Landry ML, Leland D.** 2016. Primary isolation of viruses, p 80. *In* Loeffelholz MJ, Hodinka RL, Young SA, Pinsky BA (ed), *Clinical Virology Manual*, 5th ed. ASM Press, Washington, DC.
27. **Madej RM, Davis J, Holden MJ, Kwang S, Labourier E, Schneider GJ.** 2010. International standards and reference materials for quantitative molecular infectious disease testing. *J Mol Diagn* **12:**133–143.
28. **WHO.** 2016. *1st WHO International Standard for BK Virus DNA*. National Institute for Biological Standards and Control, World Health Organization, Hertfordshire, United Kingdom.
29. **Chosewood LC, Wilson DE (ed).** 2009. *Biosafety in Microbiological and Biomedical Laboratories*. US Department of Health and Human Services, Washington, DC.
30. **Centers for Disease Control and Prevention.** 2016. Protecting healthcare personnel. https://www.cdc.gov/hai/prevent/ppe.html. Accessed 9 February 2018.
31. **Siegel JD, Rhinehart E, Jackson M, Chiarello L.** 2007. *2007 Guideline for Isolation Precautions: Preventing Transmission of Infectious Agents in Healthcare Settings*. Centers for Disease Control and Prevention, Atlanta, GA.
32. **Van de Perre P.** 2003. Transfer of antibody via mother's milk. *Vaccine* **21:**3374–3376 .
33. **Palmeira P, Quinello C, Silveira-Lessa AL, Zago CA, Carneiro-Sampaio M.** 2012. IgG placental transfer in healthy and pathological pregnancies. *Clin Dev Immunol* **2012:**985646.
34. **Waaijenborg S, Hahné SJ, Mollema L, Smits GP, Berbers GA, van der Klis FR, de Melker HE, Wallinga J.** 2013. Waning of maternal antibodies against measles, mumps, rubella, and varicella in communities with contrasting vaccination coverage. *J Infect Dis* **208:**10–16.
35. **Centers for Disease Control and Prevention.** 2015. General recommendations on immunization. *In Epidemiology and Prevention of Vaccine-Preventable Diseases*, 13th ed. Centers for Disease Control and Prevention, Atlanta, GA. https://www.cdc.gov/vaccines/pubs/pinkbook/genrec.html. Accessed 9 February 2018.
36. **Centers for Disease Control and Prevention.** 2016. List of vaccines used in United States. https://www.cdc.gov/vaccines/vpd/vaccines-list.html. Accessed 9 February 2018.
37. **Immunization Action Coalition.** 2017. Administering vaccines: dose, route, site, and needle size. http://immunize.org/catg.d/p3085.pdf. Accessed 9 February 2018.
38. **Centers for Disease Control and Prevention.** 2017. Recommended immunization schedule for children and adolescents aged 18 years or younger, United States, 2018. https://www.cdc.gov/vaccines/schedules/hcp/child-adolescent.html. Accessed 9 February 2018.
39. **Centers for Disease Control and Prevention.** 2017. Recommended immunization schedules for adults aged 19 years or older, United States 2018. https://www.cdc.gov/vaccines/schedules/hcp/adult.html. Accessed 9 February 2018.
40. **Barr Labs, Inc.** 2014. *Adenovirus type 4 and type 7 vaccine, live, oral. Package insert*. Barr Labs, Inc, Montvale, NJ.
41. **World Health Organization.** 2016. Dengue vaccine: WHO position paper— July 2016. *Wkly Epidemiol Rec* **91:**349–364.
42. **Schmaljohn CS.** 2012. Vaccines for hantaviruses: progress and issues. *Expert Rev Vaccines* **11:**511–513.
43. **World Health Organization.** 2015. Hepatitis E vaccine: WHO position paper, May 2015. *Wkly Epidemiol Rec* **90:**185–200.
44. **Fischer M, Lindsey N, Staples JE, Hills S, Centers for Disease Control and Prevention (CDC).** 2010. Japanese encephalitis vaccines: recommendations of the Advisory Committee on Immunization Practices (ACIP). *MMWR Recomm Rep* **59**(RR01)**:**1–27.
45. **Fischer M, Rabe IB, Rollin PE.** 2017. Tickborne encephalitis. *In CDC Yellow Book 2018: Health Information for International Travel*. Centers for Disease Control and Prevention, Atlanta, GA. https://wwwnc.cdc.gov/travel/yellowbook/2018/infectious-diseases-related-to-travel/tickborne-encephalitis. Accessed 9 February 2018.

46. **World Health Organization.** 2017. Yellow fever. http://www.who.int/ith/vaccines/yf/en/. Accessed 9 February 2018.
47. **Centers for Medicare and Medicaid Services.** 2017. *CLIA Program and Medicare Laboratory Services*. Centers for Medicare and Medicaid Services, Baltimore, MD.
48. **United States Food and Drug Administration.** 2007. Guidance for industry and FDA staff—commercially distributed analyte specific reagents (ASRs): frequently asked questions. https://www.fda.gov/RegulatoryInformation/Guidances/ucm078423.htm. Accessed 9 February 2018.
49. **Centers for Medicare and Medicaid Services.** 2008. *Proficiency Testing*. Centers for Medicare and Medicaid Services, Baltimore, MD.
50. **CLSI.** 2011. *Quality Management System: A Model for Laboratory Services*. Approved guideline (QMS01-A4), 4th ed. Clinical and Laboratory Standards Institute, Wayne, PA.
51. **CAP.** 2017. *All Common Checklist*. College of American Pathologists, Northfield, IL.
52. **Jennings L, Van Deerlin VM, Gulley ML, College of American Pathologists Molecular Pathology Resource Committee.** 2009. Recommended principles and practices for validating clinical molecular pathology tests. *Arch Pathol Lab Med* **133:**743–755.
53. **Burd EM.** 2010. Validation of laboratory-developed molecular assays for infectious diseases. *Clin Microbiol Rev* **23:**550–576.
54. **Clark RB, Lewinski MA, Loeffelholz MJ, Tibbetts RJ.** 2009. *Cumitech 31A, Verification and Validation of Procedures in the Clinical Microbiology Laboratory*. Coordinating ed, Sharp SE. ASM Press, Washington, DC.
55. **Wadsworth Center.** 2011. *Approval of Microbiology Nucleic Acid Amplification Assays, Microbiology Molecular Checklist*. State of New York Department of Health, Wadsworth Center, Albany, NY.
56. **Luca DC, August CZ, Weisenberg E.** 2003. Adult T-cell leukemia/lymphoma in a peripheral blood smear. *Arch Pathol Lab Med* **127:**636.
57. **Wright TC Jr.** 2006. CHAPTER 3 Pathology of HPV infection at the cytologic and histologic levels: basis for a 2-tiered morphologic classification system. *Int J Gynaecol Obstet* **94**(Suppl 1)**:**S22–S31.
58. **Leland DS, Ginocchio CC.** 2007. Role of cell culture for virus detection in the age of technology. *Clin Microbiol Rev* **20:**49–78.
59. **Weiss LM, Chen YY.** 2013. EBER in situ hybridization for Epstein-Barr virus. *Methods Mol Biol* **999:**223–230.
60. **American Society of Hematology.** 2010. Flower cells of leukemia. *Blood* **115:**1668.
61. **Centers for Disease Control and Prevention.** Direct fluorescent antibody test. https://www.cdc.gov/rabies/diagnosis/direct_fluorescent_antibody.html.

ANSWERS

CHAPTER 1

1. C
2. D
3. B
4. A
5. B
6. C
7. A.
8. D
9. C
10. A
11. D
12. C
13. C, D, A, B
14. F
15. T
16. T

CHAPTER 2

1. B
2. A
3. B
4. D
5. A
6. A
7. F
8. F
9. T
10. T

CHAPTER 3

1. A
2. A
3. E
4. C
5. A
6. B
7. B
8. A
9. A
10. D
11. T
12. F
13. T
14. F
15. T

CHAPTER 4

1. C
2. D
3. A
4. A
5. B
6. B
7. D
8. C
9. C, A, E, B, D
10. E, A, D, C, B
11. F
12. T
13. T
14. T
15. F

CHAPTER 5

1. B
2. A
3. C
4. C
5. A
6. A
7. D
8. B
9. B
10. D, C, B, A
11. F
12. T
13. T
14. F
15. F

CHAPTER 6

1. D
2. A
3. B
4. B
5. A
6. D
7. C, B, D, A
8. F
9. T
10. T
11. F

CHAPTER 7

1. D
2. A
3. A
4. B
5. C
6. D
7. A
8. T
9. T
10. T

CHAPTER 8

1. A
2. B
3. B
4. A
5. C
6. D
7. B
8. B
9. A
10. D
11. F
12. T
13. F
14. T
15. F

CHAPTER 9

1. B
2. C
3. A
4. C
5. A
6. A
7. D
8. B
9. C
10. C
11. A
12. T
13. T
14. F
15. F

CHAPTER 10

1. C
2. C
3. B
4. C
5. A
6. B, A, C, D
7. T
8. T
9. F
10. F

CHAPTER 11

1. B
2. C
3. D
4. A
5. C
6. A
7. C
8. B, E, C, A, D
9. C, D, B, E, A
10. T
11. F
12. F
13. T

CHAPTER 12

1. A
2. A
3. C
4. B
5. C
6. D
7. A
8. D
9. A
10. D
11. T
12. T
13. T
14. F
15. F

CHAPTER 13

1. C
2. B
3. A
4. D
5. B, C, D, A
6. D, C, A, B
7. A, D, B, C
8. F
9. T
10. T

CHAPTER 14

1. B
2. D
3. B
4. A
5. A
6. C
7. F
8. T
9. T
10. F
11. F

CHAPTER 15

1. A
2. B
3. B
4. D
5. A
6. B
7. C
8. A
9. C
10. F
11. T
12. T

CHAPTER 16

1. C
2. C
3. A
4. A
5. D
6. B
7. C
8. D
9. B
10. C
11. F
12. T
13. F
14. F
15. F

CHAPTER 17

1. B
2. C
3. B
4. D
5. D
6. C
7. F
8. F
9. T
10. T

CHAPTER 18

1. A
2. D
3. B
4. C
5. B
6. D
7. A
8. C
9. A
10. B
11. C
12. T
13. F
14. F
15. F

CHAPTER 19

1. A
2. C
3. D
4. C
5. A
6. A
7. D
8. D
9. D
10. A
11. C
12. T
13. F
14. T
15. F

CHAPTER 20

1. C
2. D
3. B
4. C
5. A
6. F
7. T
8. T
9. T
10. T

CHAPTER 21

1. A
2. C
3. A
4. B
5. B
6. F
7. F
8. F
9. T
10. F

INDEX

B

D

E

F

I

J

K

L

M

Q

T